THE ENDS OF THE EARTH

THE ANTARCTIC

THE ENDS OF THE EARTH

An Anthology of the Finest Writing on the Arctic and the Antarctic

THE ANTARCTIC

Edited by Francis Spufford

B L O O M S B U R Y

CONTENTS

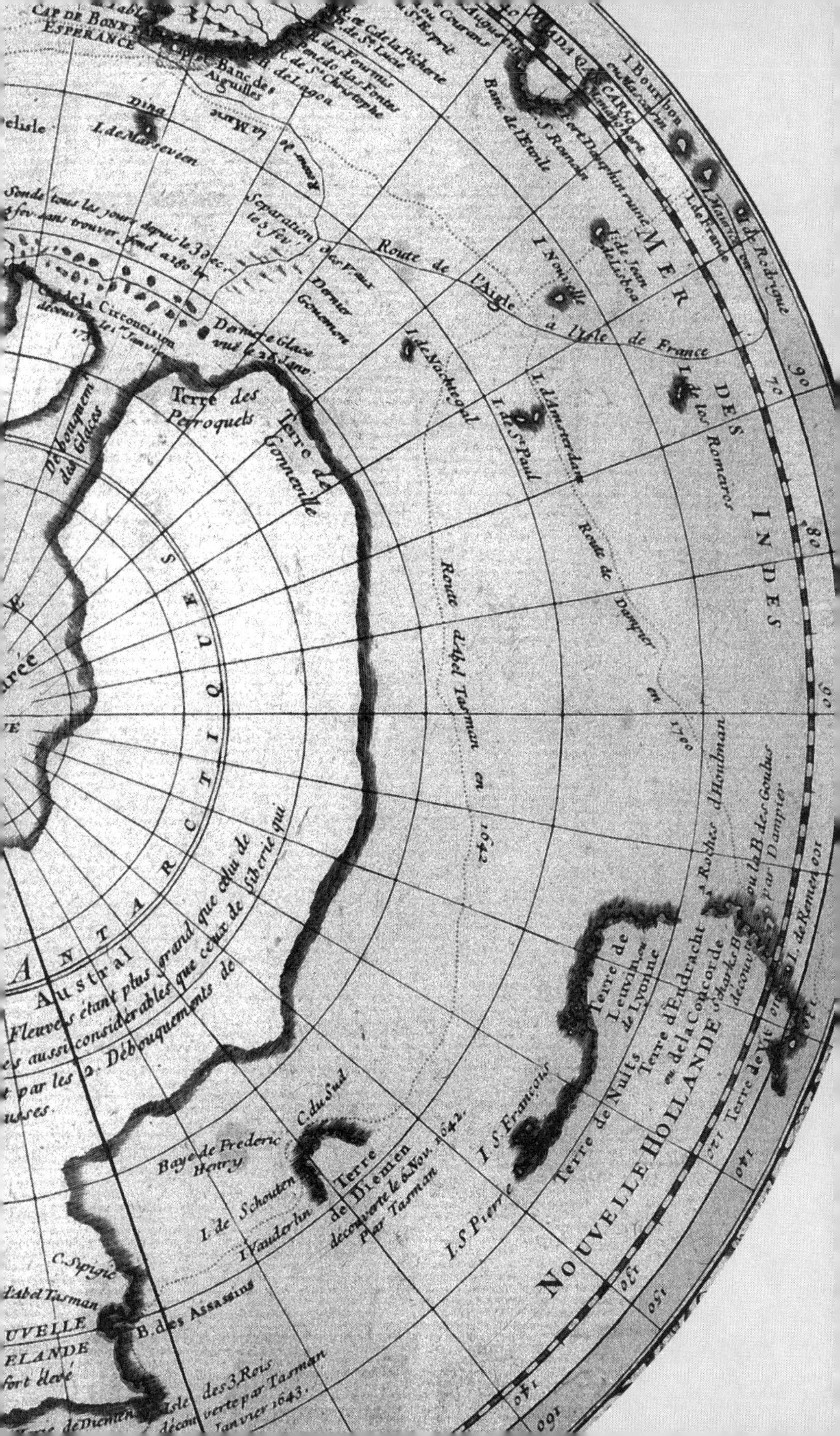

ESPERANCE
Banc des Aiguilles
I. de Marseveen
Sonde tous les jours depuis le 3 dec.
Dernière Glace vue le 23 Janv.
Route de l'Aigle a l'Isle de France
MER DES INDES
I. Bourbon ou Mascarin
I. de France
I. de Rodrigue
Banc de l'Etoile
I. Nouvelle
I. de Nachtegal
I. d'Amsterdam
I. de St. Paul
I. de los Romeros
Terre des Perroquels
Terre de Gonneville
Débouquement des Glaces
ANTARCTIQUES
Austral
Route de Dampier en 1700
Route d'Abel Tasman en 1642
Roches d'Houtman
Terre de Leuwin ou de Lyonne
Terre de Nuits
Terre d'Endracht ou de la Concorde
NOUVELLE HOLLANDE
I. S. François
I. S. Pierre
Terre de Witt
C. du Sud
Baye de Frederic Henry
I. de Schouten
I. Vanderlin
Terre de Diemen découverte le 6 Nov. 1642 par Tasman
B. des Assassins
Isle des 3 Rois découverte par Tasman Janvier 1643.
90
180
100
110
120
130
140
150
160

INTRODUCTION

FRANCIS SPUFFORD

Polar explorers don't all come to an illustrious end. Nobu Shirase was famous once. A crowd of fifty thousand people turned out in the rain when his Japanese Antarctic Expedition sailed home in 1912. They sang patriotic songs and held up paper lanterns "like wet peonies." But at the age of 85, in the hungry year after the Second World War ended, he died of malnutrition in a rented room over a fish shop, almost completely forgotten. He left behind a farewell poem, in the traditional *waka* form of five lines and exactly thirty-one syllables. It said (in blunt English paraphrase): *Study the treasures of the Antarctic, even after I am dead.*

What treasures are these? From the financial point of view, no-one got rich exploring the Antarctic in the "heroic age" at the beginning of the 20th century, despite the occasional fantasy of finding gold lying about in nuggets on some glacial moraine sealed away by ice and distance. Getting there cost so much that it took the equivalent of a space programme to land a few humans and their dogs or ponies on the shore of the southern continent. In effect, exploring the interior of Antarctica *was* a space programme, 1900-style; a venture into a white space on the map that was so inhospitable to life, you had to take along every scrap of food and fuel and equipment you needed, and every night retreat into a capsule of the warmer climate you came from. Only the air to breathe was provided, and that sometimes came choked with flying snow, or at temperatures that froze the moisture in your exhaled breaths till a crystalline mask, a beard of icicles, formed on your face. So there were no fortunes to be won at that time, no literal treasures to be had; just funding to find, through magic-lantern lecture tours, and rubber-chicken banquets with sceptical legislators, and weary presentations to patrons who might bring out the cheque-book if you thrilled them enough.

Moreover, almost all Antarctic travellers seemed to insist, in what they wrote afterwards, on the physical misery of Antarctic living. "Polar exploration," runs the famous opening line of Apsley Cherry-Garrard's *Worst Journey in the World*, "is at once the cleanest and the most isolated way of having a bad time which has been devised." Or, as Shirase's own *Nankyokuki* put it, "The whole enterprise was indescribably difficult and fraught with danger, and it was without doubt the worst of the trials and tribulations we had experienced since we left our mothers' wombs." The human body, in the classic exploration narratives, was usually in a state of discomfort at best. It was chilled, it was soaked, it was contorted, it was blinded by glare, it craved for food to the point where its owner would dream happily of deep-fried sardines dipped in chocolate. Ernest Shackleton's sailing master, Frank Worsley, said the worst of their boat journey to South Georgia was the hanks of coarse reindeer hair they had to filter out of the soup with their teeth. And beyond discomfort, bodies were subject to a collection of utterly un-treasureable catastrophes. Sensations to look forward to included having the pus in your blisters turn to ice and expand, having wounds ten years old reopen, and having the soles of your feet fall off.

Yet these same books did indeed open Antarctica to their readers as a treasurehouse. They made it as self-evident that the snow has treasures as Shirase had found it; as almost everybody finds it who has been fortunate enough to go to the Antarctic themselves, and to feel the cold brilliance of the south light, and to see the enormity of the ice. Antarctica is rarely a continent that towers, rarely a place that overwhelms you with height. Its sublimity lies in width, in horizontals that go on and on. A tabular berg flecked with blue-green will lie across a whole quadrant of the horizon. The fractured white band of an ice shelf will stretch as far as the eye can see in both directions. The frieze of a far mountain range will be coloured by tender early light in a pattern that repeats; buttercup yellow where the snow's in the sun, a blue-yellow where it's not, chilled as the cream in some ultimately shaded dairy, over and over again, horizon-long. Antarctica reaches round you, on clear days. Its bigness takes you by the heart and squeezes. Once it has you, it tends to keep you. Whether you went there as an explorer or as a tourist, as a scientist or as a climber, as an administrator or as a cook, you start scheming, the moment you leave, to get back there again.

Oh yes, there are treasures. Antarctica is a wonder and a delight. But *how* it is treasured has changed with the changing perceptions of travellers. The human history of Antarctica is very short, barely more than a hundred years, for though James Clark Ross in Her Majesty's Ships *Erebus* and *Terror* made the first Antarctic landfall in the 1840s, serious exploration did not begin until the 1890s, and the layers of memory and story which make a place into a place were all laid down thereafter. Geologically speaking, Antarctica may be a piece of Gondwanaland; imaginatively, it is an artifact of the twentieth century, about as old as jazz or the Ford Motor Company. Yet the twentieth-century was a densely packed one, for alterations in the understanding of nature, and science, and human society, and it is these that have played out, in displaced form, in the way we have seen Antarctica—in the way we've treasured it.

The first layer of Antarctic history was the heroic one. People usually put quote marks around "heroic," and there is certainly a lot to be sceptical about in the nationalistic fervour that enfolded the great expeditions of the decade leading up to the First World War. But "heroic" is the right word for the role those expeditions have played, ever after, in Antarctic memory. Together, the stories of them make up Antarctica's *Iliad*—a collective epic, in which figures not quite of human scale struggle and clash and sometimes die. Later times have endlessly reinterpreted its characters, but however much the interpretations change, the stories remain, because this barbarous beginning is the foundation of the human sense we've made of the continent, and is still the touchstone—the reference point—by which later generations of travellers understand what they themselves are doing there.

At the opening of the twentieth century, tiny gangs of human beings, all male, arrived on the margins of an unmapped space about the size of the continental United States, built themselves wooden cabins to live in, and set off inwards into the unknown. Ordinary-sized people when they were at home, they expanded, in their own conceptions of themselves, to match the giant scale of the landscape. Antarctica was an enormous stage which they were treading across virtually alone: how could they not feel as momentous, as *big* as the place, at the same time as they felt fragile, miniscule, infinitely vulnerable in relation to it? Antarctica made them big, first in the echo-chamber of their own breathing, as they sledged laboriously across the snow, then in the eager gaze of the world when they returned. Explorers had been famous figures for a long time, but mostly they had reported the strenuous discovery of populated places; they had been famous for encountering human otherness. Here, the explorers encountered only nature and themselves.

Or rather, nature and each other, because Antarctica like all deserts is an intensely gregarious place: the very few inhabitants necessarily live right on top of each other, getting to know each other's characteristics till they are either devoted friends, or maddened by homicidal irritation, or both. All sojourns in the ice accordingly generate emotions at high pressure. The difference of the Heroic Age was that, then, these were fused together with the business of dealing, for the very first time, from scratch, with the dangerous environment. The disagreements were disagreements about basic strategy for survival, about the elementary technology for moving around, about the proper way for authority to be exercised in an utterly isolated group.

Theoretically, the early explorers were engaged, in Antarctica, in a kind of romantic imperialism, and they did indeed try to claim what happened to them for various national stories, for the ongoing epics of British or Norwegian or German or Japanese or Australian identity—but, even in a world of empires, Antarctica was a domain where empire turned abstract, almost absurd; where you could clearly impose only a kind of vacuous diagram of possession on the snow if you claimed it for Queen Maud or King Edward VII or Emperor Meiji. Not surprisingly, what has survived of the Heroic Age in the modern imagination is more personal. It's a set of iconic figures, iconic events, iconic moments. Apsley Cherry-Garrard and his companions suffer through the horrors of the "Winter Journey" in a state of unearthly calm and mutual kindness; Douglas Mawson loses both his sledging partners, over on the other side of the continent, and all their food, and walks home to base alone, holding his feet together with string. We look at what was endured, and ask if we could endure it. We look at the expeditions' famous leaders and ask ourselves who we resemble. Which are you? Are you a Captain Scott, tense, anxious, man-hauling your way through the snow by main force yet describing it brilliantly afterwards, relying for your authority on military rank and on charm? Are you a Shackleton, with exactly the same prejudice against dog-sledging as Scott, having learned it with him on the same disastrous journey in 1902, but allied to a wonderfully supple gift for managing people, maternally kind when you could be, unhesitatingly ruthless when you had to be? Are you Amundsen, driven, impeccably self-educated in polar technique, yet far more of a polar performance artist than a word man, and so best appreciated ever after by skiers, mountaineers, ice athletes who can dance through the same moves he made, on his way to the Pole in 1912? Are you, far more obscurely, a Shirase, scarcely noticed by the main contenders for the Pole when he turned up in the Ross Sea in 1912, yet determined to be there, to make a start?

So the main treasures of Antarctica's first age are pieces of human behaviour, far removed from the present sometimes in manners or sensibility, urgently sympathetic because they are the actions of our lone representatives amid a vast, cold, indifferent, wilderness. But there were treasures for the senses, too. The early explorers were romantic observers as well as romantic imperialists. They saw Antarctica with the eyes for wilderness that had been developed in Europe and North America by Romantic literature. Not too long before, wild landscapes had only looked to travellers like a barren mess; but now, thanks to Wordsworth and Keats and Goethe and Thoreau, people had learned to see beauty in the fractal, unplanned complexity of stone, of snow, of water, of ice. The journeys of the heroic age were often journeys into visual delight, reverently noted down by cold hands holding colder pencils. The Australian scientist Edgeworth David led a side trip on Shackleton's *Nimrod* expedition of 1907–9 to climb the great volcano Mount Erebus. From near the summit, he found, you could see a giant shadow of the mountain projected across the sea of cloud below. "All within the shadow of Erebus was a soft bluish grey; all without was warm, bright and golden. Words fail to describe a scene of such transcendent majesty and beauty." Antarctica, the explorers learned, was a place of sensory extremes. In the winter dark, or in the strange depthless glow of a white-out, it shut down your senses with tormenting thoroughness, but at other times it overloaded them with colour and dazzle. The Antarctic light refracted shades of cobalt or emerald or lilac off snow crystals that you would never see in any of the planet's other landscapes. Extra suns danced in the sky. The moon rose as a gilded smudge, as a burning slit, as a crimson block. The aurora shimmered against black, in slow gauzy pulses. But the explorers were also noting the porridge texture of new grey ice as it formed on a still sea, and the quiddity of individual boulders, perched like abandoned game pieces on low plinths of unmelted snow. All the detail of Antarctica, not just the spectacle, flooded into their willing eyes.

In this lay the seed of what the continent would become in the future. Many of the early expeditions did substantial scientific work: Scott's 1910–13 *Terra Nova* expedition produced its leader's heroic (or incompetent) defeat in the race for the pole, and a fistful of tragic death, but it also produced six closely printed blue and gold folio volumes of scientific results, in geology, marine biology, meteorology, measurement of the earth's magnetic field. Science, though, tended to be subordinated to the language of *conquest* which almost all the early explorers used, as if they were battling against a hostile nature in the Antarctic, and needed to assault it, attack it, overcome it.

Knowledge came in as the prize, the suitably abstract prize, for victory: you struggled through the Antarctic's defences, and the unknown became known. The prize, of course, was never more abstract than in the competition to reach the South Pole itself, an unmarked spot corresponding to the geometry of human map-making, not to any physical feature a traveller could reach or see. It turned out to be located, this theoretical point where the earth's axis intersected its crust, on an ice plateau 10,000 feet high, beyond mountains you could only climb by mounting glaciers like ladders of smashed glass. And then the imaginary war cost real lives. But when the explorers of the Heroic Age were actually gazing, fascinated, at the intricacies of what was in front of them, the rhetoric of war sometimes went away. Scott has been much criticised for stopping, on the miserable homeward march from the Pole, to collect 35 lbs of rock specimens from exposed strata beside the Beardmore Glacier. But the fossil plant leaves in the samples, 170 million years old, were beautiful early evidence of the continent's Gondwanan past. Carrying them along was, perversely, among the most forward-looking things Scott ever did. It anticipated the coming time when scientists, not explorers, would be Antarctica's defining inhabitants; when understanding, not surviving, would be the most pressing human business there.

*

But first came an interlude with aeroplanes. The Heroic Age ended with the coming of the First World War. The outbreak of the war itself was less decisive than the arrival of the news of its nature. Shackleton, having been baulked of the Pole by the Scott–Amundsen race of 1912, went back to the Antarctic for a quixotic attempt to cross it side to side in 1914. War was being declared as he left, but when he next caught up with the news, in 1917—his ship having sunk, his men marooned on a speck of sub-antarctic rock, he himself having made a brilliant dash for help in an open boat across 800 miles of sea—the manager of the whaling station in South Georgia told him that "Millions are dying. Europe is mad. The world is mad." That was the age's end, in three sentences. The war spoiled public appetite for conquering nature; in Europe, damaged the ability to pay the space-programme cost of it, too. And in the 1920s and 1930s, Antarctica was quiet again—except for the planes in its sky. The two expeditions of the American explorer Richard Byrd used dog sledges for land travel as a matter of course, since the motorized alternatives were still not reliable at low temperatures, but for covering distance, for his major forays into the unknown, he flew. The route to the Pole that had taken Amundsen a season to cover, there and back, took him two days, with views down onto the crumpled iceflows streaming over the Transantarctic Mountains which, for the first time, integrated the fragmented observations of the past into one dynamic map. This made a considerable psychological difference. A human being in Antarctica could now possess, imaginatively, a much wider spread of it than was visible through the eyeholes of any one knitted helmet. From ten thousand feet, you could begin to see how Antarctica worked; you could begin to see it as having a unity in which, paradoxically, humans played very little part. The other new technologies of communications changed Antarctica too. Even when Byrd wintered by himself in an isolated weather station on the Ross Ice Shelf, as described in his 1938 bestseller *Alone*, he was linked to the world beyond as his predecessors had never been, by radio. Leaking fumes from a heater almost killed him, and he went through a long dark night of hallucinatory self-examination, but he could get friendly voices to talk to every evening, and baseball scores, and news of his investments plummeting on Wall Street. Byrd's Antarctica was halfway between times. It was still a place of adventure and danger; and the ski-mounted biplanes could always be drafted in to serve in high-speed versions of the old fantasy of imperium. In 1939 a brief German expedition to the Weddell Sea claimed a swathe of the interior by the impatient method of overflying it, and sprinkling the snow below with thousands of little metal swastikas.

War had stopped the expeditions; now war restarted them. A strategic worry about the Falkland Islands in 1944 put the British back into the long arm of Antarctica due south of the Americas, and as the Second World War segued into the Cold War, the continent became an arena for a confrontation between the two new superpowers, very much as if (to continue the space-programme analogy) the continent were a nearby planet, habitable if not very hospitable, which neither side could leave alone for fear the other one would have it. To add further complications, Latin American expeditions busily scribbled new land claims over the earlier European grabs for Antarctic territory, until the legal map of the continent resembled a logically-impossible pie, where everyone's slice overlapped with everyone else's. The International Geophysical Year in 1957 brought activity to a new level. The U.S. Navy started bulk construction of Antarctic facilities, with a giant base on Ross Island next door to Captain Scott's first hut, another logistically astonishing base at the South Pole itself, and a string of field camps and runways all over the million square miles of wilderness. Meanwhile, Soviet efforts went into building a chain of stations in the coldest, bleakest spots of the Weddell Sea sector, connected by mighty caterpillar convoys running on adapted artillery tractors. The scene seemed set for a mutually thwarting deadlock, in which humanity's primate instincts for power and territory took priority over all the actual qualities of the wilderness.

But instead, something remarkable happened. The deadlock metamorphosed into an international agreement to put Antarctica outside the reach of the global dispute. The Antarctic Treaty, signed in 1961 by the dozen nations active in Antarctica at the time, suspended all territorial claims and reserved the continent for science. Protocols added since have forbidden mining or oil drilling, protected Antarctica from human settlement except in very restricted ways, and put the continent under ecological quarantine. Cynics might say—did say—that the signatories' virtue did not cost them very much. As of 1961, Antarctica did not seem to contain much of economic value anyway. But the first indications appeared not long afterwards that there might, after all, be literal treasures under the snows, treasures of the mineral variety: and so far they have not been exploited. So far, despite it having no enforcement apparatus, the treaty has held. The United States' hasty collection of strategic assets in Antarctica became the main backbone of its transport infrastructure, with the U.S. running the flights down from New Zealand to the great gateway at McMurdo Base, and onwards from there all over the continent. The Russians set records for enduring the lowest temperatures ever recorded on the planet: minus 80 centigrade, at Vostok station, which turned out, after the Soviet Union fell, to be sitting on top of one of the strangest of scientific surprises.

To begin with, the new, permanently inhabited Antarctica was still a noticeably military place. Into the 1970s and early 1980s, the Navy ran the facilities at American bases, and still set the mood there, especially during the winters. McMurdo was officially designated as a ship, its residents ate in a galley and, like a submarine on Cold War missile patrol, it was powered by a small nuclear reactor, known affectionately to locals as "Nukey-Poo." But the patriarchal days of Antarctica were ending, both because something more routine and organised had taken over from the improvised tribal life of the Heroic Age, and also in the straightforward sense that the continent was no longer reserved exclusively for men. Women had first set foot in Antarctica in the 1950s, but only as naval wives. From the 1970s on, women were there in growing numbers in their own right, as scientists, as pilots, as support staff, as sailors or soldiers. The gender ratio on the continent remained bizarrely skewed, even in metropolitan McMurdo, but the perceptions and experiences of the other half of the human race did at last have the chance to inform Antarctic thinking and writing.

The sensibility that created the new consensual picture of Antarctica was the scientific one. After all, Antarctica was run *for* the scientists, who came to spend the polar summer searching for the trace gases of ancient climates in ice cores, or fossil-hunting in the Transantarctic Mountains, or teasing apart the molecular biology of the slow, slow bacteria of the Dry Valleys; often coming summer after summer, until they'd racked up totals of time on the ice that put the early explorers to shame. The scientists did not, of course, have a monopoly on access to Antarctica. Increasingly, as the twentieth century ended and the twenty-first century began, very expensive tourism was possible, often using the cast-off icebreakers of the former Soviet Union as cruise ships; and for even more money, adventure travellers could fly into the private airstrip at Patriot Hills in West Antarctica, and ski to the Pole "in the footsteps" of the Heroic Age explorers. But the scientists remained the dominant presence in Antarctic life. Travellers on the tour ships could gaze at the light and the wildlife, awestruck, heart-squeezed, and join what they saw to their own biographies, add it to their own treasuries of amazement. But the scientists were still the ones who got to set the meaning of the continent. It was theirs, to the point that support workers in Antarctica, flipping burgers or sorting garbage, sometimes became as pissed-off at the "beakers" as long-ago geologists had once been at the Scotts and Shackletons who kept moulding a fabulous research opportunity into a narrative of adventure.

What the scientists learned to see was a kind of amplification, continent-wide and in immense detail, of the view from the early flights. Once again, there was a connection to technology. Because the scientists now had relative immunity from the dangers of the environment, because they could move around with relative ease, they could sink into the type of concentration which distributes itself through the material world it is thinking about. (This sort of disembodied attention is a lot easier if your body is, in fact, safe and warm.) From hundreds—thousands—of separate studies, a sense built up of Antarctica's unity as a physical system, with its separate components of ice and rock and air and sea all subtly dependent on one another; and the whole system, in turn, delicately interrelated with the functioning of the rest of the planet, from the world's weather to its ecosystems. This was another great change of perceptions, another great reframing of the whole idea of Antarctica. From the beginning, it had been crucial to its imaginative impact as a place that it was a place apart, locked away at the bottom of world, beyond terrible seas patrolled by a ring of storms. It was an icy South far less continuous with the rest of the world than the icy North was. You could, if you had a mind to, hail a taxi in New York or Berlin, and be driven to the Arctic: but the Antarctic was on a separate map. Now the separateness, though not the uniqueness, was going away. First came the awareness that the industrial processes of the populated planet off to the north did, indeed, reach down and touch the ice. Cans of oven spray hissing out chlorofluorocarbons in Tokyo could bore a hole in the atmosphere in the southern polar sky. Then, more troubling, came the realization that Antarctica could act back upon the rest of the planet. In the 1990s, it became plain that the Antarctic ice shelves, by locking away vast tonnages of fresh water, effectively governed world sea levels. If the CO_2 pumped out by humans raised global temperatures enough to loosen the ice sheets, the melt water from the Ross Sea and the Weddell Sea would come pouring down my street, your street, our street at home, if home happened to be any of the world's coastal cities. Now, Antarctica no longer looked like a fortress of ice to be conquered. It was a fragile panel in the fabric of the planet. It no longer threatened to overwhelm tiny humans doing epic deeds in a harsh landscape. The balance of threat had reversed. *We* could damage *it*, without ever leaving home—and might inadvertently summon down upon ourselves a disaster out of the ice which would dwarf the cold fate that Captain Scott had gone looking for in its indifferent spaces. The new treasure of Antarctica is knowledge. Knowledge for its own sake, for our delight; severely practical knowledge, about how we may hurt ourselves, in a world whose biological infrastructure does not grow stronger just because our economies do.

Change, change, change throughout the Antarctic century. What we think we see, in the far South, shifts with our own preoccupations, like the jumbled icefloes in Antarctic waters which have seemed to travellers to morph into very different architectures, depending on who was doing the looking, and when. Captain Scott's men saw the ruins of St Paul's Cathedral going by;

Lieutenant Shirase's saw ruined temples. Look for modernist purity in the tabular bergs, and you find it. Look for tentacular, Gaudi-esque curlicues, and you find them too. Yet the ice's repertoire of shapes has been the same all century long.

When the Antarctic makes its astonishing impact on the senses, people try their damnedest to describe it. That's why there has been so much good writing about it, over the single century of its human history. People have constantly measured up the powers of language against it. But being in Antarctica is also a constant reminder of language's secondary status, of description's belated appearance on any scene. Nowhere else on Earth is it so clear that a place has an integrity apart from what we might say about it. Nowhere are words so obviously ineffectual a response to what just, massively, exists, whole and complete and in no real need of translation. Words, Antarctica teaches us, are not what the world is made of. Stop listening to me, then. Step aside, and sit down on that wind-carved boulder. Sit for a while: there are mountains in the distance to which the best response is hush. Take a long, silent look at the treasures of the snow.

But remember to get moving again while you can still feel your toes.

INTO THE NIGHT

from *Through the First Antarctic Night* (1900)

Frederick A. Cook

Frederick Cook would win fame a decade after he wrote this, as one of the two claimants locked in dispute over the discovery of the North Pole. But here the young doctor from New York State, along with an equally young Roald Amundsen, is at the other end of the earth. The exploring vessel Belgica *has frozen fast into the Antarctic pack ice. As polar summer ends, the unprepared adventurers on board are about to become the first human beings to experience an Antarctic winter. Here, with the physical and psychological challenges of the long dark, the Heroic Age opens.*

March 4. This morning a bunch of sharp rays of light pierced my port as the sun rose over the icy stillness of the north. It was like a bundle of frosted silver wire, and it served well the purpose of an eye-opener. Sleep here is an inexpressible dream. It does not matter how difficult the work, or how great the anxiety, we sink easily into prolonged restful slumbers. We awake rested, refreshed, and full of youthful vigour, always ready for the day's task. In the first days of our life in the pack we ate when we were hungry, slept when we were tired, and worked when the spirit moved us. (But later we were never hungry, always tired, and the spirit never moved us.)

[. . .]

We have wearied of pushing southward this season, and are discouraged in our ability to move in any direction, but we have tried hard to make a higher latitude. Nature frowns upon us and refuses to reward our dearly-bought venture. She guards the mysteries of the frozen south with much jealousy. She tempts us by permitting a small advance and a long look ahead, but when we have resolved to force on into the white blank, the icy gates close as if to say, "You can look, but you must not enter." A water sky, a land blink, or some other sign, indicative of land or open water, is constantly before us and these are, to the polar explorer, like the Star of Bethlehem to the children of Israel. They perpetually urge us on. We burn down the fires and wait impatiently for better success on the morrow, feeling always that we have won our success, thus far, by our own hard efforts, and by the same methods we hope to master the barriers now walled around us. Pressing ice, blasting head winds, blinding snow squalls, and all the worst elements of sea and weather combine to bewilder and defeat us.

The south polar lands are carefully shielded and fenced off by the circumpolar pack. The regions beyond the outer edge are not to be secured from the depths of mystery by a dash or an assault. The fortifications are more firmly laid than ever a human mind suggested. The prodigious depths of snow above, and the endless expanse of ensnaring sea around are mostly impregnable to man. He who contemplates an attack on this heatless undersurface of the globe will find many tempting allurements and many disheartening rebuffs. Such has been our experience. The battle, however, should be fought, though it promises to be the fiercest of all human engagements. Science demands it, modern progress calls for it, for in this age a blank upon our chart is a blur upon our prided enlightenment.

[. . .]

We are now again firmly stationed in a moving sea of ice, with no land and nothing stable on the horizon to warn us of our movements. Even the bergs, immense, mountainous masses, though apparently fixed and immovable, sail as we do, and with the same apparent ease. The astronomical positions which we obtain from the sun and from the stars indicate to us that we drift from five to ten miles per day. It is a strange sensation to know that, blown with the winds, you are moving rapidly over an unknown sea, and yet see nothing to indicate a movement. We pass no fixed point, and can see no pieces of ice stir; everything is quiet. The entire horizon drifts with us. We are part of an endless frozen sea. Our course is zigzag, but generally west—we do not know our destination, and are always conscious that we are the only human beings to be found in the entire circumpolar region at the bottom of the globe. It is a curious situation.

[. . .]

March 15. The weather is remarkably clear. There is no wind, no noise, and no motion in the ice. During the night we saw the first aurora australis. I saw it first at eight o'clock, but it was so faint then that I could not be positively certain whether it was a cloud with an unusual ice-blink upon it or an aurora;

but at ten o'clock we all saw it in a manner which was unmistakable. The first phenomenon was like a series of wavy fragments of cirrus clouds, blown by strong, high winds across the zenith. This entirely disappeared a few minutes after eight o'clock. What we saw later was a trembling lacework, draped like a curtain, on the southern sky. Various parts were now dark, and now light, as if a stream of electric sparks illuminated the fabric. The curtain seemed to move in response to these waves of light, as if driven by the wind which shook out old folds and created new ones, all of which made the scene one of new interest and rare glory.

That I might better see the new attraction and also experiment with my sleeping-bag, I resolved to try a sleep outside upon one of the floes. For several days I had promised myself the pleasure of this experience, but for one reason or another I had deferred it. At midnight I took my bag and, leaving the warmth and comfort of the cabin, I struggled out over the icy walls of the bark's embankment, and upon a floe three hundred yards east I spread out the bag. The temperature of the cabin was the ordinary temperature of a comfortable room; the temperature of the outside air was –20° C. (–4° F.) After undressing quickly, as one is apt to do in such temperatures, I slid into the fur bag and rolled over the ice until I found a depression suitable to my ideas of comfort. At first my teeth chattered and every muscle of my body quivered, but in a few minutes this passed off and there came a reaction similar to that after a cold bath. With this warm glow I turned from side to side and peeped past the fringe of accumulating frost, around my blow-hole through the bag, at the cold glitter of the stars. As I lay there alone, away from the noise of the ship, the silence and the solitude were curiously oppressive. There was not a breath of air stirring the glassy atmosphere, and not a sound from the ice-decked sea or its life to indicate movement or commotion. Only a day ago this same ice was a mass of small detached floes, moving and grinding off edges with a complaining squeak. How different it was now! Every fragment was cemented together into one heterogeneous mass and carpeted by a hard, ivory-like sheet of snow. Every move which I made in my bag was followed by a crackling complaint from the snow crust.

At about three o'clock in the morning a little wind came from the east. My blow-hole was turned in this direction, but the slow blast of air which struck my face kept my moustache and my whiskers, and every bit of fur near the opening, covered with ice. As I rolled over to face the leeward there seemed to be a misfit somewhere. The hood portion of the bag was as hard as if coated with sheet-iron, and my head was firmly encased. My hair, my face, and the under garments about my neck were frozen to the hood. With every turn I endured an agony of hair pulling. If I remained still my head became more and more fixed by the increasing condensation. In the morning my head was boxed like that of a deep sea-diver. But aside from this little discomfort I was perfectly at ease, and might have slept if the glory of the heavens and the charm of the scene about had not been too fascinating to permit restful repose.

The aurora, as the blue twilight announced the dawn, had settled into an arc of steady brilliancy which hung low on the southern sky, while directly under the zenith there quivered a few streamers; overhead was the southern cross, and all around the blue dome there were sparkling spots which stood out like huge gems. Along the horizon from south to east there was the glow of the sun, probably reflected from the unknown southern lands. This was a band of ochre tapering to gold and ending in orange red. At four o'clock the aurora was still visible but faint. The heavens were violet and the stars were now fading behind the increasing twilight. A zone of yellow extended from west around south to east, while the other half of the circle was a vivid purple. The ice was a dark blue. An hour later the highest icebergs began to glitter as if tipped with gold, and then the hummocks brightened. Finally, as the sun rose from her snowy bed, the whole frigid sea was coloured as if flooded with liquid gold. I turned over and had dropped into another slumber when I felt a peculiar tapping on the encasement of my face. I remained quiet, and presently I heard a loud chatter. It was uttered by a group of penguins who had come to interview their new companion. I hastened to respond to the call, and, after pounding my head and pulling out some bunches of hair, I jumped into my furs, bid the surprised penguins good morning, and went aboard. Here I learned that Lecointe, not knowing of my presence on the ice, had taken me for a seal, and was only waiting for better light to try his luck with the rifle.

[. . .]

April 16. In this shiftless sea of ice everything depends upon the wind. If it is south, we have steady, clear, cold weather. If it is north we have a warm, humid air with snow and unsettled weather. If it is east or west it brings a tempest with great quantities of driving snow; but it never ceases blowing. It is blow, blow, from all points of the compass. It is because of this importance of the wind, because it is the key-note to the day which follows, that our first question in the morning is "how is the wind?" To-day it is east, and has increased to a gale, in which it is absolutely impossible to take even a short walk on the pack. For recreation we have taken to mending. Racovitza is patching his pantaloons for the tenth time. This, he says, will be the last time, and I think he is right, for he has used leather to strengthen all the weak parts. Amundsen is patching boots; Lecointe is mending instruments; Danco and I are trying to repair watches. Nearly all of our good timepieces are out of commission. Our hands are better adapted for the trade of a blacksmith than that of a jeweller, but we are trying hard and have, to some extent, succeeded. Just at present it is the crystals which we wish to replace. We have no extra glasses, but we have found some small pocket compasses with crystals too small. How can we make them fit? Danco said, "Try sealing wax," which we did. We covered half of the watch and a good part of the crystal and thus made a very effective job, but in appearance it is a woeful object.

April 20. The easterly storm which has raged unceasingly for a week, and almost continuously for a month, shows some signs this morning of ceasing. At 4 A.M. the barometer began to rise, and the temperature fell to −2° C. The wind shifted to the northeast, but its force was soon spent. During the day the wind came only in intermittent puffs. The mouse-coloured clouds separated, permitting an occasional sunburst to light up the awful gloom which has so long hung over us. To-night, at ten o'clock, it is actually calm, and snow is falling lightly in huge, feathery flakes. This sudden calmness and dark unbroken silence, after the many days of boisterous gales, instill within us a curious sensation. The ship no longer quivers and groans. The ropes about the rigging have ceased their discordant music, and the floes do not utter the usual nerve-despairing screams. This sudden stillness, seemingly increased by the falling snow, brings to us a notion of impending danger.

[. . .]

May 16. The long night began at 12 o'clock last night. We did not know this until this afternoon. At 4 o'clock Lecointe got an observation by two stars which placed us in latitude 71° 34′ 30″, longitude 89° 10′. According to a careful calculation from these figures the captain announces the melancholy news that there will be no more day—no more sun for seventy days, if our position remains about the same. If we drift north the night will be shorter, if south it will be longer. Shortly before noon the long prayed-for southerly wind came, sweeping from the pack the warm, black atmosphere, and replacing it with a sharp air and a clearing sky. Exactly at noon we saw a brightening in the north. We expected to see the sun by refraction, though we knew it was actually below the horizon, but we were disappointed. The cold whiteness of our earlier surroundings has now been succeeded by a colder blackness. Even the long, bright twilight, which gladdened our hearts on first entering the pack, has been reduced to but a fraction of its earlier glory; this now takes the place of our departed day.

[. . .]

May 17. At about seven o'clock the captain went out to find two stars from which to obtain an observation for position. The sky was too hazy to give him an observation, but his eye rested upon an inexplicable speck of light in the west. He stood and looked at it for some moments. It did not change in position, but sparkled now and then like a star. The thing came suddenly, disappeared and again reappeared in exactly the same spot. It was so curious and assumed so much the nature of a surprise, that Lecointe came into the cabin and announced the news. We accused him of having had too early an eye-opener, but we went out quickly to see the mystery. It was about eight o'clock; the sky was a streaky mouse colour. The ice was gray, with a slight suggestion of lilac in the high lights, but the entire outline of the pack was vague under a very dark twilight. We looked for some time in the direction in which Lecointe pointed, but we saw only a gloomy waste of ice, lined in places by breaks in the pack from which oozed a black cloud of vapour. We were not sure that the captain's eyesight was not defective, and began to blackguard him afresh.

After we had stood on the snow-decked bridge for ten minutes, shivering and kicking about to keep our blood from freezing, we saw on a floe some distance westward a light like that of a torch. It flickered, rose and fell, as if carried by some moving object. We went forward to find if anybody was missing—for we could only explain the thing by imagining a man carrying a lantern. Everybody was found to be on board, and then the excitement ran high. Soon all hands were on deck and all seemed to think that the light was being moved towards us. Is it a human being? Is it perhaps some one from an unknown south polar race of people? For some minutes no one ventured out on the pack to meet the strange messenger. We were, indeed, not sufficiently dressed for this mission. Few had had breakfast; all were without mittens and hats, some without coats, and others without trousers. If it were a diplomatic visitor we were certainly in an uncomfortable and undignified uniform with which to receive him. Amundsen, who was the biggest, the strongest, the bravest, and generally the best dressed man for sudden emergencies, slipped into his *annorak,* jumped on his *ski* and skated rapidly over the gloomy blackness of the pack to the light. He lingered about the spot a bit, and then returned without company and without the light, looking somewhat sheepish. It proved to be a mass of phosphorescent snow which had been newly charged by sea algae, and was occasionally raised and brushed by the pressure of the ice.

May 18. During the few hours of midday dawn we made an excursion to a favourite iceberg to view the last signs of the departing day. It was a weird jaunt. I shall always remember the peculiar impression it produced upon me. When we started almost all the party were outside, standing about in groups of three or four, discussing the prospects of the long winter night and the short glory of the scene about. A thing sadder by far than the fleeing sun was the illness of our companion, Lieutenant Danco, which was emphasised to us now by his absence from all the groups, his malady confining him to the ship. We knew at this time that he would never again see a sunrise, and we felt that perhaps others might follow him. "Who will be here to greet the returning sun?" was often asked.

My companions on the excursion were Gerlache and Amundsen. Slowly and lazily we skated over the rough surface of the snow to the northward.

[. . .]

The pack, with the strange play of deflected light upon it, the subdued high lights, the softened shadows, the little speck of human and wild life, and our good ship buried under its snows, should have been interesting to us; but we were interested only in the sky and in the northern portion of it. A few moments before twelve the cream-coloured zone in the north brightened to an orange hue, and precisely at noon half of the form of the sun ascended above the ice. It was a misshapen, dull semicircle of gold heatless, rayless, and sad. It sank again in a few moments, leaving almost no colour and nothing cheerful to remember through the seventy long days of darkness which followed. We returned to the ship and during the afternoon laid out the plans for our midwinter occupation.

May 20. It is the fifth day of the long night and it certainly seems long, very long, since we have felt the heat of the sun. Since entering the pack our spirits have not improved. The quantity of food which we have consumed, individually and collectively, has steadily decreased and our relish for food has also slowly but steadily failed. There was a time when each man enjoyed some special dish and by distributing these favoured dishes at different times it was possible to have some one gastronomically happy every day. But now we are tired of everything. We despise all articles which come out of tin, and a general dislike is the normal air of the *Belgica.* The cook is entitled, through his efforts to please us, to kind consideration, but the arrangement of the menu is condemned, and the entire food store is used as a subject for bitter sarcasm. Everybody having any connection with the selection or preparation of the food, past or present, is heaped with some criticism. Some of this is merited, but most of it is the natural outcome of our despairing isolation from accustomed comforts.

I do not mean to say that we are more discontented than other men in similar conditions. This part of the life of polar explorers is usually suppressed in the narratives. An almost monotonous discontent occurs in every expedition through the polar night. It is natural that this should be so, for when men are compelled to see one another's faces, encounter the few good and the many bad traits of character for weeks, months, and years, without any outer influence to direct the mind, they are apt to remember only the rough edges which rub up against their own bumps of misconduct. If we could only get away from each other for a few hours at a time, we might learn to see a new side and take a fresh interest in our comrades; but this is not possible. The truth is, that we are at this moment as tired of each other's company as we are of the cold monotony of the black night and of the unpalatable sameness of our food. Now and then we experience affectionate moody spells and then we try to inspire each other with a sort of superficial effervescence of good cheer, but such moods are short-lived. Physically, mentally, and perhaps morally, then, we are depressed. and from my past experience in the Arctic I know that this depression will increase with the advance of the night, and far into the increasing dawn of next summer.

The mental conditions have been indicated above. Physically we are steadily losing strength, though our weight remains nearly the same, with a slight increase in some. All seem puffy about the eyes and ankles, and the muscles, which were hard earlier, are now soft, though not reduced in size. We are pale, and the skin is unusually oily. The hair grows rapidly, and the skin about the nails has a tendency to creep over them, seemingly to protect them from the cold. The heart action is failing in force and is decidedly irregular. Indeed, this organ responds to the slightest stimulation in an alarming manner. If we walk hurriedly around the ship the pulse rises to 110 beats, and if we continue for fifteen minutes it intermits, and there is also some difficulty of respiration. The observers, going only one hundred yards to the observatories, come in almost breathless after their short run. The usual pulse, too, is extremely changeable from day to day. Now it is full, regular, and vigorous; again it is soft, intermittent and feeble. In one case it was, yesterday, 43, to-day it is 98, but the man complains of nothing and does his regular work. The sun seems to supply an indescribable something which controls and steadies the heart. In its absence it goes like an engine without a governor.

There is at present no one disabled, but there are many little complaints. About half of the men complain of headaches and insomnia; many are dizzy and uncomfortable about the head, and others are sleepy at all times, though they sleep nine hours. All of the secretions are reduced, from which it follows that digestion is difficult. Acid dyspepsia and frequent gastric discomforts are often mentioned. There are also rheumatic and neuralgic pains, muscular twitchings, and an indefinite number of small complaints, but there is but one serious case on hand. This is Danco. He has an old heart lesion, a leak of one of the valves, which has been followed by an enlargement of the heart and a thickening of its walls. In ordinary conditions, when there was no need for an unusual physical or mental strain, and when liberal fresh food and bright sunshine were at hand, he felt no defect. But these conditions are now changed. The hypertrophied muscular tissue is beginning to weaken, and atrophy of the heart is the result, dilating and weakening with a sort of measured step, which, if it continues at the present rate, will prove fatal within a month.

[. . .]

May 27. The little dusk at midday is fading more and more. A feeble deflected light falls upon the elevations, the icebergs, and the hummocks, offering a faint cheerfulness, but this soon withdraws and leaves a film of blackness. The pack presents daily the same despondent surface of gray which, by contrast to the white sparkle of some time ago, makes our outlook even more melancholy. The weather is now quite clear and in general more settled. The temperature ranges from 5° to 10° C. below zero. We have frequent falls of snow, but the quantity is small and the period is short. Generally we are able to see the stars from two in the afternoon until ten in the morning. During the four hours of midday the sky is generally screened by a thick icy vapour. There are a few white petrels about daily, and in the sounding hole we have noticed a seal occasionally, but there is now no other life. All have an abundance of work, but our ambition for regular occupation, particularly anything which requires prolonged mental concentration, is wanting; even the task of keeping up the log is too much. There is nothing new to write about, nothing to excite fresh interest. There are now no auroras, and no halos; everything on the frozen sea and over it is sleeping the long sleep of the frigid night.

[. . .]

May 31. The regular routine of our work is tiresome in the extreme, not because it is difficult of execution or requires great physical exertion, but because of its monotony. Day after day, week after week, and month after month we rise at the same hour, eat the same things, talk on the same subjects, make a pretense of doing the same work, and look out upon the same icy wilderness. We try hard to introduce new topics for thought and new concoctions for the weary stomach. We strain the truth to introduce stories of home and of flowery future prospects, hoping to infuse a new cheer; but it all fails miserably. We are under the spell of the black Antarctic night, and, like the world which it darkens, we are cold, cheerless, and inactive. We have aged ten years in thirty days.

[. . .]

June 1. It is now difficult to get out of our warm beds in the morning. There is no dawn,—nothing to mark the usual division of night and morning until nearly noon. During the early part of the night it is next to impossible to go to sleep, and if we drink coffee we do not sleep at all. When we do sink into a slumber, it is so deep that we are not easily awakened. Our appetites are growing smaller and smaller, and the little food which is consumed gives much trouble. Oh, for that heavenly ball of fire! Not for the heat—the human economy can regulate that—but for the light—the hope of life.

[. . .]

June 3. The weather is unendurable, the temperature is –30° C. and an easterly gale is burying us in a huge drift of snow. With a high wind, an air thick with flying snow, and a temperature such as we have had for the past three days, ranging from –28° to –30°, it is utterly impossible to exist outside in the open blast. In calm weather such a temperature causes delight, but in a storm it gives rise to despair. I think it is Conan Doyle who says, "What companion is there like the great restless, throbbing polar sea? What human mood is there which it does not match and sympathise with?" I should like Mr. Doyle to spend one month with us on this great, restless, throbbing sea, under this dense, restless, throbbing blackness of the antarctic night. I am sure he would find conditions to drive his pen, but where is the companionship of a sea which with every heave brings a block of ice against your berth making your only hope of life, the bark, tremble from end to end? Where is the human being who will find sympathy in the howling winds under the polar night?

[. . .]

June 4. The ice is again breaking and the pressure of the floes, as they ride over each other, makes a noise converting the otherwise dark quietude into a howling scene of groans. It is again snowing and the wind keeps veering from the north-west to north-east.

Whenever we have advanced on our mysterious drift with the restless pack, either far east or far south, or both, we are arrested in our progress and the temperature falls. In the east there is also great pressure, and it is only in the far east or south that we get easterly or southerly winds. These winds have the character of land breezes—extremely dry, with a low temperature—followed by delightful, clear weather. From these facts we must conclude that the east and south are lined with land of large proportions or islands united by ice. An easy wind south or west drives us quickly; indeed, at times we drift northward without wind. The bergs now seem to press north and east.

June 5. To-day we have to record the darkest page in our log—the death of our beloved comrade, Danco. It has not been unexpected, for we have known that he could not recover, but the awful blank left by his demise is keenly felt, and the sudden gloom of despair, thus thrown over the entire party, is impossible of description. Poor fellow! in the past forty-eight hours he had been steadily improving, and, although we were not encouraged by this, he felt so much better that he was cheerful and altogether more like his former self, but it was the calm before the storm. Without any premonition of his coming death Danco passed away easily tonight; his last words to me were, "I can breathe lighter and will soon get strength." A companion with noble traits has left us. The event is too sad to note in detail. His life has steadily and persistingly sunk with the northerly setting of the sun. In ordinary health, his circulation was so nicely balanced that it needed but the unbalancing element of the prolonged darkness to disturb the equilibrium, and send him to a premature grave.

June 7. We have made a bag of sail-cloth, and into it the remains of Danco have been sewn. This morning we searched the crevasses for an opening which might serve as a grave. We found no place sufficiently open, but with axes and chisels we cut an aperture through the young ice in a recent lead, about one hundred yards from the bark. Owing to the depressing effect upon the party, we found it necessary to place the body outside on the ice upon a sledge the day after the death. At a few minutes before noon to-day the commandant, followed by the officers and scientific staff came to this sledge. The crew, dressed in an outer suit of duck, then marched out and, taking the drag rope, they proceeded over the rough drifts southerly to the lead. The day was bitterly cold, with a wind coming out of the south-west. Much snow in fine crystals was driven through the air, and it pierced the skin like needles. The surface of the ice was gray, but the sky had here and there a touch of brightness. In the north there was a feeble metallic glow, and directly overhead there were a few stratus of rose-coloured clouds. The moon, fiery, with a ragged edge, hung low on the southern sky. There was light enough to read ordinary print, but it was a weird light. Danco was a favourite among the sailors, and his departure was as keenly felt in the forecastle as among us. The men expressed this in the funeral procession. Slowly but steadily they marched over the rough surface of the ice with an air of inexpressible sadness. The sledge was brought to the freezing water. Here the commandant made a few fitting remarks, and then two heavy weights were attached to the feet, and the body was entrusted to the frosted bosom of the Antarctic ocean.

June 8. The melancholy death, and the incidents of the melancholy burial of Danco, have brought over us a spell of despondency which we seem unable to conquer. I fear that this feeling will remain with us for some time, and we can ill afford it. Though there are none among us sick at this time, we may at any moment have small complaints which will become serious under this death-dealing spell of despair. We are constantly picturing to ourselves the form of our late companion floating about in a standing position, with the weights to his feet, under the frozen surface and perhaps under the *Belgica.*

[. . .]

June 22. It is midnight and midwinter. Thirty-five long, dayless nights have passed. An equal number of dreary, cheerless days must elapse before we again see the glowing orb, the star of day. The sun has reached its greatest northern declination. We have thus passed the antarctic midnight. The winter solstice is to us the meridian day, the zenith of the night as much so as twelve o'clock is the meridian hour to those who dwell in the more favoured lands, in the temperate and tropical zones, where there is a regular day and night three hundred and sixty-five times in the yearly cycle. Yesterday was the darkest day of the night; a more dismal sky and a more depressing scene could not be imagined, but to-day the outlook is a little brighter. The sky is lined with a few touches of orange. The frozen sea of black snow is made more cheerful by the high lights, with a sort of dull phosphorescent glimmer of the projecting peaks of ice.

[. . .]

July 12. The light is daily increasing at midday, which should be a potent encouragement, but we are failing in fortitude and in physical force. From day to day we all complain of a general enfeeblement of strength, of insufficient heart action, of a mental lethargy, and of a universal feeling of discomfort. There has, however, been one exception; one among us who has not fallen into the habit of being a chronic complainer. This is Captain Lecointe. The captain has had to do the most trying work, that of making the nautical observations, which often keeps him handling delicate instruments outside and in trying positions in the open blast for an hour at a time. He has come in with frosted fingers, frozen ears, and stiffened feet, but with characteristic good humour he has passed these discomforts off. His heart action has steadily remained full and regular. The only other man in the party of equal strength is the cook, Michotte. But to-day I have to record the saddening news that Lecointe is suddenly failing. Not that he has complained of any ill-feeling, for he still maintains that he feels well; but in the usual daily examination, I notice that his pulse is intermitting, the first sign of coming debility. He is assuming a deathly pallor, does not eat, and finds it difficult to either sleep or breathe. There is a puffiness under the eyes, his ankles are swollen, and the entire skin has a dry, glossy appearance. The symptoms are all similar to those of Danco in his last stages; but Lecointe has a steady heart and sound organs, which augur in his favour.

July 14. Lecointe has given up all hope of ever recovering, and has made out his last instructions. His case seems almost hopeless to me. The unfavourable prognosis has sent another wave of despair over the entire party. Almost everybody is alarmed and coming to me for medical treatment, for real or imaginary troubles. The complaints differ considerably, but the underlying cause is the same in all. We are developing a form of anaemia peculiar to the polar regions. An anaemia which I had noticed before among the members of the first Peary Arctic edition, but our conditions are much more serious.To overcome this trouble I have devised a plan of action, which the sailors call the "baking treatment." As soon as the pulse becomes irregular and rises to one hundred beats per minute, with a puffiness of the eyes and swollen ankles, the man is stripped and placed close to a fire for one hour each day. I prohibit all food except milk, cranberry sauce, and fresh meat, either penguin or seal steaks fried in oleomargarine. The patient is not allowed to do anything which will seriously tax the heart. His bedding is dried daily, and his clothing is carefully adjusted to the needs of his occupation. Laxatives are generally necessary: and vegetable bitters, with mineral acids, are a decided help. Strychnine is the only remedy which has given me any service in regulating the heart, and this I have used as a routine. But surely one of the most important things was to raise the patient's hopes and instil a spirit of good humour. When at all seriously afflicted, the men felt that they would surely die, and to combat this spirit of abject hopelessness was my most difficult task. My comrades, however, were excellent aids, for as soon as one of our number was down, everybody made it his business to create an air of good cheer about him.

The first upon whom I tried this system of treatment systematically was Lecointe. I had urged part of it upon Danco, but he could not eat the penguin, and when I told him he must, he said he would rather die. When Lecointe came under treatment I told him that if he would follow the treatment carefully I thought he would be out of bed in a week. I did not have this faith in the treatment at that time, but I had confidence in the soundness of Lecointe's organs and I wished to boom up the man. Lecointe replied by saying, "I will sit on the stove for a month and eat penguins for the rest of my polar life if that will do me good." (He did sit beside the stove two hours daily for a month, and he ate, by his own choosing, penguin steaks for the balance of his stay in the polar circle. In a week he was about, and in a fortnight he again made his observations, and for the rest of his polar existence he was again one of the strongest men on the *Belgica.*)

[. . .]

July 15. The weather continues cold, but clear and calm, the only three qualities which make the antarctic climate endurable during the night. There is now much light. One can read ordinary print at 9 A.M., and at noon the north is flushed with a glory of green and orange and yellow. We are still very feeble. An exercise of one hour sends the pulse up to 130, but we have all learned to like and crave penguin meat. To sleep is our most difficult task, and to avoid work is the mission of everybody. Arctowski says, "We are in a mad-house," and our humour points that way.

[. . .]

July 21. The night is clear and sharp, with a brightness in the sky and a blueness on the ice which we have not seen since the first few days after sunset. An aurora of unusual brightness is arched across the southern sky. The transformation in its figure is rapid, and the wavy movement is strikingly noticeable. We are all out looking at the aurora, some by way of curiosity, but others are seriously studying the phenomenon. Arctowski, bundled in a wealth of Siberian furs, is walking up and down the deck, ascending to the bridge and passing in and out of the laboratory, as if some great event were about to transpire. Racovitza, with a pencil in his bare hand, in torn trousers, and without a coat or a hat, comes out every few minutes and, with a shiver, returns to make serious sketches of the aurora and humorous drafts of the unfortunate workers in the "cold, ladyless south." These daily touches of humour by "Raco" are bitterly sarcastic but extremely amusing. Lecointe, lost in a Nansen suit of furs, has been out on the pack in his observatory, which he calls the "Hotel," and is particularly elated because he has succeeded in getting an observation. "Now," says he, "we will know when this bloody sun will rise." Our position is latitude 70° 36′ 19″, longitude 86° 34′ 19″. If we continue to drift northerly a little, if the temperature remains low enough to give a great refraction, and if the weather remains clear, the captain promises us a peep of the sun for a few moments to-morrow. This is the happiest bit of news which has come to us, and it sends a thrill of joy from the cabin to the forecastle.

July 22. Every man on board has long since chosen a favourite elevation from which to watch the coming sight. Some are in the crow's nest, others on the ropes and spars of the rigging; but these are the men who do little travelling. The adventurous fellows are scattered over the pack upon icebergs and high hummocks. These positions were taken at about eleven o'clock. The northern sky at this time was nearly clear and clothed with the usual haze. A bright lemon glow was just changing into an even glimmer of rose. At about half-past eleven a few stratus clouds spread over the rose, and under these there was a play in colours, too complex for my powers of description. The clouds were at first violet, but they quickly caught the train of colours which was spread over the sky beyond. There were spaces of gold, orange, blue, green, and a hundred harmonious blends, with an occasional strip like a band of polished silver to set the colours in bold relief. Precisely at twelve o'clock a fiery cloud separated, disclosing a bit of the upper rim of the sun.

All this time I had been absorbed by the pyrotechnic-like display, but now I turned about to see my companions and the glory of the new sea of ice, under the first light of the new day. Looking towards the sun the fields of snow had a velvety aspect in pink. In the opposite direction the pack was noticeably flushed with a soft lavender light. The whole scene changed in colour with every direction taken by the eye, and everywhere the ice seemed veiled by a gauzy atmosphere in which the colour appeared to rest. For several minutes my companions did not speak. Indeed, we could not at that time have found words with which to express the buoyant feeling of relief, and the emotion of the new life which was sent coursing through our arteries by the hammer-like beats of our enfeebled hearts.

Lecointe and Amundsen were standing on an iceberg close to me. They faced the light, and watched the fragment of the sun slide under bergs, over hummocks, and along the even expanse of the frozen sea, with a worshipful air. Their eyes beamed with delight, but under this delight there was noticeable the accumulated suffering of seventy dayless nights. Their faces were drawn and thin, though the weight of their bodies was not reduced. The skin had a sickly, jaundiced colour, green, and yellow, and muddy. Altogether, we accused each other of appearing as if we had not been washed for months.

LANDFALL

from *To the South Polar Regions* (1901)

Louis Bernacchi

Bernacchi, an Australian geologist on the Southern Cross expedition led by the quarrelsome Norwegian-Australian Carsten Borchgrevink, was not describing an unknown landscape in this account of the party's landfall at Cape Adare. James Clark Ross's ships back in the 1840s had discovered and named the rocky angle of Antarctica that greets you first, when you sail south from Tasmania. But Bernacchi and his companions were seeing it with a difference. Now humanity had come to stay; now they had arrived to find out where the astonishing vistas led, to learn what lay beyond the mountains. Bernacchi gazes south, and the curtain rises on a continent.

We were now in the open sea to the south, for not a particle of ice was visible in any direction. Large flocks of brown-backed petrels were seen, and numbers of whales of the finner type. A sharp look-out was kept for land, and at 7 P.M. on the 15th of February it was sighted; but it was only a glimpse we caught of it through the dense canopy of clouds. Since noon the wind had increased steadily in force, until towards evening it was blowing a furious gale from the southeast and was accompanied by clouds of drifting snow. All that night and the following day the storm raged with full fury and the ship laboured heavily in the heavy seas. She lay to under half topsails, plunging fiercely into the seas and sometimes burying her whole bows beneath the waves, whilst ever and anon mighty green billows would pour over our decks and rush down into the cabins below. Our horizon was narrowly limited by the sheets of spray borne by the wind and the drifting snow, so we could see no land although we were not far from it.

The storm gradually abated towards the afternoon of the 17th, and we were able to stand in once more for the coast. The weather continued to improve and the dense mist cleared a little. At two o'clock in the afternoon land was again sighted distant some twenty-five miles, and we headed for a dark and high mass of rock which was evidently Cape Adare. It was a Cape of a very dark basaltic appearance, with scarcely any snow laying upon it, thus forming a strong contrast to the rest of the snow-covered coast. This lack of snow is principally due to the very exposed position of the Cape to the south-east winds, and, perhaps, also to the steep and smooth nature of its sides, which afford no hold for any snowfall. As we approached the coast it changed continually in aspect. Sometimes dense clouds of mist would envelop it; at other times the clouds would roll up like a great curtain, disclosing to our eyes a long chain of snow-clad mountains, the peaks of which tapered up one above the other like the tiers of an amphitheatre or those of the Great Pyramid of Cheops; but it was only a momentary vision, quickly disappearing, then all was again sombre, nothing but the heaving mass of waters, the whistle of the wind in the cordage, and the blinding snow across our decks.

Although we were certainly twenty miles distant from the land, the intervening space seemed infinitely less; in those high latitudes the eye is constantly liable to be deceived in the estimate it forms of distances. Apart from the contrast of light and shade, the great height of the mountain ranges and their bareness (they being destitute of any trees, etc., whereby to afford a point of comparison) augment this singular deception.

The wind decreased in fury as we got under the lee of the shore, but the whole heavens were still overcast with a dark mantle of tempestuous clouds, which now and then enshrouded the land in its folds, hiding it entirely from our view. The Bay (Robertson Bay) was clear of ice excepting for a huge stranded and weather beaten iceberg in its centre, into the cavities of which the seas ever and anon rushed with a great roar. As we drew closer, the coast assumed a most formidable aspect. The most striking features were the stillness and deadness and impassibility of the new world. Nothing around but ice and rock and water. No token of vitality anywhere; nothing to be seen on the steep sides of the excoriated hills. Igneous rocks and eternal ice constituted the landscape. Here and there enormous glaciers fell into the sea, the extremities of some many miles in width. Afterwards, when the mist had cleared away, we counted about a dozen of them around the Bay, rising out of the waters like great crystal walls. Approaching this sinister coast for the first time, on such a boisterous, cold and gloomy day, our decks covered with drift snow and frozen sea water, the rigging encased in ice, the heavens as black as death, was like approaching some unknown land of punishment, and struck into our hearts a feeling preciously akin to fear when calling to mind that there, on that terrible shore, we were to live isolated from all the world for many long months to come. It was a scene, terrible in its austerity, that can only be witnessed at that extremity of the globe; truly, a land of unsurpassed desolation.

The bay, into which we had entered, was about forty miles in width, and appeared to be well sheltered from the south-west, south-south-east, and east winds, but was exposed to those from the north-west. At its southern extremity, one recognised the great peak of Mount Sabine, rising up in weird majesty to some 12,000 feet. We were now close in under the shore and in smooth water. How delightful it was to be in calm water under the lee of the cold high peaks, after being so long involved in the din of the roaring elements.

[. . .]

The Commander now decided to effect a landing and requested Mr. Fougner and me to accompany him ashore in a small canvas boat. We got into the frail craft and rowed her ashore, but it took nearly a quarter of an hour to reach the land, for, although the distance from the ship appeared small, it was actually great. The place of landing was a shelving beach, formed of gravel and pebbles; slight surf was breaking upon it and the boat had to be handled carefully so as to avoid capsizing.

Thus, after many months, we had attained our destination, notwithstanding the numerous obstacles with which our path had been beset.

After having hoisted up our boat out of reach of the sea, we commenced an examination of the place. We had not walked many yards before we met the secluded and melancholy inhabitants of that South Polar land; these were the penguins scattered about it in groups of a hundred and more. They extended us but cold courtesy and gravely regarded us from a distance; but on our approaching closer they evinced more interest and commenced talking loquaciously together in their own particular vernacular. They had evidently discovered that there was something unusual about our appearance, and some were commissioned to investigate matters. These, with perfect *sang-froid,* slowly marched right up to our feet and ogled up at us in a most ludicrous fashion. Having finished this scrutiny, they returned to their fellows as sedately as they had come, and thenceforth took no more notice of us. What impressed us greatly was the general appearance of sadness prevailing amongst them; they seemed to be under the shadow of some great trouble. It is no small matter that will arouse them from their stolidity. There were many young birds among them; no doubt most of the older ones had already migrated northwards, it being late in the year for them. The effluvium from the guano was very powerful. The strong ammoniacal odour at first gave us a sensation of nausea, but we soon got used to it and never afterwards suffered any unpleasantness. There was, however, no large accumulation of guano of any commercial value, for in no place was it deeper than from three to four inches, and this only in very small patches of only a few feet in extent. The powerful winds prevent any extensive formation by sweeping all accumulations into the sea.

[. . .]

Bleached remains of thousands of penguins were scattered all over the platform, mostly young birds that had succumbed to the severity of the climate. Thousands of years hence, if the species should become extinct, those remains, frozen and buried among the débris, will be available as a proof of what once existed in these gelid regions now just habitable, then, perhaps, not at all. Stretched out on their backs along the beach were many seals enjoying their quiet, hitherto undisturbed, siesta. They were of the species *Leptonychotes Weddelli,* of which we had not met one in the ice-pack; these were of a dark colour with light spots. As we approached them they opened their beautiful and large intelligent eyes, gazed at us nonchalantly, snorted, blinked, and went to sleep again. Fear with these animals, as also with the penguins, is evidently an acquired, and not an hereditary habit. Numerous mummified carcases of these seals were observed lying about.

We also saw a few Giant Petrels *(Ossifraga gigantea)* and a great number of skua-gulls *(Megalestris Maccormicki).* These latter seemed to resent our visit, for they repeatedly darted at our heads and made a noisy outcry.

Satisfied with our preliminary survey of the place, we returned on board. The ship was brought in close to the shore and anchor let go in about eleven fathoms of water. Then, after some champagne-drinking and speech-making, we went on shore again, for all wanted to feel the rocks beneath their feet and to climb up the cliffs and get a look round.

It was nearly midnight when Mr. Hugh Evans and I commenced the ascent of Cape Adare. The light was still fairly strong at that time, nor was the temperature at all low. By following a ridge of craggy rocks we found the climbing tolerably easy and reached the top in less than an hour. We were thus the first human beings to set foot on the summit of South Victoria Land, and we felt full compensation for our climb. On the way up, we saw a few penguins, and even at the top (950 feet by aneroid) there were traces of them.

The scene before us looked inexpressibly desolate. A more barren desert cannot be conceived, but one of immense interest from a geological point of view.

From the end of the Cape to the foot of the mountain-range beyond, a great waste of hollows and ridges lay before our eyes; ridges rising beyond ridges like ocean waves, whose tumult had been suddenly frozen into stone. Beds of snow and ice filled up some of these extensive hollows, which had been scooped out by glacier action.

Never before had I seen the evidences of volcanic and of glacier action laying side by side—the hobnobbing of extreme heat and extreme cold. Great fire-scathed masses of rock rose out of the *débris* formed by the glaciers that had passed over the land. Vast convulsions must at one time have shaken the foundations of this land. But now silence and deep peace brooded over the scene that once had been so fearfully convulsed.

[. . .]

When we had set out the weather was fine, but later on the sky became overcast, as dark ominous clouds rolled up from the north-east.

The prospect from where we were was extensive, but scarcely beautiful. Down at our feet lay the sea, almost free of ice-pack. Huge stranded icebergs, defying the power of the solar beams, were visible in various directions along the coast. Behind us lay the great Antarctic Land; snow peaks rising beyond one another until by distance they dwindled away to insignificancy. The silence and immobility of the scene was impressive; not the slightest animation or vitality anywhere. It was like a mental image of our globe in its primitive state—a spectacle of Chaos.

Around us ice and snow and the remnants of internal fires; above, a sinister sky; below, the sombre sea; and over all, the silence of the sepulchre!

SLEDGE DOGS AND ENGLISHMEN

from *Diary of the Discovery Expedition, 1901–4* (1967)

Edward Wilson

The Discovery*'s voyage to the Ross Sea in 1900 to 1904 was an expedition on a different scale from any of the nimble scouting parties that had gone before. It was imperial Britain's grand, official effort, masterminded at home by a fantasizing octogenarian savant, and equipped with no expense spared. But out in the great whiteness, as a pioneer party went south across the Ross Ice Shelf, the* Discovery*'s big plans still came down to three men and a dog team, facing the cold. In this journal kept by the biologist-surgeon Edward Wilson, you won't find more than faint traces of one of the two crises that struck the travellers. Wilson's eye was too kind, perhaps too unearthly, for him to record the rivalry that unfolded between Robert Scott, the leader, and the third man, the Anglo-Irish merchant marine lieutenant Ernest Shackleton. But his journal is full of the other problem that made this trip such a disastrous influence on later British expedition, such a crucible for trouble yet to come. Here we watch all three of them struggle, and fail, to master dog-driving—the only technology that could have made their passage through the snow something better than an ordeal.*

Sunday, 16 Nov. Twenty-fifth Sunday after Trinity. A perfectly beautiful day, and we turned out at 6 A.M. We first tried the dogs with the total weights and six sledges. They were altogether too much for them on this surface, so we decided to start relay work. We did 2½ miles three times over before lunch, bringing up 3 sledges at a time, then taking the dogs back and bringing up the other three. We did the same after lunch, covering 15 geographical miles in the day and making 5 miles good to the south. Tedious work. We had most brilliant sunshine all day. After camping, there was a fine dog fight in which every one of the team succeeded in joining, as both the dog pickets were dragged out. A white low mist rises over the Barrier surface during the small hours of the night after these sunny cloudless days, and in the mist is nearly always to be seen a white fog-bow facing the sun.

[. . .]

Wed 19 Nov. Weather as before. Dead calm all day and very misty. Ice crystals falling all day, very thickly tonight. Altogether beautiful weather as the sun breaks through now and again and patches of blue sky overhead. Hard day's work. 15 miles covered to make 5 miles southing. Dogs getting very tired and very slow. We were at it from 11.30 A.M. till 9.30 P.M. and now at 11.15 P.M. we are at last in our sleeping bags. Surface worse than ever, with a thick coating of loose ice crystals like fine sand. We pray for a wind to sweep it all off and give us a hard surface again. This is wearing us out and the dogs, and yet we cover no ground. And the exertion of driving the poor beasts is something awful. Fine halo with brilliant prismatic parhelia round the sun today. There was a double halo for a while at 4 P.M., the outer one having the more marked prismatic colours and a radius about twice that of the inner. Thick fall of ice crystals again tonight. Have seen no land today, flat Barrier surface all round, very few sastrugi visible. Seriously thinking of giving the dogs a day off to rest.

[. . .]

Sat 22 Nov. Northerly breeze, sufficient to just help us with both sails set. Beautiful day, but threatening to become overcast and windy from the northwest. Turned out at 9 A.M. Made 2½ miles southing before lunch, covering 7½ to do it, pulling hard with the dogs all the way. After lunch we made 2¼ miles, 4¾ to the S.S.W. during the day, and fourteen miles covered to make that. More new land appearing to the southwest, and during the afternoon land appeared right ahead to the S.S.W., which is very satisfactory, as we can now make straight for it and leave a depot, and so reduce our weights considerably. Altogether much more promising than this slow and tedious plod to the south on an ice plain simply to beat a southern record. Now we have new land to survey, and I have the prospect of sketching, and we may find out something too which will explain this extraordinary Great Barrier, as we are so to speak getting to the back of it.

The surface today is very soft and heavy. We tried the dogs with the whole weights today, but they could hardly move them. Very tedious this relay work.

[. . .]

Tues 2 Dec. 5 A.M. really begins the day, when we are writing up our diaries in our sleeping bags after a long night's work, preparatory to turning in for the day. Slept nine hours solid. Turned out at 3 P.M. Covered 13 miles in making 4¼ to the southwest. My full day on. The Captain had the luncheon camping and in preparing our hot stuff set the tent alight, luckily just as we came up with the second loads, or the blessed thing would have been burnt. Providentially I was able to grab the thing the moment the flame came through to the outside and put it out, so that the only damage was a hole you could put your head through.

We had a heavy day and turned in late about 6 A.M. The weather from 5 P.M. till midnight was overcast and we could see nothing. Then it cleared quickly and the new land all came in sight ahead of us, more and more new land appearing to the south. One can now see a lot of detail in it, and it seems to consist of a series of fine bold mountain ranges with splendid peaks, all snow-clad to the base of course, but here and there rocky precipices, too steep to hold the snow, stood out bold and dark. It was a wonderful sight, the pale blue shadows in the white ranges standing against a greenish sky.

The dogs pulled well until the last lap when the sun was very hot and they "threw their 'ands in," and refused to do anything. The only thing then is to beat them and get them on yard by yard, sickening work. Many of them are badly chafed by the harness.

[. . .]

Fri 5 Dec. Just turned in, 4 A.M. Turned out again at 4 P.M. Nearly every night now we dream of eating and food. Very hungry always, our allowance being a very bare one. Dreams as a rule of splendid food, ball suppers, sirloins of beef, caldrons full of steaming vegetables. But one spends all one's time shouting at waiters who won't bring one a plate of anything, or else one finds the beef is only ashes when one gets it, or a pot full of honey has been poured out on a sawdusty floor. One very rarely gets a feed in one's sleep, though occasionally one does. For one night I dreamed that I ate the whole of a large cake in the hall at Westal without thinking and was horribly ashamed when I realized it had been put there to go in for drawing room tea, and everyone was asking where the cake was gone. These dreams were very vivid, I remember them now, though it is two month since I dreamed them. One night I dreamed that Sir David Gill at the Cape was examining me in Divinity and I told him I had only just come back from the farthest south journey and was frightfully hungry, so he got in a *huge* roast sirloin of beef and insisted on filling me up to the brim before he examined me.

Glorious day again, without a cloud. New plan today. We did 4 miles straight off with the first half load, then had our cold lunch as we went back for the second half with the dogs. Twelve miles we covered to make this four towards our depot. At a mile and a half from camp I went on ahead on ski and prepared the tent and got the supper under way. My eyes are a bit touched by the sun glare today. On these night marches we have the sun straight in our faces to the south and so strong is the glare that it catches one, notwithstanding all one's care. I have worn snow goggles every day since our start.

[. . .]

Tues 9 Dec. 2 A.M. Just turned in. Warm sunny morning. Beautiful range of rugged snowcovered mountains before us, with a long rounded snow hill at the foot. Did some darning. Woke up at 10 A.M. and as it was very hot in the tent, I sat outside in the hot sun and made a sketch of the panorama of new land before us. The heat was intense and there was no breeze. So one could easily sketch in bare hands and they got sun scorched. Turned in again when I had finished and slept till 4 P.M. when we had breakfast. Broiling hot sun all day. The dogs managed 2½ miles with half loads, but Snatcher died in the night and on opening him I found he had died of an acute peritonitis.

The snow today was soft, ankle deep, and the work very hot and heavy. No breeze except an occasional puff from the north. No clouds, the sun scorching down on us the whole march. We turned in at 4 A.M., no one having had a rest in camp today, as all three are needed to haul and help the dogs.

[. . .]

Thurs 11 Dec. 5 A.M. Just turned in. The skua is still with us. Turned out at 4 P.M. after another very hot night in the tent. We started seal meat with our breakfast, as well as having it for lunch today. Very heavy day's work. Covered nine miles and made 3 good towards the land. Soft and heavy snow, the dogs constantly stopping all together and refusing to pull on. The Captain had the camp shift. The dogs seem a trifle better in health for the pieces of their companion, whom I cut up and distributed among them. Not one refused to eat it, indeed most of them neglected their fish for it. There was no hesitation. "Dog don't eat dog" certainly doesn't hold down here, any more than does Ruskin's aphorism in *Modern Painters* that "A fool always wants to shorten space and time; a wise man wants to lengthen both." We must be awful fools at that rate, for our one desire is to shorten the space between us and the land. Perhaps Ruskin would agree that we are awful fools to be here at all, though I think if he saw these new mountain ranges he might think perhaps it was worth it.

The snow is soft and one sinks in at every step, making the walking very fatiguing. Supper, sleeping bag and breakfast are joys worth living for under these conditions. Only our appetites have clean outgrown our daily rations and we are always ravenously hungry, before meals painfully so for an hour or two.

[. . .]

Wed 17 Dec. 4 A.M. and a well marked fog-bow of white light. Just turned into our bags after camping and supper. I act as butcher every night now, killing a dog when necessary and cutting him up to feed the others. Slept well and woke at 11.20 when all the fog was gone and it was a brilliantly clear hot sunny day. No breeze. So I left my bag and the tent and sat sketching on the sledges for nearly three hours, when the others turned out and we had breakfast. We spent an hour upon odd jobs and started away south or a trifle east of south towards the land farthest visible.

I was leading, but after 3 hours got such a violent attack of sunglare in my eyes that I could see nothing. Yet I had worn grey glasses all the time sketching, and grey glasses as well as leather snow goggles on the march. I put in some drops and for the remaining five hours pulled behind with both eyes blindfold. They were very painful and streaming with water. By the time we camped they were better. We had lovely weather and the dogs pulled fairly well, making in all some 8 miles during the day. We passed some very splendid cliffs of rock and ice and glaciated land. The sun's heat was intense today, scorching our faces and hands. Noses constantly skinning, lips very painful, swollen and raw.

[. . .]

Sun 21 Dec. Fourth Sunday in Advent. I now save half a biscuit from supper to eat when I wake at night, otherwise I simply can't sleep again. I have never experienced such craving for more food before. We turned out about 7 A.M. and after breakfast made 2¼ miles when the dogs became so utterly rotten that we camped. It was very hot and close and we decided to start again at night time. Soft snowflakes falling, no sun, but low stratus. We eat our cold lunch and then slumbered and did odd jobs from 1 till 7 P.M. when we had some hot Bovril chocolate with somatose, and made another start.

The dogs were worse than ever and after a mile and a half we had to give it up. They simply wouldn't attempt to pull and seem as weak as kittens. We then tried some experiments—uncoupled all the dogs and took off all the food we were carrying for them, leaving only our own kit and food for the month. The surface was so heavy that we three could hardly budge it at all. We camped and had supper and turned in for a short night, intending to get back to the day routine again tomorrow.

The night was as raw and cold as the day had been oppressively hot, up to plus 28° F. by the sling thermometer. Did my butcher's job and turned in.

[. . .]

Wed 24 Dec. Christmas Eve. 2 A.M. In our bags writing up diaries or talking of food, letters and the relief ship. Turned out at 9.30 A.M. to a bright sunny morning, nice and warm. We had several jobs to do. We discarded and left the large sledge which carried all the dogs' fish and cut in two our own provision tank to take also the remains of the fish and carry the dog flesh. This meant a good bit of sewing as they are canvas tanks. Shackle and I did this, while the Captain took a round of angles and a sight, which put 81° 33½′ S.lat. We started away about 1 P.M. and made 5 miles by 6 P.M. Camped for lunch and then did 3 more by 9 P.M. when we camped for the night. As a result of today's medical examination I told the Captain that both he and Shackleton had suspicious looking gums, though hardly enough to swear to scurvy in them. No sketching today, very hazy light and excessive mirage. Surface rather better and dogs pulling better. My eyes touched by sun glare very painful during the night.

[. . .]

Fri 26 Dec. Woke up at 5 A.M. and as the left eye is still uncomfortable, made a sketch, using the right eye only. About 10 A.M. we started off and made nearly 5 miles, when my left eye got so intensely painful and watered so profusely that I could see nothing and could hardly stand the pain. I cocainized it repeatedly on the march, but the effect didn't last for more than a few minutes. For two days too I have had this eye blind-fold for a trifling grittiness and now it came to this, while the right eye, which I had been using freely, was perfectly well. The Captain decided we should camp for lunch and the pain got worse and worse. I never had such pain in the eye before, and all the afternoon it was all I could do to lie still in my sleeping bag, dropping in cocaine from time to time. We tried ice, and zinc solution as well. After supper I tried hard to sleep, but after two hours of misery I gave myself a dose of morphia and then slept soundly the whole night and woke up practically well.

*

Sat 27 Dec. Turned out at 7 A.M. Again a bright sunny day and no wind. I lay in my bag with my eyes bandaged while all the cooking was done, for fear of starting off again. The Captain and Shackle did everything for me. Nothing could have been nicer than the way I was treated. We started off at 10 A.M.

and without camping for lunch, made a good march of 10 miles by 7 P.M., through a long day with a scorching sun. We then camped for the night.

From start to finish today I went blindfold both eyes, pulling on ski. Luckily the surface was smooth and I only fell twice. I had the strangest thoughts or day dreams as I went along, all suggested by the intense heat of the sun I think. Sometimes I was in beech woods, sometimes in fir woods, sometimes in the Birdlip woods, all sorts of places connected in my mind with a hot sun. And the swish-swish of the ski was as though one's feet were brushing through dead leaves, or cranberry undergrowth or heather or juicy bluebells. One could almost see them and smell them. It was delightful. I had no pain in the eyes all day, a trifling headache. Towards evening we came in sight of a splendid new range of mountains still farther to the south.

[. . .]

Wed 31 Dec. Turned out at 6 A.M. NAO ration breakfast. Sun coming out and clouds rapidly clearing. We started off and made 4 or 5 miles by 2 P.M., watching the strait all the while to see if the head of it would clear. The mountains were perfectly beautiful today in the sunshine. As far as we could see, and apparently it was a blue horizon line, there was no land blocking the strait, so we must suppose this southernmost high land of 13,000 ft. to be insular perhaps. The strait was about 20 miles or more across and ran in due west, I should say. But all these details will appear when the Captain's map has been made out. He took a sort of running survey of the whole coast line.

All the morning we were crossing very immense pressure ridges radiating from the cape which formed the northern boundary of the strait. They were cut up in all directions by immense crevasses which were all filled in and bridged over with compacted snow. Sometimes from edge to edge the crevasse would measure 50 to 60 ft. and the whole train of sledges, dogs and all would be on the bridge at once. Only at the edges was there a risk of going through and some of the narrow crevasses too let one down suddenly.

At 2 P.M. we camped and after lunch, with some food in our pockets, we started off on ski to try and reach bare rock and bring back some specimens for the geologist. We also wanted to know whether there was anything in the way of a tide crack on shore or not, as this would practically decide whether the Barrier was afloat or aground. We ran in some 4 miles over smooth rounded pressure ridges, hills and vales and then were brought up by a perfect chaos of pressure and crevasses. We roped ourselves together, took off and left our ski at the edge and commenced to try and cross what appeared to be a gigantic tide crack, extending about a mile and a half across to the snow slopes that came off the land. Once across this, we could get our rock specimens which appeared quite close, but we had more than we could manage before us.

We started by going down steps cut in an ice slope, then by continual winding from side to side we made our way gradually across what at first looked like impassable crevasses, but in places they were filled in, though 50 to 80 ft. deep and blue ice, and in places they were bridged over. But after a while we were faced by more and more precarious bridges, and they got narrower and fewer, and were constantly giving way as we crossed them one by one on the rope. We never unroped the whole time, as there were crevasses everywhere and not a sign of some of them, till one of us went in and saw blue depths below to any extent you like. Shackleton was tied up in the middle, and the Captain and I at each end. Sometimes he led, sometimes I, if he came to an impasse and we had to go back.

As we got deeper and deeper in among this chaos of ice, the travelling became more and more difficult, and the ice all more recently broken up, so that no snow bridges had formed and we were faced by crevasses, ten, twenty, and thirty feet across, with sheer cliff ice sides to a depth of 50 or 80 ft. Unknown depths sometimes, because the bottom seemed a jumble of ice and snow and frozen pools of water and great screens of immense icicles. A very beautiful sight indeed, but an element of uncertainty about it, as one was always expecting to see someone drop in a hole, and while keeping your rope taut in case that happened, you would suddenly drop in a hole yourself. We tried hard to cross all this and reach the rock, but after covering a mile or more of it we came to impassable crevasses, and then saw that the land snow slope ended in a sheer ice cliff, a true ice foot, of some 40 ft. which decided us to retrace our steps, as even if we reached it, this ice foot would prevent our reaching rock.

The sun's heat was intense and not a breath of wind stirring. The heat has a very great power on these ice masses, as is evidenced by the immense icicles and frozen pools of what has been water. The colours, all shades of blue and pale green and shimmering lights were to be seen among these crevasses. The prismatic colours of the ice crystals were wonderful too today, forming what looked literally like a carpet of snow, glittering with gems of every conceivable colour, crimson, blue, violet, yellow, green and orange, and of a brilliance that would put any jewel in the shade. Our supper got upset in the tent sad to say, and we are so short of food that we scraped it all up off the floor cloth and cooked it up again. It was a soup so didn't suffer much. Another dog died today from sheer weakness.

Thurs 1 Jan. New Year's Day. Best wishes to all at home, and the best of good luck to all of us. We turned out at 8 A.M. to breakfast. Fine hot sunny day again, slightly overcast by cirro-stratus. Got away by 11 A.M. and made 4½ miles by 2.30. Camped for an hour for lunch and then made 4 miles more by 7.30, when we camped for supper. 8½ miles in the day, homewards now. We had sail set all the afternoon with a fair breeze. The dogs are terribly weak and of very little use. Another dog dropped on the march today, too weak even to walk. We put him on the sledge till evening, and when we had fed them and gone into the tent, he was killed by his neighbour in the trace, for his food. Our tent floor cloth, a large square of Willesden canvas, we use as a sail and it makes an excellent one.

[. . .]

What we have to consider is that we shall soon have no dogs at all and shall have to pull all our food and gear ourselves. And we don't know anything about the snow surface of the Barrier during summer. It may be quite different to what it was on the way south. One *must* leave a margin for heavy surfaces, bad travelling, and weather, difficulty in picking up depots, and of course the possibility of one of us breaking down. We have been making outwards from the coast today and find the surface improving as we do so. Close in shore it is apparently windless, and there is evidently a tremendous precipitate every night during the summer of fog-crystals, which lie inches deep, feet deep in places, forming a smooth, soft, crustless surface of flocculent snow.

[. . .]

Mon 5 Jan. Turned out 6.30 A.M. Fine sunny morning with northerly breeze and a little cirrus about. Made 4¼ miles in the morning, lunching at one. I sketched outside on my bag as the sun was warm and the breeze had dropped. Our lunch is a meagre meal—a biscuit and a half, eight lumps of sugar and a piece of sealmeat. One can hold it all in one hand, but it serves to carry us on for three hours' hauling in the afternoon. During this we covered 3¾ miles today in very deep soft ice crystals. We can now pull the sledges ourselves, so we have given up driving the dogs. They are doing no work at all.

We had yesterday the most beautiful cloud colouring round the sun, and again today much the same. There were cirrus clouds on a blue sky round the sun and these were beautifully edged with a vivid scarlet—a real vermilion, which ran into orange, yellow and pure violet on white cirrus clouds. The breeze today continued northerly, but the sun on the snow crystals made pull the sledges run very easily on the whole. One sweats very freely all day long. We have got into a routine now which balances our work and our food supply to a nicety, keeping us always hungry.

[. . .]

Wed 7 Jan. Turned out at 6 A.M. Bright warm and sunny morning. Started off with a light head wind from the northwest. Cast all the remaining dogs adrift and pulled the sledges ourselves on foot. Good surface. Made 5½ miles by 12.30. Camped for lunch. All the land partly obscured by low stratus. After lunch made up the 10 miles and camped at 5 P.M. Very hot sun, but more or less overcast sky. Very free perspiration, wet through all day. Washed our feet and hands in the snow on camping. Another dog dropped today.

The cloud effects over the land have been very fine all day, rolling cumulus clouds among the snow mountains and deep shadows, alternating with bright sunlight. It was a great relief to us today to plod along with no worry from the dogs. They all followed at their own pace. We had spells of conversation and long spells of silence, during which my thoughts wandered on ahead of me to the days that are yet to come.

FARTHEST SOUTH

from *The Heart of the Antarctic* (1909)

Ernest Shackleton

Shackleton came back to Antarctica in 1907 as the leader of his own expedition, outraging Scott, who believed Shackleton had promised to leave alone the whole Ross Sea route to the Pole. What really petrified Scott, of course, was the astonishing hair's-breadth failure to reach the Pole excerpted here. With a man-hauling technique for travel as dogged and painful as his rival's, Shackleton had nonetheless traversed the Ice Shelf, discovered a glacier leading up through the blockading mountains, and burst out into the globe's very last terrain, the high ice plateau where the South Pole lay. Frantically calculating and recalculating his supplies, Shackleton discovered that even starvation rations would not let him take the last steps—and turned for home ninety-seven miles from glory, a decision with a humane magnificence about it that still endures, even if luck then played a part in bringing him back alive from the risks he had already *taken to reach his Farthest South.*

December 21. Midsummer Day, with 28° of frost! We have frost-bitten fingers and ears, and a strong blizzard wind has been blowing from the south all day, all due to the fact that we have climbed to an altitude of over 8000 ft. above sea-level. From early morning we have been striving to the south, but six miles is the total distance gained, for from noon, or rather from lunch at 1 P.M., we have been hauling the sledges up, one after the other, by standing pulls across crevasses and over great pressure ridges. When we had advanced one sledge some distance, we put up a flag on a bamboo to mark its position, and then roped up and returned for the other. The wind, no doubt, has a great deal to do with the low temperature, and we feel the cold, as we are going on short commons. The altitude adds to the difficulties, but we are getting south all the time. We started away from camp at 6.45 A.M. to-day, and except for an hour's halt at lunch, worked on until 6 P.M. Now we are camped in a filled-up crevasse, the only place where snow to put round the tents can be obtained, for all the rest of the ground we are on is either névé or hard ice. We little thought that this particular pressure ridge was going to be such an obstacle; it looked quite ordinary, even a short way off, but we have now decided to trust nothing to eyesight, for the distances are so deceptive up here. It is a wonderful sight to look down over the glacier from the great altitude we are at, and to see the mountains stretching away east and west, some of them over 15,000 ft. in height. We are very hungry now, and it seems as cold almost as the spring sledging. Our beards are masses of ice all day long. Thank God we are fit and well and have had no accident, which is a mercy, seeing that we have covered over 130 miles of crevassed ice.

December 22. All day long, from 7 A.M., except for the hour when we stopped for lunch, we have been relaying the sledges over the pressure mounds and across crevasses. Our total distance to the good for the whole day was only four miles southward, but this evening our prospects look brighter, for we must now have come to the end of the great glacier. It is flattening out, and except for crevasses there will not be much trouble in hauling the sledges to-morrow. One sledge to-day, when coming down with a run over a pressure ridge, turned a complete somersault, but nothing was damaged, in spite of the total weight being over 400 lb. We are now dragging 400 lb. at a time up the steep slopes and across the ridges, working with the alpine rope all day, and roping ourselves together when we go back for the second sledge, for the ground is so treacherous that many times during the day we are saved only by the rope from falling into fathomless pits. Wild describes the sensation of walking over this surface, half ice and half snow, as like walking over the glass roof of a station. The usual query when one of us falls into a crevasse is: "Have you found it?" One gets somewhat callous as regards the immediate danger, though we are always glad to meet crevasses with their coats off, that is, not hidden by the snow covering. To-night we are camped in a filled-in crevasse. Away to the north down the glacier a thick cumulus cloud is lying, but some of the largest mountains are standing out clearly. Immediately behind us lies a broken sea of pressure ice. Please God, ahead of us there is a clear road to the Pole.

December 23. Eight thousand eight hundred and twenty feet up, and still steering upwards amid great waves of pressure and ice-falls, for our plateau, after a good morning's march, began to rise in higher ridges, so that it really was not the plateau after all. To-day's crevasses have been far more dangerous than any others we have crossed, as the soft snow hides all trace of them until we fall through. Constantly to-day one or another of the party has had to be hauled out from a chasm by means of his harness, which had alone saved him from death in the icy vault below. We started at 6.40 A.M. and worked on steadily until 6 P.M., with the usual lunch hour in the middle of the day. The pony maize does not swell in the water now, as the temperature is very low and the water freezes. The result is that it swells inside after we have eaten it. We are very hungry indeed, and talk a great deal of what we would like to eat. In spite of the crevasses, we have done thirteen miles to-day to the south, and we are now in latitude 85° 41′ South. The temperature at noon was plus 6° Fahr. and at 6 P.M. it was minus 1° Fahr., but it is much lower at night. There was a strong south-east to south-south-east wind blowing all day, and it was cutting to our noses and burst lips. Wild was frost-bitten. I do trust that to-morrow will see the end of this bad travelling, so that we can stretch out our legs for the Pole.

December 24. A much better day for us; indeed, the brightest we have had since entering our Southern Gateway. We started off at 7 A.M. across waves and undulations of ice, with some one or other of our little party falling through the thin crust of snow every now and then. At 10.30 A.M. I decided to steer more to the west, and we soon got on to a better surface, and covered 5 miles 250 yards in the forenoon. After lunch, as the surface was distinctly improving, we discarded the second sledge, and started our afternoon's march with one sledge. It has been blowing freshly from the south and drifting all day, and this, with over 40° of frost, has coated our faces with ice. We get superficial frost-bites every now and then. During the afternoon the surface improved greatly, and the cracks and crevasses disappeared, but we are still going uphill, and from the summit of one ridge saw some new land, which runs south-south-east down to latitude 86° South. We camped at 6 P.M., very tired and with cold feet. We have only the clothes we stand up in now, as we depoted everything else, and this continued rise means lower temperatures than I had anticipated. To-night we are 9095 ft. above sea-level, and the way before us is still rising. I trust that it will soon level out, for it is hard work pulling at this altitude. So far there is no sign of the very hard surface that Captain Scott speaks of in connection with his journey on the Northern Plateau. There seem to be just here regular layers of snow, not much windswept, but we will see better the surface conditions in a few days. To-morrow will be Christmas Day, and our thoughts turn to home and all the attendant joys of the time. One longs to hear "the hansoms slurring through the London mud." Instead of that, we are lying in a little tent, isolated high on the roof of the end of the world, far, indeed, from the ways trodden of men. Still, our thoughts can fly across the wastes of ice and snow and across the oceans to those whom we are striving for and who are thinking of us now. And, thank God, we are nearing our goal. The distance covered to-day was 11 miles 250 yards.

December 25. Christmas Day. There has been from 45° to 48° of frost, drifting snow and a strong biting south wind, and such has been the order of the day's march from 7 A.M. to 6 P.M. up one of the steepest rises we have yet done, crevassed in places. Now, as I write, we are 9500 ft. above sea-level, and our latitude at 6 P.M. was 85° 55′ South. We started away after a good breakfast, and soon came to soft snow, through which our worn and torn sledge-runners dragged heavily. All morning we hauled along, and at noon had done 5 miles 250 yards. Sights gave us latitude 85° 51′ South. We had lunch then, and I took a photograph of the camp with the Queen's flag flying and also our tent flags, my companions being in the picture. It was very cold, the temperature being minus 16° Fahr., and the wind went through us. All the afternoon we worked steadily uphill, and we could see at 6 P.M. the new land plainly trending to the south-east. This land is very much glaciated. It is comparatively bare of snow, and there are well-defined glaciers on the side of the range, which seems to end up in the south-east with a large mountain like a keep. We have called it "The Castle." Behind these the mountains have more gentle slopes and are more rounded. They seem to fall away to the south-east, so that, as we are going south, the angle opens and we will soon miss them. When we camped at 6 P.M. the wind was decreasing. It is hard to understand this soft snow with such a persistent wind, and I can only suppose that we have not yet reached the actual plateau level, and that the snow we are travelling over just now is on the slopes, blown down by the south and south-east wind. We had a splendid dinner. First came hoosh, consisting of pony ration boiled up with pemmican and some of our emergency Oxo and biscuit. Then in the cocoa water I boiled our little plum pudding, which a friend of Wild's had given him. This, with a drop of medical brandy, was a luxury which Lucullus himself might have envied; then came cocoa, and lastly cigars and a spoonful of *creme de menthe* sent us by a friend in Scotland. We are full to-night, and this is the last time we will be for many a long day. After dinner we discussed the situation, and we have decided to still further reduce our food. We have now nearly 500 miles, geographical, to do if we are to get to the Pole and back to the spot where we are at the present moment. We have one month's food, but only three weeks' biscuit, so we are going to make each week's food last ten days. We will have one biscuit in the morning, three at mid-day, and two at night. It is the only thing to do. To-morrow we will throw away everything except the most absolute necessities. Already we are, as regards clothes, down to the limit, but we must trust to the old sledge-runners and dump the spare ones. One must risk this. We are very far away from all the world, and home thoughts have been much with us to-day, thoughts interrupted by pitching forward into a hidden crevasse more than once. Ah, well, we shall see all our own people when the work here is done. Marshall took our temperatures to-night. We are all two degrees sub normal, but as fit as can be. It is a fine open-air life and we are getting south.

December 26. Got away at 7 A.M. sharp, after dumping a lot of gear. We marched steadily all day except for lunch, and we have done 14 miles 480 yards on an uphill march, with soft snow at times and a bad wind. Ridge after ridge we met, and though the surface is better and harder in places, we feel very tired at the end of ten hours' pulling. Our height to-night is 9590 ft. above sea-level according to the hypsometer. The ridges we meet with are almost similar in appearance. We see the sun shining on them in the distance, and then the rise begins very gradually. The snow gets soft, and the weight of the sledge becomes more marked. As we near the top the soft snow gives place to a hard surface, and on the summit of the ridge we find small crevasses. Every time we reach the top of a ridge we say to ourselves: "Perhaps this is the last," but it never is the last; always there appears away ahead of us another ridge. I do not think that the land lies very far below the ice-sheet, for the crevasses on the summits of the ridges suggest that the sheet is moving over land at no great depth. It would seem that the descent towards the glacier proper from the plateau is by a series of terraces. We lost sight of the land to-day, having left it all behind us, and now we have the waste of snow all around. Two more days and our maize will be finished. Then our hooshes will be more woefully thin than ever. This shortness of food is unpleasant, but if we allow ourselves what, under ordinary circumstances, would be a reasonable amount, we would have to abandon all idea of getting far south.

December 27. If a great snow plain, rising every seven miles in a steep ridge, can be called a plateau, then we are on it at last, with an altitude above the sea of 9820 ft. We started at 7 A.M. and marched till noon, encountering at 11 A.M. a steep snow ridge which pretty well cooked us, but we got the sledge up by noon and camped. We are pulling 150 lb. per man. In the afternoon we had good going till 5 P.M. and then another ridge as difficult as the previous one, so that our backs and legs were in a bad way when we reached the top at 6 P.M., having done 14 miles 930 yards for the day. Thank heaven it has been a fine day, with little wind. The temperature is minus 9° Fahr. This surface is most peculiar, showing layers of snow with little sastrugi all pointing south-south-east. Short food make us think of plum puddings, and hard half-cooked maize gives us indigestion, but we are getting south. The latitude is 86° 19′ South to-night. Our thoughts are with the people at home a great deal.

December 28. If the Barrier is a changing sea, the plateau is a changing sky. During the morning march we continued to go up hill steadily, but the surface was constantly changing. First there was soft snow in layers, then soft snow so deep that we were well over our ankles, and the temperature being well below zero, our feet were cold through sinking in. No one can say what we are going to find next, but we can go steadily ahead. We started at 6.55 A.M., and had done 7 miles 200 yards by noon, the pulling being very hard. Some of the snow is blown into hard sastrugi, some that look perfectly smooth and hard have only a thin crust through which we break when pulling; all of it is a trouble. Yesterday we passed our last crevasse, though there are a few cracks or ridges fringed with crystals shining like diamonds, warning us that the cracks are open. We are now 10,199 ft. above sea-level, and the plateau is gradually flattening out, but it was heavy work pulling this afternoon. The high altitude and a temperature of 48° of frost made breathing and work difficult. We are getting south—latitude 86° 31′ South to-night. The last sixty miles we hope to rush, leaving everything possible, taking one tent only and using the poles of the other as marks every ten miles, for we will leave all our food sixty miles off the Pole except enough to carry us there and back. I hope with good weather to reach the Pole on January 12, and then we will try and rush it to get to Hut Point by February 28. We are so tired after each hour's pulling that we throw ourselves on our backs for a three minutes' spell. It took us over ten hours to do 14 miles 450 yards to-day, but we did it all right. It is a wonderful thing to be over 10,000 ft. up, almost at the end of the world. The short food is trying, but when we have done the work we will be happy. Adams had a bad headache all yesterday, and to-day I had the same trouble, but it is better now. Otherwise we are all fit and well. I think the country is flattening out more and more, and hope to-morrow to make fifteen miles, at least.

December 29. Yesterday I wrote that we hoped to do fifteen miles to-day, but such is the variable character of this surface that one cannot prophesy with any certainty an hour ahead. A strong southerly wind, with from 44° to 49° of frost, combined with the effect of short rations, made our distance 12 miles 600 yards instead. We have reached an altitude of 10,310 ft., and an uphill gradient gave us one of the most severe pulls for ten hours that would be possible. It looks serious, for we must increase the food if we are to get on at all, and we must risk a depot at seventy miles off the Pole and dash for it then. Our sledge is badly strained, and on the abominably bad surface of soft snow is dreadfully hard to move. I have been suffering from a bad headache all day, and Adams also was worried by the cold. I think that these headaches are a form of mountain sickness, due to our high altitude. The others have bled from the nose, and that must relieve them. Physical effort is always trying at a high altitude, and we are straining at the harness all day, sometimes slipping in the soft snow that overlies the hard sastrugi. My head is very bad. The sensation is as though the nerves were being twisted up with a corkscrew and then pulled out. Marshall took our temperatures to-night, and we are all at about 94°, but in spite of this we are getting south. We are only 198 miles off our goal now. If the rise would stop the cold would not matter, but it is hard to know what is man's limit. We have only 150 lb. per man to pull, but it is more severe work than the 250 lb. per man up the glacier was. The Pole is hard to get.

December 30. We only did 4 miles 100 yards to-day. We started at 7 A.M., but had to camp at 11 A.M., a blizzard springing up from the south. It is more than annoying. I cannot express my feelings. We were pulling at last on a level surface, but very soft snow, when at about 10 A.M. the south wind and drift commenced to increase, and at 11 A.M. it was so bad that we had to camp. And here all day we have been lying in our sleeping-bags trying to keep warm and listening to the threshing drift on the tent-side. I am in the cooking-tent, and the wind comes through, it is so thin. Our precious food is going and the time also, and it is so important to us to get on. We lie here and think of how to make things better, but we cannot reduce food now, and the only thing will be to rush all possible at the end. We will do and are doing all humanly possible. It is with Providence to help us.

December 31. The last day of the old year, and the hardest day we have had almost, pushing through soft snow uphill with a strong head wind and drift all day. The temperature is minus 7° Fahr., and our altitude is 10,477 ft. above sea-level. The altitude is trying. My head has been very bad all day, and we are all feeling the short food, but still we are getting south. We are in latitude 86° 54′ South to-night, but we have only three weeks' food and two weeks' biscuit to do nearly 500 geographical miles. We can only do our best. Too tired to write more to-night. We all get iced-up about our faces, and are on the verge of frost-bite all the time. Please God the weather will be fine during the next fourteen days. Then all will be well. The distance to-day was eleven miles.

[. . .]

January 1, 1909. Head too bad to write much. We did 11 miles 900 yards (statute) to-day, and the latitude at 6 P.M. was 87° 6½′ South, so we have beaten North and South records. Struggling uphill all day in very soft snow. Every one done up and weak from want of food. When we camped at 6 P.M. fine warm weather, thank God. Only 172½ miles from the Pole. The height above sea-level, now 10,755 ft., makes all work difficult. Surface seems to be better ahead. I do trust it will be so to-morrow.

January 2. Terribly hard work to-day. We started at 6.45 A.M. with a fairly good surface, which soon became very soft. We were sinking in over our ankles, and our broken sledge, by running sideways, added to the drag. We have been going uphill all day, and to-night are 11,034 ft. above sea-level. It has taken us all day to do 10 miles 450 yards, though the weights are fairly light. A cold wind, with a temperature of minus 14° Fahr., goes right through us now, as we are weakening from want of food, and the high altitude makes every movement an effort, especially if we stumble on the march. My head is giving me trouble all the time. Wild seems the most fit of us. God knows we are doing all we can, but the outlook is serious if this surface continues and the plateau gets higher, for we are not travelling fast enough to make our food spin out and get back to our depot in time. I cannot think of failure yet. I must look at the matter sensibly and consider the lives of those who are with me. I feel that if we go on too far it will be impossible to get back over this surface, and then all the results will be lost to the world. We can now definitely locate the South Pole on the highest plateau in the world, and our geological work and meteorology will be of the greatest use to science; but all this is not the Pole. Man can only do his best, and we have arrayed against us the strongest forces of nature. This cutting south wind with drift plays the mischief with us, and after ten hours of struggling against it one pannikin of food with two biscuits and a cup of cocoa does not warm one up much. I must think over the situation carefully to-morrow, for time is going on and food is going also.

January 3. Started at 6.55 A.M., cloudy but fairly warm. The temperature was minus 8° Fahr. at noon. We had a terrible surface all the morning, and did only 5 miles 100 yards. A meridian altitude gave us latitude 87° 22′

South at noon. The surface was better in the afternoon, and we did six geographical miles. The temperature at 6 P.M. was minus 11° Fahr. It was an uphill pull towards the evening, and we camped at 6.20 P.M., the altitude being 11,220 ft. above the sea. To-morrow we must risk making a depot on the plateau, and make a dash for it, but even then, if this surface continues, we will be two weeks in carrying it through.

January 4. The end is in sight. We can only go for three more days at the most, for we are weakening rapidly. Short food and a blizzard wind from the south, with driving drift, at a temperature of 47° of frost, have plainly told us to-day that we are reaching our limit, for we were so done up at noon with cold that the clinical thermometer failed to register the temperature of three of us at 94°. We started at 7.40 A.M., leaving a depot on this great wide plateau, a risk that only this case justified, and one that my comrades agreed to, as they have to every one so far, with the same cheerfulness and regardlessness of self that have been the means of our getting as far as we have done so far. Pathetically small looked the bamboo, one of the tent poles, with a bit of bag sewn on as a flag, to mark our stock of provisions, which has to take us back to our depot, one hundred and fifty miles north. We lost sight of it in half an hour, and are now trusting to our footprints in the snow to guide us back to each bamboo until we pick up the depot again. I trust that the weather will keep clear. To-day we have done 12½ geographical miles, and with only 70 lb. per man to pull it is as hard, even harder, work than the 100 odd lb. was yesterday, and far harder than the 250 lb. were three weeks ago, when we were climbing the glacier. This, I consider, is a clear indication of our failing strength. The main thing against us is the altitude of 11,200 ft. and the biting wind. Our faces are cut, and our feet and hands are always on the verge of frost-bite. Our fingers, indeed, often go, but we get them round more or less. I have great trouble with two fingers on my left hand. They had been badly jammed when we were getting the motor up over the ice face at winter quarters, and the circulation is not good. Our boots now are pretty well worn out, and we have to halt at times to pick the snow out of the soles. Our stock of sennegrass is nearly exhausted, so we have to use the same frozen stuff day after day. Another trouble is that the lamp-wick with which we tie the finnesko is chafed through, and we have to tie knots in it. These knots catch the snow under our feet, making a lump that has to be cleared every now and then. I am of the opinion that to sledge even in the height of summer on this plateau, we should have at least forty ounces of food a day per man, and we are on short rations of the ordinary allowance of thirty-two ounces. We depoted our extra underclothing to save weight about three weeks ago, and are now in the same clothes night and day. One suit of underclothing, shirt and guernsey, and our thin Burberries, now all patched. When we get up in the morning, out of the wet bag, our Burberries become like a coat of mail at once, and our heads and beards get iced-up with the moisture when breathing on the march. There is half a gale blowing dead in our teeth all the time. We hope to reach within 100 geographical miles of the Pole; under the circumstances we can expect to do very little more. I am confident that the Pole lies on the great plateau we have discovered, miles and miles from any outstanding land. The temperature to-night is minus 24° Fahr.

January 5. To-day head wind and drift again, with 50° of frost, and a terrible surface. We have been marching through 8 in. of snow, covering sharp sastrugi, which plays havoc with our feet, but we have done 13⅓ geographical miles, for we increased our food, seeing that it was absolutely necessary to do this to enable us to accomplish anything. I realise that the food we have been having has not been sufficient to keep up our strength, let alone supply the wastage caused by exertion, and now we must try to keep warmth in us, though our strength is being used up. Our temperatures at 5 A.M. were 94° Fahr. We got away at 7 A.M. sharp and marched till noon, then from 1 P.M. sharp till 6 P.M. All being in one tent makes our camp-work slower, for we are so cramped for room, and we get up at 4.40 A.M. so as to get away by 7 A.M. Two of us have to stand outside the tent at night until things are squared up inside, and we find it cold work. Hunger grips us hard, and the food-supply is very small. My head still gives me great trouble. I began by wishing that my worst enemy had it instead of myself, but now I don't wish even my worst enemy to have such a headache; still, it is no use talking about it. Self is a subject that most of us are fluent on. We find the utmost difficulty in carrying through the day, and we can only go for two or three more days. Never once has the temperature been above zero since we got on to the plateau, though this is the height of summer. We have done our best, and we thank God for having allowed us to get so far.

January 6. This must be our last outward march with the sledge and camp equipment. To-morrow we must leave camp with some food, and push as far south as possible, and then plant the flag. To-day's story is 57° of frost, with a strong blizzard and high drift; yet we marched 13¼ geographical miles through soft snow, being helped by extra food. This does not mean full rations, but a bigger ration than we have been having lately. The pony maize is all finished. The most trying day we have yet spent, our fingers and faces being frost-bitten continually. To-morrow we will rush south with the flag. We are at 88° 7′ South to-night. It is our last outward march. Blowing hard to-night. I would fail to explain my feelings if I tried to write them down, now that the end has come. There is only one thing that lightens the disappointment, and that is the feeling that we have done all we could. It is the forces of nature that have prevented us from going right through. I cannot write more.

January 7. A blinding, shrieking blizzard all day, with the temperature ranging from 60° to 70° of frost. It has been impossible to leave the tent, which is snowed up on the lee side. We have been lying in our bags all day, only warm at food time, with fine snow making through the walls of the worn tent and covering our bags. We are greatly cramped. Adams is suffering from cramp every now and then. We are eating our valuable food without marching. The wind has been blowing eighty to ninety miles an hour. We can hardly sleep. To-morrow I trust this will be over. Directly the wind drops we march as far south as possible, then plant the flag, and turn homeward. Our chief anxiety is lest our tracks may drift up, for to them we must trust mainly to find our depot; we have no land bearings in this great plain of snow. It is a serious risk that we have taken, but we had to play the game to the utmost, and Providence will look after us.

January 8. Again all day in our bags, suffering considerably physically from cold hands and feet, and from hunger, but more mentally, for we cannot get on south, and we simply lie here shivering. Every now and then one of our party's feet go, and the unfortunate beggar has to take his leg out of the sleeping-bag and have his frozen foot nursed into life again by placing it inside the shirt, against the skin of his almost equally unfortunate neighbour. We must do something more to the south, even though the food is going, and we weaken lying in the cold, for with 72° of frost the wind cuts through our thin tent, and even the drift is finding its way in and on to our bags, which are wet enough as it is. Cramp is not uncommon every now and then, and the drift all round the tent has made it so small that there is hardly room for us at all. The wind has been blowing hard all day; some of the gusts must be over seventy or eighty miles an hour. This evening it seems as though it were going to ease down, and directly it does we shall be up and away south for a rush. I feel that this march must be our limit. We are so short of food, and at this high altitude, 11,600 ft., it is hard to keep any warmth in our bodies between the scanty meals. We have nothing to read now, having depoted our little books to save weight, and it is dreary work lying in the tent with nothing to read, and too cold to write much in the diary.

January 9. Our last day outwards. We have shot our bolt, and the tale is latitude 88° 23′ South, longitude 162° East. The wind eased down at 1 A.M., and at 2 A.M. we were up and had breakfast. At 4 A.M. started south, with the Queen's Union Jack, a brass cylinder containing stamps and documents to place at the furthest south point, camera, glasses, and compass. At 9 A.M. we were in 88° 23′ South, half running and half walking over a surface much hardened by the recent blizzard. It was strange for us to go along without the nightmare of a sledge dragging behind us. We hoisted Her Majesty's flag and the other Union Jack afterwards, and took possession of the plateau in the name of His Majesty. While the Union Jack blew out stiffly in the icy gale that cut us to the bone, we looked south with our powerful glasses, but could see nothing but the dead white snow plain. There was no break in the plateau as it extended towards the Pole, and we feel sure that the goal we have failed to reach lies on this plain. We stayed only a few minutes, and then, taking the Queen's flag and eating our scanty meal as we went, we hurried back and reached our camp about 3 P.M. We were so dead tired that we only did two hours' march in the afternoon and camped at 5.30 P.M. The temperature was minus 19° Fahr. Fortunately for us, our tracks were not obliterated by the blizzard; indeed, they stood up, making a trail easily followed. Homeward bound at last. Whatever regrets may be, we have done our best.

LT. SHIRASE'S CALLING CARD

from *Nankyokuki* (1913)

Nobu Shirase

In January 1912, the race to the Pole between Scott's unwieldy British expedition and Amundsen's sleek Norwegian team was already working itself out, deep in the interior of the continent. But these two competing heavyweights were not the only ones in Antarctica. Almost unnoticed, Nobu Shirase's Japanese "Dash Patrol" had moored near to Amundsen's ship in the Bay of Whales, and were unloading supplies onto the ice shelf for their own polar adventure. Very enthusiastic, very patriotic, and very much novices in the polar environment, Shirase's group depended on the dog-sledding skills of two Ainu he had brought along, from the indigenous community of the northern island of Hokkaido. With the guidance—not always acknowledged—of these experts, the Dash Patrol in fact managed to set a sledging speed record during their brief foray south. But the interest of Shirase's story lies not so much in what was achieved, as in how it filters the common themes of the Heroic Age through a strikingly different sensibility, and claims Antarctica as the imaginative possession of a world wider than Europe.

AT 2 A.M. we made out a faint pale grey line on the horizon to port, which we thought must be either a mountain or a cloud. Not until 4:20 A.M. did we see that it was actually the undulating wall of the Great Ice Barrier itself. As we drew nearer we could see it more and more clearly. At first sight the Barrier appeared as a sweeping crescent of ice about 150 *shaku* high; it was like a series of pure white folding screens, or perhaps a gigantic white snake at rest. The sea was fortunately clear of ice, and as we looked about us we were surrounded by the rippling greasy blue-green waters so characteristic of the Ross Sea. The ship was by now surging forward, with all sails set and engine full ahead.

We were now at a point some thirty miles east of the Bay of Whales, so this led to a suggestion that we change our plans and make a landing in King

Edward VII Land. This was decided upon, and bringing the ship to within a mile of the Barrier, we sailed eastward along it in search of a place to land.

Looking to starboard there were numerous caves and fissures in the Barrier, the nearer ones reflecting a deep blue onto the sea, and the further ones black as a scattering of brush strokes on the surface. From projections along the walls of these fissures hung icicles, like glass rods suspended in festoons, and these caught the lapis lazuli reflection of the ever-moving waters below.

Suddenly, a gunshot echoed round the deck. It was Takeda testing the strength of the ice by firing a shot at an overhanging section which looked about to fall. The results only led us to conclude that for all its delicate appearance we were looking at an extremely tough form of ice. Takeda leant on his rifle and sighed. "If only we had a 120 mm cannon! Just one shot with that would knock such a hole in this ice that we could land wherever we wanted."

At 7:30 A.M. we rounded a promontory and came upon a small bay approximately two miles wide and running inland for about a mile. At the far end of the bay the ice rose steeply in a series of ridges, on one side the Barrier formed a cliff, and on the other the shore was low and flat rather like a harbour quay. It seemed a suitable place to land, and we brought *Kainan-maru* into the bay.

As soon as the ship hove to the stern the boat was lowered and Takeda, Tsuchiya, Onitarō Watanabe and Hanamori all clambered aboard. Lt. Shirase instructed them to reconnoitre the area around the bay, and they rowed off. With perfect timing, the sun came out from behind the clouds and the silver of the Barrier on both sides of us shone brilliantly on the blue-green waves. The boat slowly neared the shore, the red of its ensign plain to see as it fluttered on high. *Kainan-maru* followed it into the bay.

On the right-hand side of the bay was a big ice cave which cast a deep blue reflection on the water, and just at that moment we spotted a large seal lying on the ice at the entrance to this cave. We called out to the men on the boat, and they made fast to a corner of the ice and went for the seal, full of courage and shouting with excitement now that they had at last found a worthy opponent. Their four poles came down on the seal again and again, but the enemy bared its fangs and fought back. Its mouth opened wide and it stuck out its blood-red tongue like some huge serpent. The four men finally subdued their foe by a concerted attack from all sides during which it was surrounded and beaten down in a veritable frenzy of blows. This battle lasted a full thirty minutes, and all the combatants were showered with blood and drenched with sweat.

No sooner was the battle over than the four warriors shouted a loud *Banzai!* and started climbing the steep slope. They disappeared behind one of the ridges and then reappeared further up, moving towards the left. Advancing in single file, frequently stumbling and getting to their feet again, but all the while going onwards and upwards, they soon reached the top of the Barrier, from where they continued to march inland.

Back on the ship we waited impatiently for a sign of the four men. Forty minutes later, we saw their distant silhouettes moving on the Barrier. We could see the red of their raised ensign as they drew up in a line beneath it. We could hear them in the distance as they shouted *Banzai!* three times, and we saw four pairs of arms raised and lowered in salute as they shouted. They had just completed this little ceremony and started back when the seal with which they had done battle earlier suddenly revived. Slowly lifting its head it took a quiet look around and seeing that its enemies had gone started crawling towards the water's edge. "Look out! The seal's come back to life! Oh no! It's going to get away!" We all shouted and gesticulated wildly from the deck, but our warriors had only just set off on their way back and couldn't possibly have got there in time to stop it.

We could only stamp our feet in frustration, and shout "Oh! What a shame!" but it was all to no avail. The seal was already nearing the water's edge and was just about to dive in when Muramatsu could stand it no longer, and raising his gun took a pot-shot at it. Unfortunately he missed. Startled by the report, the seal quickened its pace and finally disappeared into the depths of the ocean. It seems that seals recover completely the moment they are back in the sea, however badly injured they may have seemed when out of the water. The creatures would appear to enjoy some kind of divine protection.

The boat returned with the four men at around 9:20 A.M. and Takeda gave Lt. Shirase a detailed description of what they had found.

From the landing place, which had been excellent, they had seen what looked like a whole range of icy peaks in the distance. When they had gone a little further these turned out to be not distant mountains but the end of a vast glacier which attended for tens of *ri* [200 km] inland. Such terrain would make the Dash Patrol well-nigh impossible. One side of the glacier was as precipitous as the other and there were crevasses running in all directions. Many of these were covered with a thin crust of ice which made the surface look safe where in fact it was extremely dangerous. Hanamori, who was in the lead, put his foot right through this icy crust and fell into a crevasse about three *skaku* [1 m] across and stretching north-south for many *ri.* Luckily Tsuchiya was right beside him and managed to haul him out, otherwise he would certainly have perished, but it was an extremely dangerous place, and after that incident they thought it best to give up all idea of landing there. They had lowered Lt. Shirase's visiting card into a deep crack in the ice, and come straight back.

We were all very disappointed, not least Lt. Shirase. However, Takeda urged us not to lose heart. He was convinced that our efforts in exploring the bay had not been wasted, and spoke with great feeling about the solemn grandeur of the Barrier. They had seen icicles the size of elephant's tusks, and just the opportunity of studying these at close quarters had made it all worth while.

Lt. Shirase named the huge glacier they had surveyed Yonin Hyōga and decided to call this bay Kainan Bay. The position is 78° 17′ S and 162° 50′ W. When the naming ceremony was over *Kaịnan-maru* put back out to sea.

After a general discussion about where to land, it was agreed that only the Main Landing Party would go ashore at the Bay of Whales, and that *Kainan-maru* would then take a Coastal Party back to explore King Edward VII Land. We altered course at 10 A.M., and started to steam westwards.

During the afternoon the sky gradually clouded over. The wind blew mournfully, the temperature dropped, and snow squalls drove sporadically across the deck. Then suddenly we realised that the dark shape we could just make out about twenty miles ahead was a ship. "Look! Pirates!" said one of the sailors to Yoshino, who happened to be on deck.

Total panic ensued! Yoshino was so astonished he went round telling the whole ship, and everyone crowded on deck in disbelief. As we drew nearer, we could see that it was a lone sailing ship, but we were still uncertain where she was from. Just as the Japanese flag flew from the mast of *Kainan-maru,* this vessel was also flying a flag, but because of the distance and poor visibility we were unable to see it properly.

Eventually when only about five miles remained between the two ships, we managed to identify their flag. It was a blue cross on a red ground, and we were now in no doubt that she was *Fram,* the ship of the Norwegian Polar Expedition.

Soon after this we sailed into the Bay of Whales as planned. Unlike Kainan Bay which we had just left, the sea here was quite frozen over and we were unable to take *Kainan-maru* very far in. There was no helping it, so at

10 P.M. we rammed *Kainan-maru*'s bow into the ice edge to the east of *Fram* and approximately one and a half miles west of the eastern limit of the bay, and there we moored her to the ice.

On examining our surroundings we saw that the sea ice extended all the way to where the Barrier rose at the end of the bay, which was about fifteen miles from the ice edge. The bay was indeed enormous, and we had a truly panoramic view. The sea ice was thick and smooth and stretched rigid across from east to west, linking the ice cliffs of the Barrier which rose on either hand. Seals and penguins were dotted around on this vast white plain, and skuas and snow petrels swooped to left and right. Pods of whales swam about rippling the surface swell of the silent deep-blue sea, coming up close to the ship and emitting loud and eerie sounds as they blew fountains of water into the air.

The day had started snowy, but later as the dark clouds cleared it had turned into one of bright sunshine. Although it was almost midnight, the south polar sun shone tirelessly on and on, and we were blessed with light and warmth around the clock. This enabled us to carry out our work aboard according to plan.

No sooner had we made fast than Lt. Shirase gave orders for all expedition members to get ready to go ashore and explore. At this everyone's spirits rose, and we made an immediate start on all the rigorous preparations required. It was 11 P.M. when we left the ship and set off. As the crow flies it was about two miles from *Kainan-maru* to the Barrier, and our way lay over a sheet of ice which was three or four *shaku* [1 m] thick, but with only about five or six *sun* [18 cm] showing above water. Its level surface promised well for our advance.

However, it seemed that the season was approaching when this sheet of ice would float out to sea, and there were several cracks running in various directions, along the edges of which small embankments had formed which reminded one of the dykes separating paddy fields back home. It looked as if a sudden gust of wind from the south would blow it all straight out into the ocean. Moreover, some of the cracks were about seven or eight *sun* [22 cm] wide and we could see the deep blue of the sea and in it a large number of minnow-like fish swimming around. We had to exercise the greatest caution as we advanced and all felt that we really were walking on thin ice.

Our group advanced in a straight line towards the Barrier ahead, our feet making strange and ghostly squeaks as we trod on the half-frozen snow. We spied two or three seals along the way. Their coats, spotted but dark and glossy as lacquer, together with their general corpulence, were sufficient for us to surmise that there were plenty of fish in the surrounding seas. These seals seemed to be of a somewhat different species to those found in northern waters, and also to those we had caught so far.

We were all dressed alike with three shirts and two pairs of long underpants, over which we wore our uniforms complete with hoods, snow goggles, ear-muffs and gloves. On our feet we wore soft felt boots with metal crampons, and we each carried a long bamboo pole to steady us as we walked. As we advanced further and further onto the ice, the heat of the sun beat down on us from above and reflected up at us from the snow, and we began to feel uncomfortably hot. Soon the sweat was streaming off us and we were completely soaked. Those with overcoats took them off and walked on panting, with the coats slung over their shoulders. To make matters worse, the steam rising from our bodies condensed so as to form an extremely disagreeable mist on the dark lenses of the goggles we wore as protection against snowblindness. However, as we couldn't take them off, we just had to carry on walking in increasing discomfort, occasionally mopping the lenses as we went.

Eventually, after marching across the sea ice for about an hour, we reached the foot of the Barrier. A great wall of ice towered steeply more than two hundred *shaku* [60 m] above us with blue and purple lights rising like flames overhead, and as we gazed up in horror our skin turned to goose flesh at the sight that met our eyes. We could see that anything much stronger than a breeze would bring some of the steepest parts crashing straight down. Protruding blocks of ice, avalanches past and yet to come, precariously jutting chunks and lumps, all these contrasted with strangely carved shapes of a polished chalky whiteness. The blocks of ice continually falling from this cliff had smashed the sea ice below to such an extent that in places you could even see the water beneath it. Some of the floes were ratted on top of one another and as the great swells of the ocean surged to and fro beneath them they moved slowly up and down, sending strange rending noises like ripping silk echoing across the silent land and the sky. From holes in the ice seals occasionally showed their heads, baring their cruel fangs as they came up for air. Such awesome sights and stirrings are unique to the polar regions.

As you will have gathered from this description, the climb up the Barrier looked like being an extremely difficult one. However, our position as we attempted to walk along beneath the cliffs was no less dangerous.

We therefore decided to press on. Muramatsu, Yoshino and Hanamori took the lead and started to advance. They struck out with caution, their poles in their hands, picking their way across the rotten sea ice, scrambling on their bellies to climb atop a huge block of fallen ice, then leaping a yawning chasm. Picking their way carefully to left and right they slowly advanced until at last they reached the foot of the cliff itself. Ahead of them a recent mighty avalanche had left a terrifying scar, and this was where their labours really started. The ice ahead was steep, smooth and slippery, but before they could even start their assault on the cliff they had to cross a deep crevasse which barred their way. Glancing up they saw above them an overhanging section of the Barrier which showed every sign of imminent collapse, and they knew they were in mortal danger. To put it bluntly, the slightest lapse of attention and their fates would be sealed, either crushed beneath a block of ice or sent plummeting into the depths of a crevasse. However, only by surmounting the obstacle of the ice cliff would there be any further advance, and as they had already pledged their very lives to this venture they gathered their courage and determination in both hands and fought with all their might, brandishing their snow-shovels as weapons. The method was for one man to forge ahead and dig a pathway through the snow and ice, while the others waited behind him holding the rope attached to his waist as a safety precaution in case he fell. They did not dare to raise their voices. They took their turns to lead the way with apprehensive awe, as if looking into the depths of a bottomless pool, or walking on tip-toe across very thin ice.

However, their patient labours brought them step by step towards the top. On reaching a smooth plateau of ice they looked down, and saw the narrow path which they had cut snaking endlessly through the snow like a sheep's intestines. Then, turning around, they suddenly realised that they had at last arrived on the Barrier itself! Without a moment's pause to catch their breath or mop away the sweat which now drenched their bodies, they shouted loud *Banzais!,* raising both hands high in triumphal salute. The second and third units echoed their *Banzais!* from below, and soon to a succession of *Banzai! Banzai!* their fellow explorers emerged to join them at the top. Glancing at their watches they saw it was precisely midnight.

A minute later it was the morning of the following day, the 17th. As we all looked back at the way we had come, we saw the blue sea lying in an almost flat calm, the white ice floes scattered on its surface, and the two ships, *Kainan-maru* and *Fram,* floating in lonely isolation alongside the expanse of sea ice which covered the entire bay. This was a *sumie* world painted in Indian ink on white paper. On the ice around *Kainan-maru* we could just make out the black shapes of people dotted about and moving hither and thither to the sound of occasional gun shots, and surmised that the ship's crew were out and about on the ice, hunting penguins, seals and suchlike to dissipate the weariness of the long voyage, like little birds let out of their cage.

Turning to look in the other direction, we saw a boundless plain of white ice stretching undisturbed into infinity, meeting the blue sky and continuing beyond. Though we could sense the many secrets hidden in its depths, there was not a shadow to be seen. The sun was reflected off the white snow with dazzling brightness, and we were all struck to the very heart by a feeling of awe.

THE WINTER JOURNEY

from *The Worst Journey in the World* (1922)

Apsley Cherry-Garrard

Probably the greatest piece of writing in all exploration literature, Cherry-Garrard's Worst Journey *is an account of the whole of Scott's catastrophic return to Antarctica, the* Terra Nova *expedition of 1910–13, as witnessed by a shy, impressionable, intensely short-sighted "gentleman volunteer," whose experiences left him grief-struck almost to the point of madness. But the core of the book is the "worst journey" itself, this unforgettable description of a side trip made in the subzero stillness of the winter before Scott started for the Pole. Cherry-Garrard and his companions, Edward Wilson ("Bill") and Herbert Bowers ("Birdie"), set off to collect the eggs of the Emperor Penguin, which only nests at the darkest point of the polar year. The near-impossibility of the route, given their primitive polar equipment, made the sixty-mile walk into a miniature epic of endurance. Wilson and Bowers would both die with Scott on the polar journey; but here, Cherry-Garrard remembers his friends at a time of paradoxical triumph, rising to the occasion with a devoted, graceful stoicism. No one in this very British trio ever curses, snaps or grumbles. Understatement does all their screaming for them.*

THE HORROR OF the nineteen days it took us to travel from Cape Evans to Cape Crozier would have to be re-experienced to be appreciated; and any one would be a fool who went again: it is not possible to describe it. The weeks which followed them were comparative bliss, not because later our conditions were better—they were far worse—but because we were callous. I for one had come to that point of suffering at which I did not really care if only I could die without much pain. They talk of the heroism of the dying—they little know—it would be so easy to die, a dose of morphia, a friendly crevasse, and blissful sleep. The trouble is to go on . . .

It was the darkness that did it. I don't believe minus seventy temperatures would be bad in daylight, not comparatively bad, when you could see where you were going, where you were stepping, where the sledge straps were, the cooker, the primus, the food; could see your footsteps lately trodden deep into the soft snow that you might find your way back to the rest of your load; could see the lashings of the food bags; could read a compass without striking three or four different boxes to find one dry match; could read your watch to see if the blissful moment of getting out of your bag was come without groping in the snow all about; when it would not take you five minutes to lash up the door of the tent, and five hours to get started in the morning . . .

But in these days we were never less than four hours from the moment when Bill cried "Time to get up" to the time when we got into our harness. It took two men to get one man into his harness, and was all they could do, for the canvas was frozen and our clothes were frozen until sometimes not even two men could bend them into the required shape.

The trouble is sweat and breath. I never knew before how much of the body's waste comes out through the pores of the skin. On the most bitter days, when we had to camp before we had done a four-hour march in order to nurse back our frozen feet, it seemed that we must be sweating. And all this sweat, instead of passing away through the porous wool of our clothing and gradually drying off us, froze and accumulated. It passed just away from our flesh and then became ice: we shook plenty of snow and ice down from inside our trousers every time we changed our foot-gear, and we could have shaken it from our vests and from between our vests and shirts, but of course we could not strip to this extent. But when we got into our sleeping-bags, if we were fortunate, we became warm enough during the night to thaw this ice: part remained in our clothes, part passed into the skins of our sleeping-bags, and soon both were sheets of armour-plate.

As for our breath—in the daytime it did nothing worse than cover the lower parts of our faces with ice and solder our balaclavas tightly to our heads. It was no good trying to get your balaclava off until you had had the primus going quite a long time, and then you could throw your breath about if you wished. The trouble really began in your sleeping-bag, for it was far too cold to keep a hole open through which to breathe. So all night long our breath froze into the skins, and our respiration became quicker and quicker as the air in our bags got fouler and fouler: it was never possible to make a match strike or burn inside our bags!

Of course we were not iced up all at once: it took several days of this kind of thing before we really got into big difficulties on this score. It was not until I got out of the tent one morning fully ready to pack the sledge that I realized the possibilities ahead. We had had our breakfast, struggled into our foot-gear, and squared up inside the tent, which was comparatively warm. Once outside, I raised my head to look round and found I could not move it back. My clothing had frozen hard as I stood—perhaps fifteen seconds. For four hours I had to pull with my head stuck up, and from that time we all took care to bend down into a pulling position before being frozen in.

By now we had realized that we must reverse the usual sledging routine and do everything slowly, wearing when possible the fur mitts which fitted over our woollen mitts, and always stopping whatever we were doing, directly we felt that any part of us was getting frozen, until the circulation was restored. Henceforward it was common for one or other of us to leave the other two to continue the camp work while he stamped about in the snow, beat his arms, or nursed some exposed part. But we could not restore the circulation of our feet like this—the only way then was to camp and get some hot water into ourselves before we took our foot-gear off. The difficulty was to know whether our feet were frozen or not, for the only thing we knew for certain was that we had lost all feeling in them. Wilson's knowledge as a doctor came in here: many a time he had to decide from our descriptions of our feet whether to camp or to go on for another hour. A wrong decision meant disaster, for if one of us had been crippled the whole party would have been placed in great difficulties. Probably we should all have died.

[. . .]

We were now getting into that cold bay which lies between the Hut Point Peninsula and Terror Point. In consequence of the lack of high winds the surface of the snow is never swept and hardened and polished as elsewhere: it was now a mass of the hardest and smallest snow crystals, to pull through which in cold temperatures was just like pulling through sand. I have spoken elsewhere of Barrier surfaces, and how, when the cold is very great, sledge runners cannot melt the crystal points but only advance by rolling them over and over upon one another. That was the surface we met on this journey, and in soft snow the effect is accentuated. Our feet were sinking deep at every step.

And so when we tried to start on June 30 we found we could not move both sledges together. There was nothing for it but to take one on at a time and come back for the other. This has often been done in daylight when the only risks run are those of blizzards which may spring up suddenly and obliterate tracks. Now in darkness it was more complicated. From 11 A.M. to 3 P.M. there was enough light to see the big holes made by our feet, and we took on one sledge, trudged back in our tracks, and brought on the second. Bowers used to toggle and untoggle our harnesses when we changed sledges. Of course in this relay work we covered three miles in distance for every one mile forward, and even the single sledges were very hard pulling. When we lunched the temperature was –61°. After lunch the little light had gone, and we carried a naked lighted candle back with us when we went to find our second sledge. It was the weirdest kind of procession, three frozen men and a little pool of light. Generally we steered by Jupiter, and I never see him now without recalling his friendship in those days.

We were very silent, it was not very easy to talk: but sledging is always a silent business. I remember a long discussion which began just now about cold snaps—was this the normal condition of the Barrier, or was it a cold snap?—what constituted a cold snap? The discussion lasted about a week. Do things slowly, always slowly, that was the burden of Wilson's leadership: and every now and then the question, Shall we go on? and the answer Yes. "I think we are all right as long as our appetites are good," said Bill. Always patient, self-possessed, unruffled, he was the only man on earth, as I believe, who could have led this journey.

That day we made 3¼ miles, and travelled 10 miles to do it. The temperature was –66° when we camped, and we were already pretty badly iced up. That was the last night I lay (I had written slept) in my big reindeer bag without the lining of eider-down which we each carried. For me it was a very bad night: a succession of shivering fits which I was quite unable to stop, and which took possession of my body for many minutes at a time until I thought my back would break, such was the strain placed upon it. They talk of chattering teeth: but when your body chatters you may call yourself cold. I can only compare the strain to that which I have been unfortunate enough to see in a case of lock-jaw.

We did the same relay work on July 1, but found the pulling still harder; and it was all that we could do to move the one sledge forward. From now onwards Wilson and I, but not to the same extent Bowers, experienced a curious optical delusion when returning in our tracks for the second sledge. I have said that we found our way back by the light of a candle, and we found it necessary to go back in our same footprints. These holes became to our tired brains not depressions but elevations: hummocks over which we stepped, raising our feet painfully and draggingly. And then we remembered, and said what fools we were, and for a while we compelled ourselves to walk through these phantom hills. But it was no lasting good, and as the days passed we realized that we must suffer this absurdity, for we could not do anything else. But of course it took it out of us.

During these days the blisters on my fingers were very painful. Long before my hands were frost-bitten, or indeed anything but cold, which was of course a normal thing, the matter inside these big blisters, which rose all down my fingers with only a skin between them, was frozen into ice. To handle the cooking gear or the food bags was agony, to start the primus was worse and when, one day, I was able to prick six or seven of the blisters after supper and let the liquid matter out, the relief was very great. Every night after that I treated such others as were ready in the same way until they gradually disappeared. Sometimes it was difficult not to howl.

I *did* want to howl many times every hour of these days and nights, but I invented a formula instead, which I repeated to myself continually. Especially, I remember, it came in useful when at the end of the march with my feet frost-bitten, my heart beating slowly, my vitality at its lowest ebb, my body solid with cold, I used to seize the shovel and go on digging snow on to the tent skirting while the cook inside was trying to light the primus. "You've got it in the neck—stick it—stick it—you've got it in the neck," was the refrain, and I wanted every little bit of encouragement it would give me: then I would find myself repeating "Stick it—stick it—stick it—stick it," and then "You've got it in the neck." One of the joys of summer sledging is that you can let your mind wander thousands of miles away for weeks and weeks. Oates used to provision his little yacht (there was a pickled herring he was going to have): I invented the compactest little revolving bookcase which was going to hold not books, but pemmican and chocolate and biscuit and cocoa and sugar, and have a cooker on the top, and was going to stand always ready to quench my hunger when I got home: and we visited restaurants and theatres and grouse moors, and we thought of a pretty girl, or girls, and . . . But now that was all impossible. Our conditions forced themselves upon us without pause: it was not possible to think of anything else. We got no respite. I found it best to refuse to let myself think of the past or the future—to live only for the job of the moment, and to compel myself to think only how to do it most efficiently. Once you let yourself imagine . . .

[. . .]

I have met with amusement people who say, "Oh, we had minus fifty temperatures in Canada; they didn't worry *me*," or "I've been down to minus sixty something in Siberia." And then you find that they had nice dry clothing, a nice night's sleep in a nice aired bed, and had just walked out after lunch for a few minutes from a nice warm hut or an overheated train. And they look back upon it as an experience to be remembered. Well! of course as an experience of cold this can only be compared to eating a vanilla ice with hot chocolate cream after an excellent dinner at Claridge's. But in our present state we began to look upon minus fifties as a luxury which we did not often get.

That evening, for the first time, we discarded our naked candle in favour of the rising moon. We had started before the moon on purpose, but as we shall see she gave us little light. However, we owed our escape from a very sticky death to her on one occasion.

It was a little later on when we were among crevasses, with Terror above us, but invisible, somewhere on our left, and the Barrier pressure on our right. We were quite lost in the darkness, and only knew that we were running downhill, the sledge almost catching our heels. There had been no light all day, clouds obscured the moon, we had not seen her since yesterday. And quite suddenly a little patch of clear sky drifted, as it were, over her face, and she showed us three paces ahead a great crevasse with just a shining icy lid not much thicker than glass. We should all have walked into it, and the sledge would certainly have followed us down. After that I felt we had a chance of pulling through: God could not be so cruel as to have saved us just to prolong our agony.

But at present we need not worry about crevasses; for we had not reached the long stretch where the moving Barrier, with the weight of many hundred miles of ice behind it, comes butting up against the slopes of Mount Terror, itself some eleven thousand feet high. Now we were still plunging ankle-deep in the mass of soft sandy snow which lies in the windless area. It seemed to have no bottom at all, and since the snow was much the same temperature as the air, our feet, as well as our bodies, got colder and colder the longer we marched: in ordinary sledging you begin to warm up after a quarter of an hour's pulling, here it was just the reverse. Even now I find myself unconsciously kicking the toes of my right foot against the heel of my left: a habit I picked up on this journey by doing it every time we halted. Well no. Not always. For there was one halt when we just lay on our backs and gazed up into the sky, where, so the others said, there was blazing the most wonderful aurora they had ever seen. I did not see it, being so near-sighted and unable to wear spectacles owing to the cold. The aurora was always before us as we travelled east, more beautiful than any seen by previous expeditions wintering in McMurdo Sound, where Erebus must have hidden the most brilliant displays.

Now most of the sky was covered with swinging, swaying curtains which met in a great whirl overhead: lemon yellow, green and orange.

We got forward only 2½ miles, and by this time I had silently made up my mind that we had not the ghost of a chance of reaching the penguins. I am sure that Bill was having a very bad time these nights, though it was an impression rather than anything else, for he never said so. We knew we did sleep, for we heard one another snore, and also we used to have dreams and nightmares; but we had little consciousness of it, and we were now beginning to drop off when we halted on the march.

Our sleeping-bags were getting really bad by now, and already it took a long time to thaw a way down into them at night. Bill spread his in the middle, Bowers was on his right, and I was on his left. Always he insisted that I should start getting my legs into mine before *he* started: we were rapidly cooling down after our hot supper, and this was very unselfish of him. Then came seven shivering hours and first thing on getting out of our sleeping-bags in the morning we stuffed our personal gear into the mouth of the bag before it could freeze: this made a plug which when removed formed a frozen hole for us to push into as a start in the evening.

We got into some strange knots when trying to persuade our limbs into our bags, and suffered terribly from cramp in consequence. We would wait and rub, but directly we tried to move again down it would come and grip our legs in a vice. We also, especially Bowers, suffered agony from cramp in the stomach. We let the primus burn on after supper now for a time—it was the only thing which kept us going—and when one who was holding the primus was seized with cramp we hastily took the lamp from him until the spasm was over. It was horrible to see Birdie's stomach cramp sometimes: he certainly got it much worse than Bill or I. I suffered a lot from heartburn, especially in my bag at nights: we were eating a great proportion of fat and this was probably the cause. Stupidly I said nothing about it for a long time. Later when Bill found out, he soon made it better with the medical case.

Birdie always lit the candle in the morning—so called, and this was an heroic business. Moisture collected on our matches if you looked at them. Partly I suppose it was bringing them from outside into a comparatively warm tent; partly from putting boxes into pockets in our clothing. Sometimes it was necessary to try four or five boxes before a match struck. The temperature of the boxes and matches was about a hundred degrees of frost, and the smallest touch of the metal on naked flesh caused a frost-bite. If you wore mitts you could scarcely feel anything—especially since the tips of our fingers were already very callous. To get the first light going in the morning was a beastly cold business, made worse by having to make sure that it was at last time to get up. Bill insisted that we must lie in our bags seven hours every night.

In civilization men are taken at their own valuation because there are so many ways of concealment, and there is so little time, perhaps even so little understanding. Not so down South. These two men went through the Winter Journey and lived: later they went through the Polar Journey and died. They were gold, pure, shining, unalloyed. Words cannot express how good their companionship was.

Through all these days, and those which were to follow, the worst I suppose in their dark severity that men have ever come through alive, no single hasty or angry word passed their lips. When, later, we were sure, so far as we can be sure of anything, that we must die, they were cheerful, and so far as I can judge their songs and cheery words were quite unforced. Nor were they ever flurried, though always as quick as the conditions would allow in moments of emergency. It is hard that often such men must go first when others far less worthy remain.

There are those who write of Polar Expeditions as though the whole thing was as easy as possible. They are trusting, I suspect, in a public who will say, "What a fine fellow this is! we know what horrors he has endured, yet see, how little he makes of all his difficulties and hard-ships." Others have gone to the opposite extreme. I do not know that there is any use in trying to make a −18° temperature appear formidable to an uninitiated reader by calling it fifty degrees of frost. I want to do neither of these things. I am not going to pretend that this was anything but a ghastly journey, made bearable and even pleasant to look back upon by the qualities of my two companions who have gone. At the same time I have no wish to make it appear more horrible than it actually was: the reader need not fear that I am trying to exaggerate.

[. . .]

Luckily we were spared wind. Our naked candle burnt steadily as we trudged back in our tracks to fetch our other sledge, but if we touched metal for a fraction of a second with naked fingers we were frost-bitten. To fasten the strap buckles over the loaded sledge was difficult: to handle the cooker, or mugs, or spoons, the primus or oil can was worse. How Bowers managed with the meteorological instruments I do not know, but the meteorological log is perfectly kept. Yet as soon as you breathed near the paper it was covered with a film of ice through which the pencil would not bite. To handle rope was always cold and in these very low temperatures dreadfully cold work. The toggling up of our harnesses to the sledge we were about to pull, the untoggling at the end of the stage, the lashing up of our sleeping-bags in the morning, the fastening of the cooker to the top of the instrument box, were bad, but not nearly so bad as the smaller lashings which were now strings of ice. One of the worst was round the weekly food bag, and those round the pemmican, tea and butter bags inside were thinner still. But the real devil was the lashing of the tent door: it was like wire, and yet had to be tied tight. If you had to get out of the tent during the seven hours spent in our sleeping-bags you must tie a string as stiff as a poker, and re-thaw your way into a bag already as hard as a board. Our paraffin was supplied at a flash point suitable to low temperatures and was only a little milky: it was very difficult to splinter bits off the butter.

The temperature that night was −75.8°, and I will not pretend that it did not convince me that Dante was right when he placed the circles of ice below the circles of fire. Still we slept sometimes, and always we lay for seven hours. Again and again Bill asked us how about going back, and always we said no. Yet there was nothing I should have liked better: I was quite sure that to dream of Cape Crozier was the wildest lunacy. That day we had advanced 1½ miles by the utmost labour, and the usual relay work. This was quite a good march—and Cape Crozier is 67 miles from Cape Evans!

More than once in my short life I have been struck by the value of the man who is blind to what appears to be a common-sense certainty: he achieves the impossible. We never spoke our thoughts: we discussed the Age of Stone which was to come, when we built our cosy warm rock hut on the slopes of Mount Terror, and ran our stove with penguin blubber, and pickled little Emperors in warmth and dryness. We were quite intelligent people, and we must all have known that we were not going to see the penguins and that it was folly to go forward. And yet with quiet perseverance, in perfect friendship, almost with gentleness those two men led on. I just did what I was told.

[. . .]

Bill was anxious. It seems that Scott had twice gone for a walk with him during the Winter, and tried to persuade him not to go, and only finally consented on condition that Bill brought us all back unharmed: we were Southern Journey men. Bill had a tremendous respect for Scott, and later when we were about to make an effort to get back home over the Barrier, and our case was very desperate, he was most anxious to leave no gear behind at Cape Crozier, even the scientific gear which could be of no use to us and of which we had plenty more at the hut. "Scott will never forgive me if I leave gear behind," he said. It is a good sledging principle, and the party which does not follow it, or which leaves some of its load to be fetched in later is seldom a good one: but it is a principle which can be carried to excess.

And now Bill was feeling terribly responsible for both of us. He kept on saying that he was sorry, but he had never dreamed it was going to be as bad as this. He felt that having asked us to come he was in some way chargeable with our troubles. When leaders have this kind of feeling about their men they get much better results, if the men are good: if men are bad or even moderate they will try and take advantage of what they consider to be softness.

[. . .]

In the early morning of the next day snow began to fall and the fog was dense: when we got up we could see nothing at all anywhere. After the usual four hours to get going in the morning we settled that it was impossible to relay, for we should never be able to track ourselves back to the second sledge. It was with very great relief that we found we could move both sledges together, and I think this was mainly due to the temperature which had risen to −36°.

This was our fourth day of fog in addition to the normal darkness, and we knew we must be approaching the land. It would be Terror Point, and the fog is probably caused by the moist warm air coming up from the sea through the pressure cracks and crevasses; for it is supposed that the Barrier here is afloat.

I wish I could take you on to the great Ice Barrier some calm evening when the sun is just dipping in the middle of the night and show you the autumn tints on Ross Island. A last look round before turning in, a good day's march behind, enough fine fat pemmican inside you to make you happy, the homely smell of tobacco from the tent, a pleasant sense of soft fur and the deep sleep to come. And all the softest colours God has made are in the snow; on Erebus to the west, where the wind can scarcely move his cloud of smoke; and on Terror to the east, not so high, and more regular in form. How peaceful and dignified it all is.

That was what you might have seen four months ago had you been out on the Barrier plain. Low down on the extreme right or east of the land there was a black smudge of rock peeping out from great snow-drifts: that was the Knoll, and close under it were the cliffs of Cape Crozier, the Knoll looking quite low and the cliffs invisible, although they are eight hundred feet high, a sheer precipice falling to the sea.

It is at Cape Crozier that the Barrier edge, which runs for four hundred miles as an ice-cliff up to 200 feet high, meets the land. The Barrier is moving against this land at a rate which is sometimes not much less than a mile in a year. Perhaps you can imagine the chaos which it piles up: there are pressure ridges compared to which the waves of the sea are like a ploughed field.

[. . .]

We had therefore to find our way along the pressure to the Knoll, and thence penetrate *through* the pressure to the Emperors' Bay. And we had to do it in the dark.

Terror Point, which we were approaching in the fog, is a short twenty miles from the Knoll, and ends in a long snow-tongue running out into the Barrier. The way had been travelled a good many times in daylight, and Wilson knew there was a narrow path, free from crevasses, which skirted along between the mountain and the pressure ridges running parallel to it. But it is one thing to walk along a corridor by day, and quite another to try to do so at night, especially when there are no walls by which you can correct your course—only crevasses. Anyway, Terror Point must be somewhere close to us now, and vaguely in front of us was that strip of snow, neither Barrier nor mountain, which was our only way forward.

We began to realize, now that our eyes were more or less out of action, how much we could do with our feet and ears. The effect of walking in finnesko is much the same as walking in gloves, and you get a sense of touch which nothing else except bare feet could give you. Thus we could feel every small variation in surface, every crust through which our feet broke, every hardened patch below the soft snow. And soon we began to rely more and more upon the sound of our footsteps to tell us whether we were on crevasses or solid ground. From now onwards we were working among crevasses fairly constantly. I loathe them in full daylight when much can be done to avoid them, and when if you fall into them you can at any rate see where the sides are, which way they run and how best to scramble out; when your companions can see how to stop the sledge to which you are all attached by your harness; how most safely to hold the sledge when stopped; how, if you are dangling fifteen feet down in a chasm, to work above you to get you up to the surface again. And then our clothes were generally something like clothes. Even under the ideal conditions of good light, warmth and no wind, crevasses are beastly, whether you are pulling over a level and uniform snow surface, never knowing what moment will find you dropping into some bottomless pit, or whether you are rushing for the Alpine rope and the sledge, to help some companion who has disappeared. I dream sometimes now of bad days we had on the Beardmore and elsewhere, when men were dropping through to be caught up and hang at the full length of the harnesses and toggles many times in an hour. On the same sledge as myself on the Beardmore one man went down once head first, and another eight times to the length of his harness in 25 minutes. And always you wondered whether your harness was going to hold when the jerk came. But those days were a Sunday School treat compared to our days of blind-man's buff with the Emperor penguins among the crevasses of Cape Crozier.

Our troubles were greatly increased by the state of our clothes. If we had been dressed in lead we should have been able to move our arms and necks and heads more easily than we could now. If the same amount of icing had extended to our legs I believe we should still be there, standing unable to move: but happily the forks of our trousers still remained movable. To get into our canvas harnesses was the most absurd business. Quite in the early days of our journey we met with this difficulty, and somewhat foolishly decided not to take off our harness for lunch. The harnesses thawed in the tent, and froze back as hard as boards. Likewise our clothing was hard as boards and stuck out from our bodies in every imaginable fold and angle. To fit one board over the other required the united efforts of the would-be wearer and his two companions, and the process had to be repeated for each one of us twice a day. Goodness knows how long it took; but it cannot have been less than five minutes' thumping at each man.

As we approached Terror Point in the fog we sensed that we had risen and fallen over several rises. Every now and then we felt hard slippery snow under our feet. Every now and then our feet went through crusts in the surface. And then quite suddenly, vague, indefinable, monstrous, there loomed a something ahead. I remember having a feeling as of ghosts about as we untoggled our harnesses from the sledge, tied them together, and thus roped walked upwards on that ice. The moon was showing a ghastly ragged mountainous edge above us in the fog, and as we rose we found that we were on a pressure ridge. We stopped, looked at one another, and then *bang*—right under our feet. More bangs, and creaks and groans; for that ice was moving and splitting like glass. The cracks went off all round us, and some of them ran along for hundreds of yards. Afterwards we got used to it, but at first the effect was very jumpy. From first to last during this journey we had plenty of variety and none of that monotony which is inevitable in sledging over long distances of Barrier in summer. Only the long shivering fits following close one after the other all the time we lay in our dreadful sleeping-bags, hour after hour and night after night in those temperatures—they were as monotonous as could be. Later we got frost-bitten even as we lay in our sleeping-bags. Things are getting pretty bad when you get frost-bitten in your bag.

There was only a glow where the moon was; we stood in a moonlit fog, and this was sufficient to show the edge of another ridge ahead, and yet another on our left. We were utterly bewildered. The deep booming of the ice continued, and it may be that the tide has something to do with this, though we were many miles from the ordinary coastal ice. We went back, toggled up to our sledges again and pulled in what we thought was the right direction, always with that feeling that the earth may open underneath your feet which you have in crevassed areas. But all we found were more mounds and banks of snow and ice, into which we almost ran before we saw them. We were clearly lost. It was near midnight, and I wrote, "it may be the pressure ridges or it may be Terror, it is impossible to say,—and I should think it is impossible to move till it clears. We were steering N.E. when we got here and returned S. W. till we seemed to be in a hollow and camped."

The temperature had been rising from −36° at 11 A.M. and it was now −27°; snow was falling and nothing whatever could be seen. From under the tent came noises as though some giant was banging a big empty tank. All the signs were for a blizzard, and indeed we had not long finished our supper and were thawing our way little by little into our bags when the wind came away from the south. Before it started we got a glimpse of black rock, and knew we must be in the pressure ridges where they nearly join Mount Terror.

[. . .]

I will not say that I was entirely easy in my mind as we lay out that blizzard somewhere off Terror Point; I don't know how the others were feeling. The unearthly banging going on underneath us may have had something to do with it. But we were quite lost in the pressure and it might be the deuce and all to get out in the dark. The wind eddied and swirled quite out of its usual straightforward way, and the tent got badly snowed up: our sledge had disappeared long ago. The position was not altogether a comfortable one.

Tuesday night and Wednesday it blew up to force 10, temperature from −7° to +2°. And then it began to modify and get squally. By 3 A.M. on Thursday (July 13) the wind had nearly ceased, the temperature was falling and the stars were shining through detached clouds. We were soon getting our breakfast, which always consisted of tea, followed by pemmican. We soaked our biscuits in both. Then we set to work to dig out the sledges and tent, a big job taking several hours. At last we got started. In that jerky way in which I was still managing to jot a few sentences down each night as a record, I wrote:

"Did 7½ miles during day—seems a marvellous run—rose and fell over several ridges of Terror—in afternoon suddenly came on huge crevasse on one of these—we were quite high on Terror—moon saved us walking in—it might have taken sledge and all."

To do seven miles in a day, a distance which had taken us nearly a week in the past, was very heartening. The temperature was between −20° and −30° all day, and that was good too. When crossing the undulations which ran down out of the mountain into the true pressure ridges on our right we found that the wind which came down off the mountain struck along the top of the undulation, and flowing each way, caused a N.E. breeze on one side and a N.W. breeze on the other. There seemed to be wind in the sky, and the blizzard had not cleared as far away as we should have wished.

During the time through which we had come it was by burning more oil than is usually allowed for cooking that we kept going at all. After each meal was cooked we allowed the primus to burn on for a while and thus warmed up the tent. Then we could nurse back our frozen feet and do any necessary little odd jobs. More often we just sat and nodded for a few minutes, keeping one another from going too deeply to sleep. But it was running away with the oil. We started with 6 one-gallon tins (those tins Scott had criticized), and we had now used four of them. At first we said we must have at least two one-gallon tins with which to go back; but by now our estimate had come down to one full gallon tin, and two full primus lamps. Our sleeping-bags were awful. It took me, even as early in the journey as this, an hour of pushing and thumping and cramp every night to thaw out enough of mine to get into it at all. Even that was not so bad as lying in them when we got there.

Only −35° but "a very bad night" according to my diary. We got away in good time, but it was a ghastly day and my nerves were quivering at the end, for we could not find that straight and narrow way which led between the crevasses on either hand. Time after time we found we were out of our course by the sudden fall of the ground beneath our feet—in we went and then—"are we too far right?"—nobody knows—"well let's try nearer in to the mountain," and so forth! "By hard slogging 2¾ miles this morning—then on in thick gloom which suddenly lifted and we found ourselves under a huge great mountain of pressure ridge looking black in shadow. We went on, bending to the left, when Bill fell and put his arm into a crevasse. We went over this and another, and some time after got somewhere up to the left, and both Bill and I put a foot into a crevasse. We sounded all about and everywhere was hollow, and so we ran the sledge down over it and all was well." Once we got right into the pressure and took a longish time to get out again. Bill lengthened his trace out with the Alpine rope now and often afterwards, so he found the crevasses well ahead of us and the sledge: nice for us but not so nice for Bill. Crevasses in the dark *do* put your nerves on edge.

When we started next morning (July 15) we could see on our left front and more or less on top of us the Knoll, which is a big hill whose precipitous cliffs to seaward form Cape Crozier. The sides of it sloped down towards us, and pressing against its ice-cliffs on ahead were miles and miles of great pressure ridges, along which we had travelled, and which hemmed us in. Mount Terror rose ten thousand feet high on our left, and was connected with the Knoll by a great cup-like drift of wind-polished snow. The slope of this in one place runs gently out on to the corridor along which we had sledged, and here we turned and started to pull our sledges up. There were no crevasses, only the great drift of snow, so hard that we used our crampons just as though we had been on ice, and as polished as the china sides of a giant cup which it resembled. For three miles we slogged up, until we were only 150 yards from the moraine shelf where we were going to build our hut of rocks and snow. This moraine was above us on our left, the twin peaks of the Knoll were across the cup on our right; and here, 800 feet up the mountain side, we pitched our last camp.

We had arrived.

[. . .]

The view from eight hundred feet up the mountain was magnificent and I got my spectacles out and cleared the ice away time after time to look. To the east a great field of pressure ridges below, looking in the moonlight as if giants had been ploughing with ploughs which made furrows fifty or sixty feet deep: these ran right up to the Barrier edge, and beyond was the frozen Ross Sea, lying flat, white and peaceful as though such things as blizzards were unknown. To the north and north-east the Knoll. Behind us Mount Terror on which we stood, and over all the grey limitless Barrier seemed to cast a spell of cold immensity, vague, ponderous, a breeding-place of wind and drift and darkness. God! What a place!

TOPSY-TURVY

from *The South Pole* (1912)

Roald Amundsen

It was the Norwegian explorer Roald Amundsen's misfortune as a writer that he tended to make his adventures sound, perversely, a little too easy. They were not, of course. His victorious advance to the Pole in December 1911 was a wonder of tenacity, judgment, physical courage and intelligent risk-taking. Here, with the dangerous ice-falls of the Axel Heiberg glacier successfully navigated, we see him at full stretch, accelerating across the last bleak miles of the polar plateau to reach a place which (he engagingly admits) was not *the goal of all his dreams. But as an explorer he always valued skilful actions above skilful descriptions of those actions; and so he does not quite know now how to realize on the page—for us, reading—his own state of quizzical, physical exhilaration.*

THE WEATHER DID NOT continue fine for long. Next day (December 5) there was a gale from the north, and once more the whole plain was a mass of drifting snow. In addition to this there was thick falling snow, which blinded us and made things worse, but a feeling of security had come over us and helped us to advance rapidly and without hesitation, although we could see nothing. That day we encountered new surface conditions—big, hard snow-waves (*sastrugi*). These were anything but pleasant to work among, especially when one could not see them. It was of no use for us "forerunners" to think of going in advance under these circumstances, as it was impossible to keep on one's feet. Three or four paces was often the most we managed to do before falling down. The *sastrugi* were very high, and often abrupt; if one came on them unexpectedly, one required to be more than an acrobat to keep on one's feet. The plan we found to work best in these conditions was to let Hanssen's dogs go first; this was an unpleasant job for Hanssen, and for his dogs too, but it succeeded, and succeeded well. An upset here and there was, of course, unavoidable, but with a little patience the sledge was always righted again. The drivers had as much as they could do to support their sledges among these *sastrugi,* but while supporting the sledges, they had at the same time a support for themselves. It was worse for us who had no sledges, but by keeping in the wake of them we could see where the irregularities lay, and thus get over them. Hanssen deserves a special word of praise for his driving on this surface in such weather. It is a difficult matter to drive Eskimo dogs forward when they cannot see; but Hanssen managed it well, both getting the dogs on and steering his course by compass. One would not think it possible to keep an approximately right course when the uneven ground gives such violent shocks that the needle flies several times round the compass, and is no sooner still again than it recommences the same dance; but when at last we got an observation, it turned out that Hanssen had steered to a hair, for the observations and dead reckoning agreed to a mile. In spite of all hindrances, and of being able to see nothing, the sledge-meters showed nearly twenty-five miles.

December 6 brought the same weather: thick snow, sky and plain all one, nothing to be seen. Nevertheless we made splendid progress. The *sastrugi* gradually became levelled out, until the surface was perfectly smooth; it was a relief to have even ground to go upon once more. These irregularities that one was constantly falling over were a nuisance; if we had met with them in our usual surroundings it would not have mattered so much; but up here on the high ground, where we had to stand and gasp for breath every time we rolled over, it was certainly not pleasant.

That day we passed 88° S., and camped in 88° 9′ S. A great surprise awaited us in the tent that evening. I expected to find, as on the previous evening, that the boiling-point had fallen somewhat; in other words, that it would show a continued rise of the ground, but to our astonishment this was not so. The water boiled at exactly the same temperature as on the preceding day. I tried it several times, to convince myself that there was nothing wrong, each time with the same result. There was great rejoicing among us all when I was able to announce that we had arrived on the top of the plateau.

December 7 began like the 6th, with absolutely thick weather, but, as they say, you never know what the day is like before sunset. Possibly I might have chosen a better expression than this last—one more in agreement with the natural conditions—but I will let it stand. Though for several weeks now the sun had not set, my readers will not be so critical as to reproach me with inaccuracy. With a light wind from the north-east, we now went southward at a good speed over the perfectly level plain, with excellent going. The uphill work had taken it out of our dogs, though not to any serious extent. They had turned greedy—there is no denying that—and the half kilo of pemmican they got each day was not enough to fill their stomachs. Early and late they were looking for something—no matter what—to devour. To begin with they contented themselves with such loose objects as ski-bindings, whips, boots, and the like; but as we came to know their proclivities, we took such care of everything that they found no extra meals lying about. But that was not the end of the matter. They then went for the fixed lashings of the sledges, and—if we had allowed it—would very quickly have resolved the various sledges into their component parts. But we found a way of stopping that: every evening, on halting, the sledges were buried in the snow, so as to hide all the lashings. That was successful; curiously enough, they never tried to force the "snow rampart." I may mention as a curious thing that these ravenous animals, that devoured everything they came across, even to the ebonite points of our ski-sticks, never made any attempt to break into the provision cases. They lay there and went about among the sledges with their noses just on a level with the split cases, seeing and scenting the pemmican, without once making a sign of taking any. But if one raised a lid, they were not long in showing themselves. Then they all came in a great hurry and flocked about the sledges in the hope of getting a little extra bit. I am at a loss to explain this behaviour; that bashfulness was not at the root of it, I am tolerably certain.

During the forenoon the thick, grey curtain of cloud began to grow thinner on the horizon, and for the first time for three days we could see a few miles about us. The feeling was something like that one has on waking from a good nap, rubbing one's eyes and looking around. We had become so accustomed to the grey twilight that this positively dazzled us. Meanwhile, the upper layer of air seemed obstinately to remain the same and to be doing its best to prevent the sun from showing itself. We badly wanted to get a meridian altitude, so that we could determine our latitude. Since 86° 47′ S. we had had no observation, and it was not easy to say when we should get one. Hitherto, the weather conditions on the high ground had not been particularly favourable. Although the prospects were not very promising, we halted at 11 A.M. and made ready to catch the sun if it should be kind enough to look out. Hassel and Wisting used one sextant and artificial horizon, Hanssen and I the other set.

I don't know that I have ever stood and absolutely pulled at the sun to get it out as I did that time. If we got an observation here which agreed with our reckoning, then it would be possible, if the worst came to the worst, to go to the Pole on dead reckoning; but if we got none now, it was a question whether our claim to the Pole would be admitted on the dead reckoning we should be able to produce. Whether my pulling helped or not, it is certain that the sun appeared. It was not very brilliant to begin with, but, practised as we now were in availing ourselves of even the poorest chances, it was good enough. Down it came, was checked by all, and the altitude written down. The curtain of cloud was rent more and more, and before we had finished our work—that is to say, caught the sun at its highest, and convinced ourselves that it was descending again—it was shining in all its glory. We had put away our instruments and were sitting on the sledges, engaged in the calculations. I can safely say that we were excited. What would the result be, after marching blindly for so long and over such impossible ground, as we had been doing? We added and subtracted, and at last there was the result. We looked at each other in sheer incredulity: the result was as astonishing as the most consummate conjuring trick—88° 16′ S., precisely to a minute the same as our reckoning, 88° 16′ S. If we were forced to go to the Pole on dead reckoning, then surely the most exacting would admit our right to do so. We put away our observation books, ate one or two biscuits, and went at it again.

We had a great piece of work before us that day, nothing less than carrying our flag farther south than the foot of man had trod. We had our silk flag ready; it was made fast to two ski-sticks and laid on Hanssen's sledge. I had given him orders that as soon as we had covered the distance to 88° 23′ S., which was Shackleton's farthest south, the flag was to be hoisted on his sledge. It was my turn as forerunner, and I pushed on. There was no longer any difficulty in holding one's course; I had the grandest cloud-formations to steer by, and everything now went like a machine. First came the forerunner for the time being, then Hanssen, then Wisting, and finally Bjaaland. The forerunner who was not on duty went where he liked; as a rule he accompanied one or other of the sledges. I had long ago fallen into a reverie—far removed from the scene in which I was moving; what I thought about I do not remember now, but I was so preoccupied that I had entirely forgotten my surroundings. Then suddenly I was roused from my dreaming by a jubilant shout, followed by ringing cheers. I turned round quickly to discover the reason of this unwonted occurrence, and stood speechless and overcome.

I find it impossible to express the feelings that possessed me at this moment. All the sledges had stopped, and from the foremost of them the Norwegian flag was flying. It shook itself out, waved and flapped so that the silk rustled; it looked wonderfully well in the pure, clear air and the shining white surroundings. 88° 23′ was past; we were farther south than any human being had been. No other moment of the whole trip affected me like this. The tears forced their way to my eyes; by no effort of will could I keep them back. It was the flag yonder that conquered me and my will. Luckily I was some way in advance of the others, so that I had time to pull myself together and master my feelings before reaching my comrades. We all shook hands, with mutual congratulations; we had won our way far by holding together, and we would go farther yet—to the end.

We did not pass that spot without according our highest tribute of admiration to the man, who—together with his gallant companions—had planted his country's flag so infinitely nearer to the goal than any of his precursors. Sir Ernest Shackleton's name will always be written in the annals of Antarctic exploration in letters of fire. Pluck and grit can work wonders, and I know of no better example of this than what that man has accomplished.

The cameras of course had to come out, and we got an excellent photograph of the scene which none of us will ever forget. We went on a couple of miles more, to 88° 25′, and then camped. The weather had improved, and kept on improving all the time. It was now almost perfectly calm, radiantly clear, and, under the circumstances, quite summer-like: –0.4° F. Inside the tent it was quite sultry. This was more than we had expected.

After much consideration and discussion we had come to the conclusion that we ought to lay down a depot—the last one—at this spot. The advantages of lightening our sledges were so great that we should have to risk it. Nor would there be any great risk attached to it, after all, since we should adopt a system of marks that would lead even a blind man back to the place. We had determined to mark it not only at right angles to our course—that is, from east to west—but by snow beacons at every two geographical miles to the south.

We stayed here on the following day to arrange this depot. Hanssen's dogs were real marvels, all of them; nothing seemed to have any effect on them. They had grown rather thinner, of course, but they were still as strong as ever. It was therefore decided not to lighten Hanssen's sledge, but only the two others; both Wisting's and Bjaaland's teams had suffered, especially the latter's. The reduction in weight that was effected was considerable—nearly 110 pounds on each of the two sledges; there was thus about 220 pounds in the depot. The snow here was ill-adapted for building, but we put up quite a respectable monument all the same. It was dogs' pemmican and biscuits that were left behind; we carried with us on the sledges provisions for about a month. If, therefore, contrary to expectation, we should be so unlucky as to miss this depot, we should nevertheless be fairly sure of reaching our depot in 86° 21′ before supplies ran short. The cross-marking of the depot was done with sixty splinters of black packing-case on each side, with 100 paces between each. Every other one had a shred of black cloth on the top. The splinters on the east side were all marked, so that on seeing them we should know instantly that we were to the east of the depot. Those on the west had no marks.

The warmth of the past few days seemed to have matured our frost-sores, and we presented an awful appearance. It was Wisting, Hanssen, and I who had suffered the worst damage in the last south-east blizzard; the left side of our faces was one mass of sore, bathed in matter and serum. We looked like the worst type of tramps and ruffians, and would probably not have been recognized by our nearest relations. These sores were a great trouble to us during the latter part of the journey. The slightest gust of wind produced a sensation as if one's face were being cut backwards and forwards with a blunt knife. They lasted a long time, too; I can remember Hanssen removing the last scab when we were coming into Hobart—three months later. We were very lucky in the weather during this depot work; the sun came out all at once, and we had an excellent opportunity of taking some good azimuth observations, the last of any use that we got on the journey.

December 9 arrived with the same fine weather and sunshine. True, we felt our frost-sores rather sharply that day, with −18.4° F. and a little breeze dead against us, but that could not be helped. We at once began to put up beacons—a work which was continued with great regularity right up to the Pole. These beacons were not so big as those we had built down on the Barrier; we could see that they would be quite large enough with a height of about 3 feet, as it was very easy to see the slightest irregularity on this perfectly flat surface. While thus engaged we had an opportunity of becoming thoroughly acquainted with the nature of the snow. Often—very often indeed—on this part of the plateau, to the south of 88° 25′, we had difficulty in getting snow good enough—that is, solid enough for cutting blocks. The snow up here seemed to have fallen very quietly, in light breezes or calms. We could thrust the tent-pole, which was 6 feet long, right down without meeting resistance, which showed that there was no hard layer of snow. The surface was also perfectly level; there was not a sign of *sastrugi* in any direction.

Every step we now took in advance brought us rapidly nearer the goal; we could feel fairly certain of reaching it on the afternoon of the 14th. It was very natural that our conversation should be chiefly concerned with the time of arrival. None of us would admit that he was nervous, but I am inclined to think that we all had a little touch of that malady. What should we see when we got there? A vast, endless plain, that no eye had yet seen and no foot yet trodden; or—No, it was an impossibility; with the speed at which we had travelled, we must reach the goal first, there could be no doubt about that. And yet—and yet—Wherever there is the smallest loophole, doubt creeps in and gnaws and gnaws and never leaves a poor wretch in peace. "What on earth is Uroa scenting?" It was Bjaaland who made this remark, on one of these last days, when I was going by the side of his sledge and talking to him. "And the strange thing is that he's scenting to the south. It can never be—" Mylius, Ring, and Suggen showed the same interest in the southerly direction; it was quite extraordinary to see how they raised their heads, with every sign of curiosity, put their noses in the air, and sniffed due south. One would really have thought there was something remarkable to be found there.

From 88° 25′ S. the barometer and hypsometer indicated slowly but surely that the plateau was beginning to descend towards the other side. This was a pleasant surprise to us; we had thus not only found the very summit of the plateau, but also the slope down on the far side. This would have a very important bearing for obtaining an idea of the construction of the whole plateau. On December 9 observations and dead reckoning agreed within a mile. The same result again on the 10th: observation 2 kilometres behind reckoning. The weather and going remained about the same as on the preceding days: light south-easterly breeze, temperature −18.4° F. The snow surface was loose, but ski and sledges glided over it well. On the 11th, the same weather conditions. Temperature −13° F. Observation and reckoning again agreed exactly. Our latitude was 89° 15′ S. On the 12th we reached 89° 30′, reckoning 1 kilometre behind observation. Going and surface as good as ever. Weather splendid—calm with sunshine. The noon observation on the 13th gave 89° 37′ S. Reckoning 89° 38.5′ S. We halted in the afternoon, after going eight geographical miles, and camped in 89° 45′, according to reckoning.

The weather during the forenoon had been just as fine as before; in the afternoon we had some snow-showers from the south-east. It was like the eve of some great festival that night in the tent. One could feel that a great event was at hand. Our flag was taken out again and lashed to the same two ski-sticks as before. Then it was rolled up and laid aside, to be ready when the time came. I was awake several times during the night, and had the same feeling that I can remember as a little boy on the night before Christmas Eve—an intense expectation of what was going to happen. Otherwise I think we slept just as well that night as any other.

On the morning of December 14 the weather was of the finest, just as if it had been made for arriving at the Pole. I am not quite sure, but I believe we despatched our breakfast rather more quickly than usual and were out of the tent sooner, though I must admit that we always accomplished this with all reasonable haste. We went in the usual order—the forerunner, Hanssen, Wisting, Bjaaland, and the reserve forerunner. By noon we had reached 89° 53′ by dead reckoning, and made ready to take the rest in one stage. At 10 A.M. a light breeze had sprung up from the south-east, and it had clouded over, so that we got no noon altitude; but the clouds were not thick, and from time to time we had a glimpse of the sun through them. The going on that day was rather different from what it had been; sometimes the ski went over it well, but at others it was pretty bad. We advanced that day in the same mechanical way as before; not much was said, but eyes were used all the more. Hanssen's neck grew twice as long as before in his endeavour to see a few inches farther. I had asked him before we started to spy out ahead for all he was worth, and he did so with a vengeance. But, however keenly he stared, he could not descry anything but the endless flat plain ahead of us. The dogs had dropped their scenting, and appeared to have lost their interest in the regions about the earth's axis.

At three in the afternoon a simultaneous "Halt!" rang out from the drivers. They had carefully examined their sledge-meters, and they all showed the full distance—our Pole by reckoning. The goal was reached, the journey ended. I cannot say—though I know it would sound much more effective—that the object of my life was attained. That would be romancing rather too barefacedly. I had better be honest and admit straight out that I have never known any man to be placed in such a diametrically opposite position to the goal of his desires as I was at that moment. The regions around the North Pole—well, yes, the North Pole itself—had attracted me from childhood, and here I was at the South Pole. Can anything more topsy-turvy be imagined?

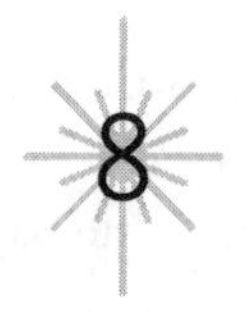

TRAGEDY ALL ALONG THE LINE

from *Scott's Last Expedition* (1913)

Robert Falcon Scott

Robert Scott perceived the Pole very differently than Amundsen when his party of five arrived there a month later, already exhausted and hungry, already suffering from weaknesses which would be exacerbated by the extreme cold they would encounter on the long trek back toward safety. What followed was a slow-motion calamity, which generations have been able to enter into, imaginatively, by reading Scott's own chronicle of it in his journal—each time hoping, as one reader put it, that this time the terrible tale may come out differently. This is Antarctica's passion story, its famous founding tragedy. Even if you blame Scott himself for what happened, it isn't easy to look away.

Wednesday, January 17. Camp 69. T. –22° at start. Night –21°. The Pole. Yes, but under very different circumstances from those expected. We have had a horrible day—add to our disappointment a head wind 4 to 5, with a temperature –22°, and companions labouring on with cold feet and hands.

We started at 7.30, none of us having slept much after the shock of our discovery. We followed the Norwegian sledge tracks for some way; as far as we make out there are only two men. In about three miles we passed two small cairns. Then the weather overcast, and the tracks being increasingly drifted up and obviously going too far to the west, we decided to make straight for the Pole according to our calculations. At 12.30 Evans had such cold hands we camped for lunch—an excellent "week-end one." We had marched 7.4 miles. Lat. sight gave 89° 53′ 37″. We started out and did 6½ miles due south. To-night little Bowers is laying himself out to get sights in terrible difficult circumstances; the wind is blowing hard, T. –21°, and there is that curious damp, cold feeling in the air which chills one to the bone in no time. We have been descending again, I think, but there looks to be a rise ahead; otherwise there is very little that is different from the awful monotony of past days. Great God! this is an awful place and terrible enough for us to have laboured to it without the reward of priority. Well, it is something to have got here, and the wind may be our friend to-morrow. We have had a fat Polar hoosh in spite of our chagrin, and feel comfortable inside—added a small stick of chocolate and the queer taste of a cigarette brought by Wilson. Now for the run home and a desperate struggle. I wonder if we can do it.

[. . .]

Wednesday, January 24. Lunch Temp. −8°. Things beginning to look a little serious. A strong wind at the start has developed into a full blizzard at lunch, and we have had to get into our sleeping-bags. It was a bad march, but we covered 7 miles. At first Evans, and then Wilson went ahead to scout for tracks. Bowers guided the sledge alone for the first hour, then both Oates and he remained alongside it; they had a fearful time trying to make the pace between the soft patches. At 12.30 the sun coming ahead made it impossible to see the tracks further, and we had to stop. By this time the gale was at its height and we had the dickens of a time getting up the tent, cold fingers all round. We are only 7 miles from our depôt, but I made sure we should be there to-night. This is the second full gale since we left the Pole. I don't like the look of it. Is the weather breaking up? If so, God help us, with the tremendous summit journey and scant food. Wilson and Bowers are my standby. I don't like the easy way in which Oates and Evans get frostbitten.

[. . .]

Saturday, January 27. R. 10. Temp. −16° (lunch), −14.3° (evening). Minimum −19°. Height 9900. Barometer low? Called the hands half an hour late, but we got away in good time. The forenoon march was over the belt of storm-tossed sastrugi; it looked like a rough sea. Wilson and I pulled in front on ski, the remainder on foot. It was very tricky work following the track, which pretty constantly disappeared, and in fact only showed itself by faint signs anywhere—a foot or two of raised sledge-track, a dozen yards of the trail of the sledge-meter wheel, or a spatter of hard snow-flicks where feet had trodden. Sometimes none of these were distinct, but one got an impression of lines which guided. The trouble was that on the outward track one had to shape course constantly to avoid the heaviest mounds, and consequently there were many zig-zags. We lost a good deal over a mile by these halts, in which we unharnessed and went on the search for signs. However, by hook or crook, we managed to stick on the old track. Came on the cairn quite suddenly, marched past it, and camped for lunch at 7 miles. In the afternoon the sastrugi gradually diminished in size and now we are on fairly level ground to-day, the obstruction practically at an end, and, to our joy, the tracks showing up much plainer again. For the last two hours we had no difficulty at all in following them. There has been a nice helpful southerly breeze all day, a clear sky and comparatively warm temperature. The air is dry again, so that tents and equipment are gradually losing their icy condition imposed by the blizzard conditions of the past week.

Our sleeping-bags are slowly but surely getting wetter and I'm afraid it will take a lot of this weather to put them right. However, we all sleep well enough in them, the hours allowed being now on the short side. We are slowly getting more hungry, and it would be an advantage to have a little more food, especially for lunch. If we get to the next depôt in a few marches (it is now less than 60 miles and we have a full week's food) we ought to be able to open out a little, but we can't look for a real feed till we get to the pony food depôt. A long way to go, and, by Jove, this is tremendous labour.

[. . .]

Tuesday, January 30. R. 13. 9860. Lunch Temp. –25°, Supper Temp. –24.5°. Thank the Lord, another fine march—19 miles. We have passed the last cairn before the depôt, the track is clear ahead, the weather fair, the wind helpful, the gradient down—with any luck we should pick up our depôt in the middle of the morning march. This is the bright side; the reverse of the medal is serious. Wilson has strained a tendon in his leg; it has given pain all day and is swollen to-night. Of course, he is full of pluck over it, but I don't like the idea of such an accident here. To add to the trouble Evans has dislodged two finger-nails to-night; his hands are really bad, and to my surprise he shows signs of losing heart over it. He hasn't been cheerful since the accident. The wind shifted from S.E. to S. and back again all day, but luckily it keeps strong. We can get along with bad fingers, but it [will be] a mighty serious thing if Wilson's leg doesn't improve.

[. . .]

Tuesday, February 6. Lunch 7900; Supper 7210. Temp. –15°. We've had a horrid day and not covered good mileage. On turning out found sky overcast; a beastly position amidst crevasses. Luckily it cleared just before we started. We went straight for Mt. Darwin, but in half an hour found ourselves amongst huge open chasms, unbridged, but not very deep, I think. We turned to the north between two, but to our chagrin they converged into chaotic disturbance. We had to retrace our steps for a mile or so, then struck to the west and got on to a confused sea of sastrugi, pulling very hard; we put up the sail, Evans' nose suffered, Wilson very cold, everything horrid. Camped for lunch in the sastrugi; the only comfort, things looked clearer to the west and we were obviously going downhill. In the afternoon we struggled on, got out of sastrugi and turned over on glazed surface, crossing many crevasses—very easy work on ski. Towards the end of the march we realised the certainty of maintaining a more or less straight course to the depôt, and estimate distance 10 to 15 miles.

Food is low and weather uncertain, so that many hours of the day were anxious; but this evening, though we are not as far advanced as I expected, the outlook is much more promising. Evans is the chief anxiety now; his cuts and wounds suppurate, his nose looks very bad, and altogether he shows considerable signs of being played out. Things may mend for him on the glacier, and his wounds get some respite under warmer conditions. I am indeed glad to think we shall so soon have done with plateau conditions. It took us 27 days to reach the Pole and 21 days back—in all 48 days—nearly 7 weeks in low temperature with almost incessant wind.

[. . .]

Thursday, February 8. R. 22. Height 6260. Start Temp. −11°; Lunch Temp. −5°; Supper, zero. 9.2 miles. Started from the depôt rather late owing to weighing biscuit, &c., and rearranging matters. Had a beastly morning. Wind very strong and cold. Steered in for Mt. Darwin to visit rock. Sent Bowers on, on ski, as Wilson can't wear his at present. He obtained several specimens, all of much the same type, a close-grained granite rock which weathers red. Hence the pink limestone. After he rejoined we skidded downhill pretty fast, leaders on ski, Oates and Wilson on foot alongside sledge—Evans detached. We lunched at 2 well down towards Mt. Buckley, the wind half a gale and everybody very cold and cheerless. However, better things were to follow. We decided to steer for the moraine under Mt. Buckley and, pulling with crampons, we crossed some very irregular steep slopes with big crevasses and slid down towards the rocks. The moraine was obviously so interesting that when we had advanced some miles and got out of the wind, I decided to camp and spend the rest of the day geologising. It has been extremely interesting. We found ourselves under perpendicular cliffs of Beacon sandstone, weathering rapidly and carrying veritable coal seams. From the last Wilson, with his sharp eyes, has picked several plant impressions, the last a piece of coal with beautifully traced leaves in layers, also some excellently preserved impressions of thick stems, showing cellular structure. In one place we saw the cast of small waves on the sand. To-night Bill has got a specimen of limestone with archeo-cyathus—the trouble is one cannot imagine where the stone comes from; it is evidently rare, as few specimens occur in the moraine. There is a good deal of pure white quartz. Altogether we have had a most interesting afternoon, and the relief of being out of the wind and in a warmer temperature is inexpressible. I hope and trust we shall all buck up again now that the conditions are more favourable. We have been in shadow all the afternoon, but the sun has just reached us, a little obscured by night haze. A lot could be written on the delight of setting foot on rock after 14 weeks of snow and ice and nearly 7 out of sight of aught else. It is like going ashore after a sea voyage. We deserve a little good bright weather after all our trials, and hope to get a chance to dry our sleeping-bags and generally make our gear more comfortable.

[. . .]

Wednesday, February 14. Lunch Temp. 0°; Supper Temp. −1°. A fine day with wind on and off down the glacier, and we have done a fairly good march. We started a little late and pulled on down the moraine. At first I thought of going right, but soon, luckily, changed my mind and decided to follow the curving lines of the moraines. This course has brought us well out on the glacier. Started on crampons; one hour after, hoisted sail; the combined efforts produced only slow speed, partly due to the sandy snowdrifts similar to those on summit, partly to our torn sledge runners. At lunch these were scraped and sand-papered. After lunch we got on snow, with ice only occasionally showing through. A poor start, but the gradient and wind improving, we did 6½ miles before night camp.

There is no getting away from the fact that we are not going strong. Probably none of us: Wilson's leg still troubles him and he doesn't like to trust himself on ski; but the worst case is Evans, who is giving us serious anxiety. This morning he suddenly disclosed a huge blister on his foot. It delayed us on the march, when he had to have his crampon readjusted. Sometimes I fear he is going from bad to worse, but I trust he will pick up again when we come to steady work on ski like this afternoon. He is hungry and so is Wilson. We can't risk opening out our food again, and as cook at present I am serving something under full allowance. We are inclined to get slack and slow with our camping arrangements, and small delays increase. I have talked of the matter to-night and hope for improvement. We cannot do distance without the ponies. The next depôt some 30 miles away and nearly 3 days' food in hand.

[. . .]

Friday, February 16. 12.5 m. Lunch Temp. –6.1°; Supper Temp. –7°. A rather trying position. Evans has nearly broken down in brain, we think. He is absolutely changed from his normal self-reliant self. This morning and this afternoon he stopped the march on some trivial excuse. We are on short rations with not very short food; spin out till to-morrow night. We cannot be more than 10 or 12 miles from the depôt, but the weather is all against us. After lunch we were enveloped in a snow sheet, land just looming. Memory should hold the events of a very troublesome march with more troubles ahead. Perhaps all will be well if we can get to our depôt to-morrow fairly early, but it is anxious work with the sick man. But it's no use meeting troubles half way, and our sleep is all too short to write more.

Saturday, February 17. A very terrible day. Evans looked a little better after a good sleep, and declared, as he always did, that he was quite well. He started in his place on the traces, but half an hour later worked his ski shoes adrift, and had to leave the sledge. The surface was awful, the soft recently fallen snow clogging the ski and runners at every step, the sledge groaning, the sky overcast, and the land hazy. We stopped after about one hour, and Evans came up again, but very slowly. Half an hour later he dropped out again on the same plea. He asked Bowers to lend him a piece of string. I cautioned him to come on as quickly as he could, and he answered cheerfully as I thought. We had to push on, and the remainder of us were forced to pull very hard, sweating heavily. Abreast the Monument Rock we stopped, and seeing Evans a long way astern, I camped for lunch. There was no alarm at first, and we prepared tea and our own meal, consuming the latter. After lunch, and Evans still not appearing, we looked out, to see him still afar off. By this time we were alarmed, and all four started back on ski. I was first to reach the poor man and shocked at his appearance; he was on his knees with clothing disarranged, hands uncovered and frostbitten, and a wild look in his eyes. Asked what was the matter, he replied with a slow speech that he didn't know, but thought he must have fainted. We got him on his feet, but after two or three steps he sank down again. He showed every sign of complete collapse. Wilson, Bowers, and I went back for the sledge, whilst Oates remained with him. When we returned he was practically unconscious, and when we got him into the tent quite comatose. He died quietly at 12.30 A.M. On discussing the symptoms we think he began to get weaker just before we reached the Pole, and that his downward path was accelerated first by the shock of his frostbitten fingers, and later by falls during rough travelling on the glacier, further by his loss of all confidence in himself. Wilson thinks it certain he must have injured his brain by a fall. It is a terrible thing to lose a companion in this way, but calm reflection shows that there could not have been a better ending to the terrible anxieties of the past week. Discussion of the situation at lunch yesterday shows us what a desperate pass we were in with a sick man on our hands at such a distance from home.

At 1 A.M. we packed up and came down over the pressure ridges, finding our depôt easily.

Sunday, February 18. R. 32. Temp. –5.5°. At Shambles Camp. We gave ourselves 5 hours' sleep at the lower glacier depôt after the horrible night, and came on at about 3 to-day to this camp, coming fairly easily over the divide. Here with plenty of horsemeat we have had a fine supper, to be followed by others such, and so continue a more plentiful era if we can keep good marches up. New life seems to come with greater food almost immediately, but I am anxious about the Barrier surfaces.

[. . .]

Tuesday, February 28. Lunch. Thermometer went below –40° last night; it was desperately cold for us, but we had a fair night. I decided to slightly increase food; the effect is undoubtedly good. Started marching in –32° with a slight north-westerly breeze—blighting. Many cold feet this morning; long time over foot gear, but we are earlier. Shall camp earlier and get the chance of a good night, if not the reality. Things must be critical till we reach the depôt, and the more I think of matters, the more I anticipate their remaining so after that event. Only 24½ miles from the depôt. The sun shines brightly, but there is little warmth in it. There is no doubt the middle of the Barrier is a pretty awful locality.

[. . .]

Friday, March 2. Lunch. Misfortunes rarely come singly. We marched to the (Middle Barrier) depôt fairly easily yesterday afternoon, and since that have suffered three distinct blows which have placed us in a bad position. First we found a shortage of oil; with most rigid economy it can scarce carry us to the next depôt on this surface (71 miles away). Second, Titus Oates disclosed his feet, the toes showing very bad indeed, evidently bitten by the late temperatures. The third blow came in the night, when the wind, which we had hailed with some joy, brought dark overcast weather. It fell below –40° in the night, and this morning it took 1½ hours to get our foot gear on, but we got away before eight. We lost cairn and tracks together and made as steady as we could N. by W., but have seen nothing. Worse was to come—the surface is simply awful. In spite of strong wind and full sail we have only done 5½ miles. We are in a *very* queer street since there is no doubt we cannot do the extra marches and feel the cold horribly.

[. . .]

Monday, March 5. Lunch. Regret to say going from bad to worse. We got a slant of wind yesterday afternoon, and going on 5 hours we converted our wretched morning run of 3½ miles into something over 9. We went to bed on a cup of cocoa and pemmican solid with the chill off. (R. 47.) The result is telling on all, but mainly on Oates, whose feet are in a wretched condition. One swelled up tremendously last night and he is very lame this morning. We started march on tea and pemmican as last night—we pretend to prefer the pemmican this way. Marched for 5 hours this morning over a slightly better surface covered with high moundy sastrugi. Sledge capsized twice; we pulled on foot, covering about 5½ miles. We are two pony marches and 4 miles about from our depôt. Our fuel dreadfully low and the poor Soldier nearly done. It is pathetic enough because we can do nothing for him; more hot food might do a little, but only a little, I fear. We none of us expected these terribly low temperatures, and of the rest of us Wilson is feeling them most; mainly, I fear, from his self-sacrificing devotion in doctoring Oates' feet. We cannot help each other, each has enough to do to take care of himself. We get cold on the march when the trudging is heavy, and the wind pierces our warm garments. The others, all of them, are unendingly cheerful when in the tent. We mean to see the game through with a proper spirit, but it's tough work to be pulling harder than we ever pulled in our lives for long hours, and to feel that the progress is so slow. One can only say "God help us!" and plod on our weary way, cold and very miserable, though outwardly cheerful. We talk of all sorts of subjects in the tent, not much of food now, since we decided to take the risk of running a full ration. We simply couldn't go hungry at this time.

[. . .]

Wednesday, March 7. A little worse I fear. One of Oates' feet *very* bad this morning; he is wonderfully brave. We still talk of what we will do together at home.

We only made 6½ miles yesterday. (R. 49.) This morning in 4½ hours we did just over 4 miles. We are 16 from our depôt. If we only find the correct proportion of food there and this surface continues, we may get to the next depôt [Mt. Hooper, 72 miles farther] but not to One Ton Camp. We hope against hope that the dogs have been to Mt. Hooper; then we might pull through. If there is a shortage of oil again we can have little hope. One feels that for poor Oates the crisis is near, but none of us are improving, though we are wonderfully fit considering the really excessive work we are doing. We are only kept going by good food. No wind this morning till a chill northerly air came ahead. Sun bright and cairns showing up well. I should like to keep the track to the end.

[. . .]

Saturday, March 10. Things steadily downhill. Oates' foot worse. He has rare pluck and must know that he can never get through. He asked Wilson if he had a chance this morning, and of course Bill had to say he didn't know. In point of fact he has none. Apart from him, if he went under now, I doubt whether we could get through. With great care we might have a dog's chance, but no more. The weather conditions are awful, and our gear gets steadily more icy and difficult to manage. At the same time of course poor Titus is the greatest handicap. He keeps us waiting in the morning until we have partly lost the warming effect of our good breakfast, when the only wise policy is to be up and away at once; again at lunch. Poor chap! it is too pathetic to watch him; one cannot but try to cheer him up.

[. . .]

Sunday, March 11. Titus Oates is very near the end, one feels. What we or he will do, God only knows. We discussed the matter after breakfast; he is a brave fine fellow and understands the situation, but he practically asked for advice. Nothing could be said but to urge him to march as long as he could. One satisfactory result to the discussion; I practically ordered Wilson to hand over the means of ending our troubles to us, so that anyone of us may know how to do so. Wilson had no choice between doing so and our ransacking the medicine case. We have 30 opium tabloids apiece and he is left with a tube of morphine. So far the tragical side of our story. (R. 53.)

[. . .]

Friday, March 16 or Saturday 17. Lost track of dates, but think the last correct. Tragedy all along the line. At lunch, the day before yesterday, poor Titus Oates said he couldn't go on; he proposed we should leave him in his sleeping-bag. That we could not do, and induced him to come on, on the afternoon march. In spite of its awful nature for him he struggled on and we made a few miles. At night he was worse and we knew the end had come.

Should this be found I want these facts recorded. Oates' last thoughts were of his Mother, but immediately before he took pride in thinking that his regiment would be pleased with the bold way in which he met his death. We can testify to his bravery. He has borne intense suffering for weeks without complaint, and to the very last was able and willing to discuss outside subjects. He did not—would not—give up hope to the very end. He was a brave soul. This was the end. He slept through the night before last, hoping not to wake; but he woke in the morning—yesterday. It was blowing a blizzard. He said, "I am just going outside and may be some time." He went out into the blizzard and we have not seen him since.

I take this opportunity of saying that we have stuck to our sick companions to the last. In case of Edgar Evans, when absolutely out of food and he lay insensible, the safety of the remainder seemed to demand his abandonment, but Providence mercifully removed him at this critical moment. He died a natural death, and we did not leave him till two hours after his death. We knew that poor Oates was walking to his death, but though we tried to dissuade him, we knew it was the act of a brave man and an English gentleman. We all hope to meet the end with a similar spirit, and assuredly the end is not far.

I can only write at lunch and then only occasionally. The cold is intense, –40° at midday. My companions are unendingly cheerful, but we are all on the verge of serious frostbites, and though we constantly talk of fetching through I don't think any one of us believes it in his heart.

We are cold on the march now, and at all times except meals. Yesterday we had to lay up for a blizzard and to-day we move dreadfully slowly. We are at No. 14 pony camp, only two pony marches from One Ton Depôt. We leave here our theodolite, a camera, and Oates' sleeping-bags. Diaries, &c., and geological specimens carried at Wilson's special request, will be found with us or on our sledge.

[. . .]

Monday, March 19. Lunch. We camped with difficulty last night, and were dreadfully cold till after our supper of cold pemmican and biscuit and a half a pannikin of cocoa cooked over the spirit. Then, contrary to expectation, we got warm and all slept well. To-day we started in the usual dragging manner. Sledge dreadfully heavy. We are 15½ miles from the depôt and ought to get there in three days. What progress! We have two days' food but barely a day's fuel. All our feet are getting bad—Wilson's best, my right foot worst, left all right. There is no chance to nurse one's feet till we can get hot food into us. Amputation is the least I can hope for now, but will the trouble spread? That is the serious question. The weather doesn't give us a chance—the wind from N. to N.W. and –40° temp. to-day.

Wednesday, March 21. Got within 11 miles of depôt Monday night; had to lay up all yesterday in severe blizzard. To-day forlorn hope, Wilson and Bowers going to depôt for fuel.

Thursday, March 22 and 23. Blizzard bad as ever—Wilson and Bowers unable to start—to-morrow last chance—no fuel and only one or two of food left—must be near the end. Have decided it shall be natural—we shall march for the depôt with or without our effects and die in our tracks.

Thursday, March 29. Since the 21st we have had a continuous gale from W.S.W. and S.W. We had fuel to make two cups of tea apiece and bare food for two days on the 20th. Every day we have been ready to start for our depôt *11 miles* away, but outside the door of the tent it remains a scene of whirling drift. I do not think we can hope for any better things now. We shall stick it out to the end, but we are getting weaker, of course, and the end cannot be far.

It seems a pity, but I do not think I can write more.

R. SCOTT.

For God's sake look after our people.

SCOTT DIES

from *I May Be Some Time* (1996)

Francis Spufford

In contrast to Roald Amundsen, Scott was a very literary explorer. Some might say that he put altogether too much of his effort into his sentences, and not nearly enough into his practical skills. Writing almost to the last, he was able, to an astonishing extent, to set the scene of his own extinction. But a point must have come when the words ran out for him; a point, of course, after which there could also be no more documentary evidence. This is a recent attempt to imagine the very last hours, the ones Scott couldn't put on paper. It is fiction, not fact.

SOMETIMES YOU WAKE from a dream of guilt or horror that has filled your whole sleeping mind, a dream that feels final, as if it held a truth about you that you cannot hope to evade, and the kind day dislodges it bit by bit, showing you exits where you had thought there were none, reminding you of a world where you still move among choices. Day has always done this for you. It seems unfair that it should not, today. Scott's eyes open. Green canvas wall of tent, rush of snow outside seen only as a tireless spatter of dark. The canvas rustles. He has not been sleeping. He has been trying to drift, but the habit of self-command cuts him off, calls him back over and over to the realisation that it is all true. This irrevocable position *is* the whole, waking truth, and the tent is his life's last scene, beyond any possibility of alteration. He can make no effort that would change anything. If he had taken Oates' advice last autumn and pushed One Ton Depot further south, he might not be lying eleven miles short of it now. If he had left different instructions about the dog-teams, even now help might be on its way, rather than receding through the Barrier blizzard as Cherry-Garrard, unknowing, drives for Cape Evans. But these are ironies that have lost their power to torment, through many repetitions. Edgar

Evans is dead under a shallow mound of snow on the Beardmore: brain haemorrhage, Wilson thought. Oates "left us the other day," as it says in Birdie's letter to his mother, neatly folded on the groundsheet. Oates is a white hummock now somewhere a little to the side of the line of march. And Wilson and Bowers lie one each side of Scott in the tent, their sleeping bags pulled over their faces. How many hours ago he does not know, the breathing first of one and then of the other turned briefly ragged and then stopped. The breath sighed out and never drew in again. Except for the silence they might be sleeping. Scott has a terrible desire that he must keep quelling, to reach and shake them, to try and summon again their company. He can imagine all too well the way the illusion of sleep would break if he did; and the moment when he asked for an answer and got none would be beyond bearing. So he must not break down and ask. He must not touch them at all. He is entirely alone, beyond all hope. For who knows what scoured and whirling distance all about, he is the only living thing. There is nothing left to do but die. But he is still here. He composes himself as best he can (it is difficult to want to stop, your mind is not adjusted to it) but nothing happens. The greater nothing which he supposes will replace this tiny green space when he goes—still unimaginable—does not arrive. His heart beats in his chest with stupid strength.

It was better when he was writing. Twelve days ago Scott's feet froze at last and crippled him. Eleven days ago, the immobilising blizzard began. Ten days ago they ran out of fuel. Eight days ago they ate the last of the food, and soon the thing became absolutely certain. Everyone wrote, though the pencil was hard to hold, and the paper glazed over with ice if you exhaled on it. Wilson wrote a letter to Oriana, a letter to his parents in Cheltenham (carefully adding the address anew on the second page in case it should be separated) and a note to his friends the Smiths. Bowers apologised to his mother that his letter should be "such a short scribble," and for other things. "It will be splendid however to pass with such companions as I have . . . Oh how I do feel for you when you hear all, you will know that for me the end was peaceful as it is only sleep in the cold. Your ever loving son to the end of this life and the next when God shall wipe away all tears from our eyes—H. R. Bowers." But Scott wrote and wrote and wrote. He paid his professional debts. He told Mrs Bowers that her son had been magnificent to the end, and Mrs Wilson that her husband had the "comfortable blue look of hope" in his eyes. He told his mother that "the Great God has called me." There are twelve letters by him in the tent, besides his diary and a Message to the Public. His teeth are loose in his gums from scurvy, his feet would be gangrenous if the cold were not slowing up the bacteria, his face is cracked with snow-burn and marked with unhealed red and purple sores where the frost bit at the points of the bones; but while he wrote he commanded the kingdom of words. He was *making*. He could see the story of the expedition as a parabola that descended to earth at its completion, and might be made to do so with a power and a grace that justified the whole; that gave the whole an inevitable fall, like any good story whose end is latent in its middle and beginning. He knew exactly what to do. A century and more of expectations were to hand, anonymous and virtually instinctive to him: he shaped them. Scarcely a word needed crossing out. One, inside the cover of his diary: "Send this diary to my wife." Correction: "widow." With the authority of death he insisted "The causes of the disaster are not due to faulty organisation but to misfortune in all risks which had to be undertaken." Into the syntax of his best sentences, he wove appeals to the practical charity of the nation, so that—like a politician on television taking care his soundbite cannot be edited into smaller units—the emotion and the appeal should be indivisible. "These rough notes and our dead bodies must tell the tale, but surely, surely, a great rich country like ours will see that those who are dependent on us are properly provided for." Grand sombre cadences, funeral music in words, came to him; long sentences running parallel in sound to each other, inviting a voice to work its way through the scored heights and depths of the phrasing.

> We are weak, writing is difficult, but for my own sake I do not regret this journey, which has shown that Englishmen can endure hardships, help one another, and meet death with as great a fortitude as ever in the past. We took risks, we knew we took them; things have come out against us, and therefore we have no cause for complaint, but bow to the will of Providence, determined still to do our best to the last.

But he had to stop. "It seems a pity but I do not think I can write more." When the writing stopped, so did all that words can do to give this situation meaning. His words are exhausted. The tale is told; but he is still here, in the silence afterwards, waiting. Some people wear their roles so closely they become their skin. There is nothing left of them besides, no residue that does not fit the proper emotions of a judge, or a salesman, or an explorer. Scott is wonderfully good at his role, but he is not one of these; he has always been self-conscious. Tucked beneath him, "I have taken my place throughout, haven't I?" says the letter to Kathleen—whether with pride, or anxiety, or final bitterness at the explorer's place and its mortal demands, he hardly knows. "What tales you would have had for the boy, but oh, what a price to pay." After the storied death it seems there remains all of you to die that you had only glimpsed sidelong as you subdued yourself to the part. You cannot die in a story; you have to die in your body. He wonders if the other two travelled, invisibly to him, sometime in the night hours, past the end of their belief, their belief not quite stretching all the way to the fact of death, and faced this horrible vacancy. He thinks not, and is glad. They were certain enough, so far as he could judge, that the eyes they closed here would open again elsewhere. They still looked forward, "slept" only to wake. Sleep, sleep: all at once he hates this lulling metaphor for the disappearance of every slightest speck of forty-three years of thinking and feeling. Such a lie. It is not sleep, this formless prospect from which his mind recoils helplessly though it is imminent.

But whatever he thinks, whatever he wants, here he is still, both holding death at bay, he cannot *help* it, and wishing it would hurry. He wants to see Kathleen again. He wants the world to expand again from this narrow trap to the proportions you learn to trust, living. He thinks that he would very much like to go indoors. The tent is a feeble cone perched beneath a huge sky on a bed of ice sustained by black, black water. He left Cape Evans five months ago: for almost a hundred and fifty nights he's been in the open, or within this portable fiction of shelter. You never realise until you come out here that the world divides so absolutely into outdoors and indoors. It is almost metaphysical. It seems marvellous to him now that people take open space, and floor it and wall it and roof it, and transform it utterly. He thinks of doors opening, and himself passing through. The door at Cape Evans, of course; but also he stands on the steps by the railings in Buckingham Palace Road knocking at the coloured front door of his own house. He waits at the bigger door of the Geographical Society, and through the glass panes he can see the porter in the vestibule coming to let him in, quite unflustered by the balaclava and the drip of melting frost off his windproof smock. Thresholds: the thick metal door in the corridor of a destroyer, whose rivets are cool bulges in a skin of paint, whose foot-high sill is shiny steel in the centre where feet touch it. A screen-door in the verandah of an American house on a hot day, which has an aquarium cool you are glad of on the inside of it, and a remote buzz of insects and traffic. Doors squeaking, grating, gliding ajar with huge solidity. He thumps for entry at the doors of St Paul's, not on the little gateway inset in the greater but on the vast sculpted panels of the great door itself, which swing wide on a chessboard floor where his footsteps fall echoey yet distinct. The gates of ivory and of horn in the *Odyssey,* from whose parted leaves stream out true visions and false dreams . . . *Have the gates of death been opened unto thee? or hast thou seen the doors of the shadow of death? Hast thou perceived the breadth of the earth? declare if thou knowest it all. Where is the way where light dwelleth? and as for darkness, where is the place thereof, that thou shouldest take it to the bound thereof, and that thou shouldest know the paths to the house thereof? . . . Hast thou entered into the treasures of the snow? or hast thou seen the treasures of the hail . . . Out of whose womb came the ice? and the hoary frost of heaven who hath gendered it? The waters are hid as with a stone, and the face of the deep is frozen. Canst thou bind the sweet influences of Pleiades, or loose the bands of Orion?* Drift . . . Drift . . . Stop! Get a grip, man. Or, no, he supposes perhaps he ought not to take hold again. But it is already done. He has tightened whatever it is in him that lashes a crumbly fear together into a block strong enough to face things. And the tent returns. The tent, the place, the two corpses, the bottle of opium tablets that would dissolve him away irresistibly if he once chose to swallow them. He is still here.

Scott kicks out suddenly, like an insomniac angry with the bedclothes. Yes, alright, but *quickly* then, without thinking. He pulls open the sleeping bag as far down as he can reach, wrenches his coat right open too, lays his arm deliberately around the cold lump of the body of his friend Edward Wilson (who is not sleeping, no, but dead) and holds tight. It is forty below in the tent. The cold comes into him. Oh how it hurts. His skin, which was the frontier of him this whole long time past, is breached: he is no longer whole: the ice is inside his chest, a spearing and dreadful presence turning the cavities of him to blue glass. His lips pull back from his teeth in an enormous snarl; but Scott has left the surface of his face, and does not know. At its tip the cold moves inside him like a key searching for a lock. An impersonal tenderness seems to be watching as it finds the latch of a box, a box of memories, and spills them out, the most private images, one by one, some that would never have been expected because they were scarcely remembered and it was never known that they had been diligently stored here all the time; one by one, each seen complete and without passion, until the last of them is reached, and flutters away, and is gone.

MAWSON LIVES

from *Home of the Blizzard* (1915; 1930)

Douglas Mawson

At the same time as the race for the Pole was dominating the Ross Sea sector, an Australian expedition was at work far away on the other side of Antarctica, led by Douglas Mawson, a scientific veteran of Shackleton's 1907 crew. The Australians were based on a shore of Wilkes Land so steadily wind-scoured that, around their hut, they learned to walk tilted right over to the diagonal. Inland, as ice streams from the interior poured down to the coast, the terrain was heavily crevassed. Mawson was out on a survey mission with two companions when one, Ninnis, abruptly fell down a crevasse, never to be seen again: most of the food went with him. Then his other sledgemate, Xavier Mertz, died of food poisoning. Mawson was alone, virtually without supplies or equipment, and seriously ill himself, 200 miles and more from base. His death was the overwhelmingly likely outcome. Instead, step by tottering step, he walked home.

Outside the bowl of chaos was brimming with drift-snow and as I lay in the sleeping-bag beside my dead companion I wondered how, in such conditions, I would manage to break and pitch camp single-handed. There appeared to be little hope of reaching the Hut, still one hundred miles away. It was easy to sleep in the bag, and the weather was cruel outside. But inaction is hard to bear and I braced myself together determined to put up a good fight.

Failing to reach the Hut it would be something done if I managed to get to some prominent point likely to catch the eye of a search-party, where a cairn might be erected and our diaries cached. So I commenced to modify the sledge and camping gear to meet fresh requirements.

The sky remained clouded, but the wind fell off to a calm which lasted several hours. I took the opportunity to set to work on the sledge, sawing it in halves with a pocket tool and discarding the rear section. A mast was made out of one of the rails no longer required, and a spar was cut from the other. Finally, the load was cut down to a minimum by the elimination of all but the barest necessities, the abandoned articles including, sad to relate, all that remained of the exposed photographic films.

Late that evening, the 8th, I took the body of Mertz, still toggled up in his bag, outside the tent, piled snow blocks around it and raised a rough cross made of the two discarded halves of the sledge runners.

On January 9 the weather was overcast and fairly thick drift was flying in a gale of wind, reaching about fifty miles an hour. As certain matters still required attention and my chances of re-erecting the tent were rather doubtful, if I decided to move on, the start was delayed.

Part of the time that day was occupied with cutting up a waterproof clothes-bag and Mertz's burberry jacket and sewing them together to form a sail. Before retiring to rest in the evening I read through the burial service and put the finishing touches on the grave.

January 10 arrived in a turmoil of wind and thick drift. The start was still further delayed. I spent part of the time in reckoning up the food remaining and in cooking the rest of the dog meat, this latter operation serving the good object of lightening the load, in that the kerosene for the purpose was consumed there and then and had not to be dragged forward for subsequent use. Late in the afternoon the wind fell and the sun peered amongst the clouds just as I was in the middle of a long job riveting and lashing the broken shovel.

The next day, January 11, a beautiful, calm day of sunshine, I set out over a good surface with a slight down grade.

From the start my feet felt curiously lumpy and sore. They had become so painful after a mile of walking that I decided to examine them on the spot, sitting in the lee of the sledge in brilliant sunshine. I had not had my socks off for some days for, while lying in camp, it had not seemed necessary. On taking off the third and inner pair of socks the sight of my feet gave me quite a shock, for the thickened skin of the soles had separated in each case as a complete layer, and abundant watery fluid had escaped saturating the sock. The new skin beneath was very much abraded and raw. Several of my toes had commenced to blacken and fester near the tips and the nails were puffed and loose.

I began to wonder if there was ever to be a day without some special disappointment. However, there was nothing to be done but make the best of it.

I smeared the new skin and the raw surfaces with lanoline, of which there was fortunately a good store, and then with the aid of bandages bound the old skin casts back in place, for these were comfortable and soft in contact with the abraded surface. Over the bandages were slipped six pairs of thick woollen socks, then fur boots and finally crampon over-shoes. The latter, having large stiff soles, spread the weight nicely and saved my feet from the jagged ice encountered shortly afterwards.

So glorious was it to feel the sun on one's skin after being without it for so long that I next removed most of my clothing and bathed my body in the rays until my flesh fairly tingled—a wonderful sensation which spread throughout my whole person, and made me feel stronger and happier.

Then on I went, treading rather like a cat on wet ground endeavouring to save my feet from pain. By 5.30 P.M. I was quite worn out—nerve-worn—though having covered but six and a quarter miles. Had it not been a delightful evening I should not have found strength to erect the tent.

The day following passed in a howling blizzard and I could do nothing but attend to my feet and other raw patches, festering finger-nails and inflamed frost-bitten nose. Fortunately there was a good supply of bandages and antiseptic. The tent, spread about with dressings and the meagre surgical appliances at hand, was suggestive of a casualty hospital.

Towards noon the following day, January 13, the wind subsided and the snow cleared off. It turned out a beautifully fine afternoon. Soon after I had got moving the slope increased, unfolding a fine view of the Mertz Glacier ahead. My heart leapt with joy, for all was like a map before me and I knew that over the hazy blue ice ridge in the far distance lay the Hut. I was heading to traverse the depression of the glacier ahead at a point many miles above our crossing of the outward journey and some few miles below gigantic ice cascades. My first impulse was to turn away to the west and avoid crossing the fifteen miles of hideously broken ice that choked the valley before me, but on second thought, in view of the very limited quantity of food left, the right thing seemed to be to make an air-line for the Hut and chance what lay between. Accordingly, having taken an observation of the sun for position and selected what appeared to be the clearest route across the valley, I started downhill. The névé gave way to rough blue ice and even wide crevasses made their appearance. The rough ice jarred my feet terribly and altogether it was a most painful march.

So unendurable did it become that, finding a bridged crevasse extending my way, I decided to march along the snow bridge and risk an accident. It was from fifteen to twenty feet wide and well packed with winter snow. The march continued along it down slopes for over a mile with great satisfaction as far as my feet were concerned. Eventually it became irregular and broke up, but others took its place and served as well; in this way the march was made possible. At 8 P.M. after covering a distance of nearly six miles a final halt for the day was made.

About 11 P.M. as the sun skimmed behind the ice slopes to the south I was startled by loud reports like heavy gun shots. They commenced up the valley to the south and trailed away down the southern side of the glacier towards the sea. The fusillade of shots rang out without interruption for about half an hour, then all was silent. It was hard to believe it was not caused by some human agency, but I learnt that it was due to the cracking of the glacier ice.

A high wind which blew on the morning of the 14th diminished in strength by noon and allowed me to get away. The sun came out so warm that the rough ice surface underfoot was covered with a film of water and in some places small trickles ran away to disappear into crevasses.

Though the course was downhill, the sledge required a good deal of pulling owing to the wet runners. At 9 P.M., after travelling five miles, I pitched camp in the bed of the glacier. From about 9.30 P.M. until 11 P.M. "cannonading" continued like that heard the previous evening.

January 15—the date on which all the sledging parties were due at the Hut! It was overcast and snowing early in the day, but in a few hours the sun broke out and shone warmly. The travelling was so heavy over a soft snowy surface, partly melting, that I gave up, after one mile, and camped.

At 7 P.M. the surface had not improved, the sky was thickly obscured and snow fell. At 10 P.M. a heavy snowstorm was in progress, and, since there were many crevasses in the vicinity, I resolved to wait.

On the 16th at 2 A.M. the snow was falling as thick as ever, but at 5 A.M. the atmosphere lightened and the sun appeared. Camp was broken without delay. A favourable breeze sprang up, and with sail set I managed to proceed in short stages through the deep newly-fallen blanket of snow. It clung in lumps to the runners, which had to be scraped frequently. Riven ice ridges as much as eighty feet in height passed on either hand. Occasionally I got a start as a foot or a leg sank through into space, but, on the whole, all went unexpectedly well for several miles. Then the sun disappeared and the disabilities of a snow-blind light had to be faced.

After laboriously toiling up one long slope, I had just taken a few paces over the crest, with the sledge running freely behind, when it dawned on me that the surface fell away unusually steeply. A glance ahead, even in that uncertain light, flashed the truth upon me—I was on a snow cornice, rimming the brink of a great blue chasm like a quarry, the yawning mouth of an immense and partly filled crevasse. Already the sledge was gaining speed as it slid past me towards the gaping hole below. Mechanically, I bedded my feet firmly in the snow and, exerting every effort, was just able to take the weight and hold up the sledge as it reached the very brink of the abyss. There must have been an interval of quite a minute during which I held my ground without being able to make it budge. It seemed an interminable time; I found myself reckoning the odds as to who would win, the sledge or I. Then it slowly came my way, and the imminent danger was passed.

The day's march was an extremely heavy five miles; so before turning in I treated myself to an extra supper of jelly soup made from dog sinews. I thought at the time that the acute enjoyment of eating compensated in some measure for the sufferings of starvation.

January 17 was another day of overcast sky and steady falling snow. Everything from below one's feet to the sky above was one uniform ghostly glare. The irregularities in the surfaces not obliterated by the deep soft snow blended harmoniously in colour and in the absence of shadows faded into invisibility. These were most unsuitable conditions for the crossing of such a dangerous crevassed valley, but delay meant a reduction of the ration and that was out of the question, so nothing remained but to go on.

A start was made at 8 A.M. and the pulling proved more easy than on the previous day. Some two miles had been negotiated in safety when an event occurred which, but for a miracle, would have terminated the story then and there. Never have I come so near to an end; never has anyone more miraculously escaped.

I was hauling the sledge through deep snow up a fairly steep sloop when my feet broke through into a crevasse. Fortunately as I fell I caught my weight with my arms on the edge and did not plunge in further than the thighs. The outline of the crevasse did not show through the blanket of snow on the surface, but an idea of the trend was obtained with a stick. I decided to try a crossing about fifty yards further along, hoping that there it would be better bridged. Alas! it took an unexpected turn catching me unawares. This time I shot through the centre of the bridge in a flash, but the latter part of the fall was decelerated by the friction of the harness ropes which, as the sledge ran up, sawed back into the thick compact snow forming the margin of the lid. Having seen my comrades perish in diverse ways and having lost hope of ever reaching the Hut, I had already many times speculated on what the end would be like. So it happened that as I fell through into the crevasse the thought "so this is the end" blazed up in my mind, for it was to be expected that the next moment the sledge would follow through, crash on my head and all go to the unseen bottom. But the unexpected happened and the sledge held, the deep snow acting as a brake.

In the moment that elapsed before the rope ceased to descend, delaying the issue, a great regret swept through my mind, namely, that after having stinted myself so assiduously in order to save food, I should pass on now to eternity without the satisfaction of what remained—to such an extent does food take possession of one under such circumstances. Realizing that the sledge was holding I began to look around. The crevasse was somewhat over six feet wide and sheer walled, descending into blue depths below. My clothes, which, with a view to ventilation, had been but loosely secured, were now stuffed with snow broken from the roof, and very chilly it was. Above at the other end of the fourteen-foot rope, was the daylight seen through the hole in the lid.

In my weak condition, the prospect of climbing out seemed very poor indeed, but in a few moments the struggle was begun. A great effort brought a knot in the rope within my grasp, and, after a moment's rest, I was able to draw myself up and reach another, and, at length, hauled my body on to the overhanging snow-lid. Then, when all appeared to be well and before I could get to quite solid ground, a further section of the lid gave way, precipitating me once more to the full length of the rope.

There, exhausted, weak and chilled, hanging freely in space and slowly turning round as the rope twisted one way and the other, I felt that I had done my utmost and failed, that I had no more strength to try again and that all was over except the passing. Below was a black chasm; it would be but the work of a moment to slip from the harness, then all the pain and toil would be over. It was a rare situation—a chance to quit small things for great—to pass from the petty exploration of a planet to the contemplation of vaster worlds beyond. But there was all eternity for the last and, at its longest, the present would be but short. I felt better for the thought.

My strength was fast ebbing; in a few minutes it would be too late. It was the occasion for a supreme attempt. Fired by the passion that burns the blood in the act of strife, new power seemed to come as I applied myself to one last tremendous effort. The struggle occupied some time, but I slowly worked upward to the surface. This time emerging feet first, still clinging to the rope, I pushed myself out extended at full length on the lid and then shuffled safely on to the solid ground at the side. Then came the reaction from the great nerve strain and lying there alongside the sledge my mind faded into a blank.

When consciousness returned it was a full hour or two later, for I was partly covered with newly fallen snow and numb with the cold. I took at least three hours to erect the tent, get things snugly inside and clear the snow from my clothes. Between each movement, almost, I had to rest. Then reclining in luxury in the sleeping-bag I ate a little food and thought matters over. It was a time when the mood of the Persian philosopher appealed to me:

> "Unborn To-morrow and dead Yesterday,
> Why fret about them if To-day be sweet?"

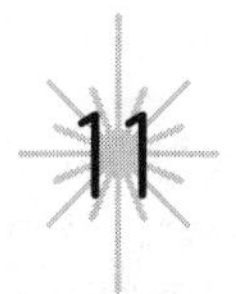

THE BLOW

from *Alone* (1938)

Richard Byrd

Midway between the Antarctica of the Heroic Age and the modern continent, the U.S. explorer who pioneered Antarctic aviation here experiences the continent at ground level; at its most icebound and solitary, in fact. In an experiment that would not have been safe without recently invented technologies, Richard Byrd overwintered solo in 1934 in an isolated weather station on the Ross Ice Shelf. It was not so very safe even so: shortly after the passage excerpted here, his heater would start pumping carbon monoxide into the air, giving him for two grim months a daily choice between heat and life and hallucinations, on the one hand, and cold and clarity of mind and death on the other. At this point, though, the inside of his shelter is still a safe haven—if he can just find his way back in.

May was a round boulder sinking before a tide. Time sloughed off the last implication of urgency, and the days moved imperceptibly one into the other. The few world news items which Dyer read to me from time to time seemed almost as meaningless and blurred as they might to a Martian. My world was insulated against the shocks running through distant economies. Advance Base was geared to different laws. On getting up in the morning, it was enough for me to say to myself: To-day is the day to change the barograph sheet, or, To-day is the day to fill the stove tank. The night was settling down in earnest. By May 17th, one month after the sun had sunk below the horizon, the noon twilight was dwindling to a mere chink in the darkness, lit by a cold reddish glow. Days when the wind brooded in the north or east, the Barrier became a vast stagnant shadow surmounted by swollen masses of clouds, one layer of darkness piled on top of the other. This was the polar night, the morbid countenance of the Ice Age. Nothing moved; nothing was visible. This was the soul of inertness. One could almost hear a distant creaking as if a great weight were settling.

Out of the deepening darkness came the cold. On May 19th, when I took the usual walk, the temperature was 65° below zero. For the first time the canvas boots failed to protect my feet. One heel was nipped, and I was forced to return to the hut and change to reindeer mukluks. That day I felt miserable; my body was racked by shooting pains—exactly as if I had been gassed. Very likely I was; in inspecting the ventilator pipes next morning I discovered that the intake pipe was completely clogged with rime and that the outlet pipe was two-thirds full. Next day—Sunday the 20th—was the coldest yet. The minimum thermometer dropped to 72° below zero; the inside thermograph, which always read a bit lower than the instruments in the shelter, stood at –74°; and the thermograph in the shelter was stopped dead—the ink, though well laced with glycerine, and the lubricant were both frozen. So violently did the air in the fuel tank expand after the stove was lit that oil went shooting all over the place; to insulate the tank against similar temperature spreads I wrapped around it the rubber air cushion which by some lucky error had been included among my gear. In the glow of a flashlight the vapor rising from the stovepipe and the outlet ventilator looked like the discharge from two steam engines. My fingers agonized over the thermograph, and I was hours putting it to rights. The fuel wouldn't flow from the drums; I had to take one inside and heat it near the stove. All day long I kept two primus stoves burning in the tunnel.

Sunday the 20th also brought a radio schedule; I had the devil's own time trying to meet it. The engine balked for an hour; my fingers were so brittle and frostbitten from tinkering with the carburetor that, when I actually made contact with Little America, I could scarcely work the key. "Ask Haines come on," was my first request. While Hutcheson searched the tunnels of Little America for the Senior Meteorologist, I chatted briefly with Charlie Murphy. Little America claimed only –60°. "But we're moving the brass monkeys below," Charlie advised. "Seventy-one below here now," I said. "You can have it," was the closing comment from the north.

Then Bill Haines's merry voice sounded in the earphones. I explained the difficulty with the thermograph. "Same trouble we've had," Bill said. "It's probably due to frozen oil. I'd suggest you bring the instrument inside, and try soaking it in gasoline, to cut whatever oil traces remain. Then rinse it in ether. As for the ink's freezing, you might try adding more glycerine." Bill was in a jovial mood. "Look at me, Admiral," he boomed. "I never have any trouble with the instruments. The trick is in having an ambitious and docile assistant." I really chuckled over that because I knew, from the first expedition, what Grimminger, the Junior Meteorologist, was going through; Bill, with his back to the fire and blandishment on his tongue, persuading the recruit that duty and the opportunity for self-improvement required him to go up into the blizzard to fix a balky trace; Bill humming to himself in the warmth of a shack while the assistant in an open pit kept a theodolite trained on the sounding balloon soaring into the night, and stuttered into a telephone the different vernier readings from which Bill was calculating the velocities and directions of the upper air currents. That day I rather wished that I, too, had an assistant. He would have taken his turn on the anemometer pole, no mistake. The frost in the iron cleats went through the fur soles of the mukluks, and froze the balls of my feet. My breath made little explosive sounds on the wind; my lungs, already sore, seemed to shrivel when I breathed.

Seldom had the aurora flamed more brilliantly. For hours the night danced to its frenetic excitement. And at times the sound of Barrier quakes was like that of heavy guns. My tongue was swollen and sore from drinking scalding hot tea, and the tip of my nose ached from frostbite. A big wind, I guessed, would come out of this still cold; it behooved me to look at my roof. I carried gallons of water topside, and poured it around the edges of the shack. It froze almost as soon as it hit. The ice was an armor plating over the packed drift.

At midnight, when I clambered topside for an auroral "ob," a wild sense of suffocation came over me the instant I pushed my shoulders through the trapdoor. My lungs gasped, but no air reached them. Bewildered and perhaps a little frightened, I slid down the ladder and lunged into the shack. In the warm air the feeling passed as quickly as it had come. Curious but cautious, I again made my way up the ladder. And again the same thing happened; I lost my breath, but I perceived why. A light air was moving down from eastward; and its bitter touch, when I faced into it, was constricting the breathing passages. So I turned my face away from it, breathing into my glove; and in that attitude finished the "ob." Before going below, I made an interesting experiment. I put a thermometer on the snow, let it lie there awhile, and discovered that the temperature at the surface was actually 5° colder than at the level of the instrument shelter, four feet higher. Reading in the sleeping bag afterwards, I froze one finger, although I shifted the book steadily from one hand to the other, slipping the unoccupied hand into the warmth of the bag.

*

Out of the cold and out of the east came the wind. It came on gradually, as if the sheer weight of the cold were almost too much to be moved. On the night of the 21st the barometer started down. The night was black as a thunderhead when I made my first trip topside; and a tension in the wind, a bulking of shadows in the night indicated that a new storm center was forming. Next morning, glad of an excuse to stay underground, I worked a long time on the Escape Tunnel by the light of a red candle standing in a snow recess. That day I pushed the emergency exit to a distance of twenty-two feet, the farthest it was ever to go. My stint done, I sat down on a box, thinking how beautiful was the red of the candle, how white the rough-hewn snow. Soon I became aware of an increasing clatter of the anemometer cups. Realizing that the wind was picking up, I went topside to make sure that everything was secured. It is a queer experience to watch a blizzard rise. First there is the wind, rising out of nowhere. Then the Barrier unwrenches itself from quietude; and the surface, which just before had seemed as hard and polished as metal, begins to run like a making sea. Sometimes, if the wind strikes hard, the drift comes across the Barrier like a hurrying white cloud, tossed hundreds of feet in the air. Other times the growth is gradual. You become conscious of a general slithering movement on all sides. The air fills with tiny scraping and sliding and rustling sounds as the first loose crystals stir. In a little while they are moving as solidly as an incoming tide, which creams over the ankles, then surges to the waist, and finally is at the throat. I have walked in drift so thick as not to be able to see a foot ahead of me; yet, when I glanced up, I could see the stars shining through the thin layer just overhead.

Smoking tendrils were creeping up the anemometer pole when I finished my inspection. I hurriedly made the trapdoor fast, as a sailor might batten down a hatch; and knowing that my ship was well secured, I retired to the cabin to ride out the storm. It could not reach me, hidden deep in the Barrier crust; nevertheless the sounds came down. The gale sobbed in the ventilators, shook the stovepipe until I thought it would be jerked out by the roots, pounded the roof with sledge-hammer blows. I could actually feel the suction effect through the pervious snow. A breeze flickered in the room and the tunnels. The candles wavered and went out. My only light was the feeble storm lantern.

Even so, I didn't have any idea how really bad it was until I went aloft for an observation. As I pushed back the trapdoor, the drift met me like a moving wall. It was only a few steps from the ladder to the instrument shelter, but it seemed more like a mile. The air came at me in snowy rushes; I breasted it as I might a heavy surf. No night had ever seemed so dark. The beam from the flashlight was choked in its throat; I could not see my hand before my face.

My windproofs were caked with drift by the time I got below. I had a vague feeling that something had changed while I was gone, but what, I couldn't tell. Presently I noticed that the shack was appreciably colder. Raising the stove lid, I was surprised to find that the fire was out, though the tank was half full. I decided that I must have turned off the valve unconsciously before going aloft; but, when I put a match to the burner, the draught down the pipe blew out the flame. The wind, then, must have killed the fire. I got it going again, and watched it carefully.

The blizzard vaulted to gale force. Above the roar the deep, taut thrumming note of the radio antenna and the anemometer guy wires reminded me of wind in a ship's rigging. The wind direction trace turned scratchy on the sheet; no doubt drift had short-circuited the electric contacts, I decided. Realizing that it was hopeless to attempt to try to keep them clear, I let the instrument be. There were other ways of getting the wind direction. I tied a handkerchief to a bamboo pole and ran it through the outlet ventilator; with a flashlight I could tell which way the cloth was whipped. I did this at hourly intervals, noting any change of direction on the sheet. But by 2 o'clock in the morning I had had enough of this periscope sighting. If I expected to sleep and at the same time maintain the continuity of the records, I had no choice but to clean the contact points.

The wind was blowing hard then. The Barrier shook from the concussions overhead; and the noise was as if the entire physical world were tearing itself to pieces. I could scarcely heave the trapdoor open. The instant it came clear I was plunged into a blinding smother. I came out crawling, clinging to the handle of the door until I made sure of my bearings. Then I let the door fall shut, not wanting the tunnel filled with drift. To see was impossible. Millions of tiny pellets exploded in my eyes, stinging like BB shot. It was even hard to breathe, because snow instantly clogged the mouth and nostrils. I made my way toward the anemometer pole on hands and knees, scared that I might be bowled off my feet if I stood erect; one false step and I should be lost forever.

I found the pole all right; but not until my head collided with a cleat. I managed to climb it, too, though ten million ghosts were tearing at me, ramming their thumbs into my eyes. But the errand was useless. Drift as thick as this would mess up the contact points as quickly as they were cleared; besides, the wind cups were spinning so fast that I stood a good chance of losing a couple of fingers in the process. Coming down the pole, I had a sense of being whirled violently through the air, with no control over my movements. The trapdoor was completely buried when I found it again, after scraping around for some time with my mittens. I pulled at the handle, first with one hand, then with both. It did not give. It's a tight fit, anyway, I mumbled to myself. The drift has probably wedged the corners. Standing astride the hatch, I braced myself and heaved with all my strength. I might just as well have tried hoisting the Barrier.

Panic took me then, I must confess. Reason fled. I clawed at the three-foot square of timber like a madman. I beat on it with my fists, trying to shake the snow loose; and, when that did no good, I lay flat on my belly and pulled until my hands went weak from cold and weariness. Then I crooked my elbow, put my face down, and said over and over again, You damn fool, you damn fool. Here for weeks I had been defending myself against the danger of being penned inside the shack; instead, I was now locked out; and nothing could be worse, especially since I had only a wool parka and pants under my windproofs. Just two feet below was sanctuary—warmth, food, tools, all the means of survival. All these things were an arm's length away, but I was powerless to reach them.

There is something extravagantly insensate about an Antarctic blizzard at night. Its vindictiveness cannot be measured on an anemometer sheet. It is more than just wind; it is a solid wall of snow moving at gale force, pounding like surf.* The whole malevolent rush is concentrated upon you as upon a personal enemy. In the senseless explosion of sound you are reduced to a crawling thing on the margin of a disintegrating world; you can't see, you can't hear, you can hardly move. The lungs gasp after the air sucked out of them, and the brain is shaken. Nothing in the world will so quickly isolate a man.

Half-frozen, I stabbed toward one of the ventilators, a few feet away. My mittens touched something round and cold. Cupping it in my hands, I pulled myself up. This was the outlet ventilator. Just why, I don't know—but instinct

* Because of this blinding, suffocating drift, in the Antarctic winds of only moderate velocity have the punishing force of full-fledged hurricanes elsewhere.

made me kneel and press my face against the opening. Nothing in the room was visible, but a dim patch of light illuminated the floor, and warmth rose up to my face. That steadied me.

Still kneeling, I turned my back to the blizzard and considered what might be done. I thought of breaking in the windows in the roof, but they lay two feet down in hard crust, and were reinforced with wire besides. If I only had something to dig with, I could break the crust and stamp the windows in with my feet. The pipe cupped between my hands supplied the first inspiration; maybe I could use that to dig with. It, too, was wedged tight; I pulled until my arms ached, without budging it; I had lost all track of time, and the despairing thought came to me that I was lost in a task without an end. Then I remembered the shovel. A week before, after levelling drift from the last light blow, I had stabbed a shovel handle up in the crust somewhere to leeward. That shovel would save me. But how to find it in the avalanche of the blizzard?

I lay down and stretched out full length. Still holding the pipe, I thrashed around with my feet, but pummeled only empty air. Then I worked back to the hatch. The hard edges at the opening provided another grip, and again I stretched out and kicked. Again no luck. I dared not let go until I had something else familiar to cling to. My foot came up against the other ventilator pipe. I edged back to that, and from the new anchorage repeated the maneuver. This time my ankle struck something hard. When I felt it and recognized the handle, I wanted to caress it.

Embracing this thrice-blessed tool, I inched back to the trapdoor. The handle of the shovel was just small enough to pass under the little wooden bridge which served as a grip. I got both hands on the shovel and tried to wrench the door up; my strength was not enough, however. So I lay down flat on my belly and worked my shoulders under the shovel. Then I heaved, the door sprang open, and I rolled down the shaft. When I tumbled into the light and warmth of the room, I kept thinking, How wonderful, how perfectly wonderful.

THE BLASPHEMOUS CITY

from *At the Mountains of Madness* (1936)

H. P. Lovecraft

Richard Byrd had a fervent reader in the horror novelist H. P. Lovecraft, who saw that the plane journeys across titanic landscapes, and even the sense of a terrain quivering with some not-quite-sayable message, could be recruited into his private cosmos. With purple gusto At the Mountains of Madness *expands the Transantarctic Range into something truly outsized, a fit location for another of the secrets which—as ever in Lovecraft—make the sanity of the beholder snap like overstretched elastic. Here, Lovecraft's use of Antarctica can stand in for the whole vein of paranoid fantasy in which the continent hides away lost tribes, mysterious temples, and swastika-painted UFOs popping out of holes in the ice-cap like Nazi party-favors.*

IN SPITE OF ALL the prevailing horrors, we were left with enough sheer scientific zeal and adventurousness to wonder about the unknown realm beyond those mysterious mountains. As our guarded messages stated, we rested at midnight after our day of terror and bafflement—but not without a tentative plan for one or more range-crossing altitude flights in a lightened plane with aerial camera and geologist's outfit, beginning the following morning. It was decided that Danforth and I try it first, and we awaked at 7 A.M. intending an early flight; however, heavy winds—mentioned in our brief bulletin to the outside world—delayed our start till nearly nine o'clock.

I have already repeated the noncommittal story we told the men at camp—and relayed outside—after our return sixteen hours later. It is now my terrible duty to amplify this account by filling in the merciful blanks with hints of what we really saw in the hidden transmontane world—hints of the revelations which have finally driven Danforth to a nervous collapse. I wish he would add a really frank word about the thing which he thinks he alone saw—even though it was probably a nervous delusion—and which was perhaps the last straw that put him where he is; but he is firm against that. All I can do is to repeat his later disjointed whispers about what set him shrieking as the plane soared back through the wind-tortured mountain pass after that real and tangible shock which I shared. This will form my last word. If the plain signs of surviving elder horrors in what I disclose be not enough to keep others from meddling with the inner antarctic—or at least from prying too deeply beneath the surface of that ultimate waste of forbidden secrets and inhuman, aeon-cursed desolation—the responsibility for unnamable and perhaps immeasurable evils will not be mine.

Danforth and I, studying the notes made by Pabodie in his afternoon flight and checking up with a sextant, had calculated that the lowest available pass in the range lay somewhat to the right of us, within sight of camp, and about twenty-three thousand or twenty-four thousand feet above sea level. For this point, then, we first headed in the lightened plane as we embarked on our flight of discovery. The camp itself, on foothills which sprang from a high continental plateau, was some twelve thousand feet in altitude; hence the actual height increase necessary was not so vast as it might seem. Nevertheless we were acutely conscious of the rarefied air and intense cold as we rose; for, on account of visibility conditions, we had to leave the cabin windows open. We were dressed, of course, in our heaviest furs.

As we drew near the forbidding peaks, dark and sinister above the line of crevasse-riven snow and interstitial glaciers, we noticed more and more the curiously regular formations clinging to the slopes; and thought again of the strange Asian paintings of Nicholas Roerich. The ancient and wind-weathered rock strata fully verified all of Lake's bulletins, and proved that these pinnacles had been towering up in exactly the same way since a surprisingly early time in earth's history—perhaps over fifty million years. How much higher they had once been, it was futile to guess; but everything about this strange region pointed to obscure atmospheric influences unfavorable to change, and calculated to retard the usual climatic processes of rock disintegration.

But it was the mountainside tangle of regular cubes, ramparts, and cave mouths which fascinated and disturbed us most. I studied them with a field glass and took aerial photographs while Danforth drove; and at times I relieved him at the controls—though my aviation knowledge was purely an amateur's—in order to let him use the binoculars. We could easily see that much of the material of the things was a lightish Archaean quartzite, unlike any formation visible over broad areas of the general surface; and that their regularity was extreme and uncanny to an extent which poor Lake had scarcely hinted.

As he had said, their edges were crumbled and rounded from untold aeons of savage weathering; but their preternatural solidity and tough material had saved them from obliteration. Many parts, especially those closest to the slopes, seemed identical in substance with the surrounding rock surface. The whole arrangement looked like the ruins of Macchu Picchu in the Andes, or the primal foundation walls of Kish as dug up by the Oxford Field Museum Expedition in 1929; and both Danforth and I obtained that occasional impression of separate Cyclopean blocks which Lake had attributed to his flight-companion Carroll. How to account for such things in this place was frankly beyond me, and I felt queerly humbled as a geologist. Igneous formations often have strange regularities—like the famous Giants' Causeway in Ireland—but this stupendous range, despite Lake's original suspicion of smoking cones, was above all else nonvolcanic in evident structure.

The curious cave mouths, near which the odd formations seemed most abundant, presented another albeit a lesser puzzle because of their regularity of outline. They were, as Lake's bulletin had said, often approximately square or semicircular; as if the natural orifices had been shaped to greater symmetry by some magic hand. Their numerousness and wide distribution were remarkable, and suggested that the whole region was honeycombed with tunnels dissolved out of limestone strata. Such glimpses as we secured did not extend far within the caverns, but we saw that they were apparently clear of stalactites and stalagmites. Outside, those parts of the mountain slopes adjoining the apertures seemed invariably smooth and regular; and Danforth thought that the slight cracks and pittings of the weathering tended toward unusual patterns. Filled as he was with the horrors and strangenesses discovered at the camp, he hinted that the pittings vaguely resembled those baffling groups of dots sprinkled over the primeval greenish soapstones, so hideously duplicated on the madly conceived snow mounds above those six buried monstrosities.

We had risen gradually in flying over the higher foothills and along toward the relatively low pass we had selected. As we advanced we occasionally looked down at the snow and ice of the land route, wondering whether we could have attempted the trip with the simpler equipment of earlier days. Somewhat to our surprise we saw that the terrain was far from difficult as such things go; and that despite the crevasses and other bad spots it would not have been likely to deter the sledges of a Scott, a Shackleton, or an Amundsen. Some of the glaciers appeared to lead up to wind-bared passes with unusual continuity, and upon reaching our chosen pass we found that its case formed no exception.

Our sensations of tense expectancy as we prepared to round the crest and peer out over an untrodden world can hardly be described on paper; even though we had no cause to think the regions beyond the range essentially different from those already seen and traversed. The touch of evil mystery in these barrier mountains, and in the beckoning sea of opalescent sky glimpsed betwixt their summits, was a highly subtle and attenuated matter not to be explained in literal words. Rather was it an affair of vague psychological symbolism and aesthetic association—a thing mixed up with exotic poetry and paintings, and with archaic myths lurking in shunned and forbidden volumes. Even the wind's burden held a peculiar strain of conscious malignity; and for a second it seemed that the composite sound included a bizarre musical whistling or piping over a wide range as the blast swept in and out of the omnipresent and resonant cave mouths. There was a cloudy note of reminiscent repulsion in this sound, as complex and unplaceable as any of the other dark impressions.

We were now, after a slow ascent, at a height of twenty-three thousand, five hundred and seventy feet according to the aneroid; and had left the region of clinging snow definitely below us. Up here were only dark, bare rock slopes and the start of rough-ribbed glaciers—but with those provocative cubes, ramparts, and echoing cave mouths to add a portent of the unnatural, the fantastic, and the dreamlike. Looking along the line of high peaks, I thought I could see the one mentioned by poor Lake, with a rampart exactly on top. It seemed to be half lost in a queer antarctic haze—such a haze, perhaps, as had been responsible for Lake's early notion of volcanism. The pass loomed directly before us, smooth and windswept between its jagged and malignly frowning pylons. Beyond it was a sky fretted with swirling vapors and lighted by the low polar sun—the sky of that mysterious farther realm upon which we felt no human eye had ever gazed.

A few more feet of altitude and we would behold that realm. Danforth and I, unable to speak except in shouts amidst the howling, piping wind that raced through the pass and added to the noise of the unmuffled engines, exchanged eloquent glances. And then, having gained those last few feet, we did indeed stare across the momentous divide and over the unsampled secrets of an elder and utterly alien earth.

I think that both of us simultaneously cried out in mixed awe, wonder, terror, and disbelief in our own senses as we finally cleared the pass and saw what lay beyond. Of course, we must have had some natural theory in the back of our heads to steady our faculties for the moment. Probably we thought of such things as the grotesquely weathered stones of the Garden of the Gods in Colorado, or the fantastically symmetrical wind-carved rocks of the Arizona desert. Perhaps we even half thought the sight a mirage like that we had seen the morning before on first approaching those mountains of madness. We must have had some such normal notions to fall back upon as our eyes swept that limitless, tempest-scarred plateau and grasped the almost endless labyrinth of colossal, regular, and geometrically eurythmic stone masses which reared their crumbled and pitted crests above a glacial sheet not more than forty or fifty feet deep at its thickest, and in places obviously thinner.

The effect of the monstrous sight was indescribable, for some fiendish violation of known natural law seemed certain at the outset. Here, on a hellishly ancient table-land fully twenty thousand feet high, and in a climate deadly to habitation since a prehuman age not less than five hundred thousand years ago, there stretched nearly to the vision's limit a tangle of orderly stone which only the desperation of mental self-defense could possibly attribute to any but conscious and artificial cause. We had previously dismissed, so far as serious thought was concerned, any theory that the cubes and ramparts of the mountainsides were other than natural in origin. How could they be otherwise, when man himself could scarcely have been differentiated from the great apes at the time when this region succumbed to the present unbroken reign of glacial death?

Yet now the sway of reason seemed irrefutably shaken, for this Cyclopean maze of squared, curved, and angled blocks had features which cut off all comfortable refuge. It was, very clearly, the blasphemous city of the mirage in stark, objective, and ineluctable reality. That damnable portent had had a material basis after all—there had been some horizontal stratum of ice dust in the upper air, and this shocking stone survival had projected its image across the mountains according to the simple laws of reflection. Of course, the phantom had been twisted and exaggerated, and had contained things which the real source did not contain; yet now, as we saw that real source, we thought it even more hideous and menacing than its distant image.

Only the incredible, unhuman massiveness of these vast stone towers and ramparts had saved the frightful things from utter annihilation in the hundreds of thousands—perhaps millions—of years it had brooded there amidst the blasts of a bleak upland. "Corona Mundi—Roof of the World—" All sorts of fantastic phrases sprang to our lips as we looked dizzily down at the unbelievable spectacle. I thought again of the eldritch primal myths that had so persistently haunted me since my first sight of this dead antarctic world—of the demoniac plateau of Leng, of the Mi-Go, or abominable Snow Men of the Himalayas, of the Pnakotic Manuscripts with their prehuman implications, of the Cthulhu cult, of the Necronomicon, and of the Hyperborean legends of formless Tsathoggua and the worse than formless star spawn associated with that semientity.

For boundless miles in every direction the thing stretched off with very little thinning; indeed, as our eyes followed it to the right and left along the base of the low, gradual foothills which separated it from the actual mountain rim, we decided that we could see no thinning at all except for an interruption at the left of the pass through which we had come. We had merely struck, at random, a limited part of something of incalculable extent. The foothills were more sparsely sprinkled with grotesque stone structures, linking the terrible city to the already familiar cubes and ramparts which evidently formed its mountain outposts. These latter, as well as the queer cave mouths, were as thick on the inner as on the outer sides of the mountains.

The nameless stone labyrinth consisted, for the most part, of walls from ten to one hundred and fifty feet in ice-clear height, and of a thickness varying from five to ten feet. It was composed mostly of prodigious blocks of dark primordial slate, schist, and sandstone—blocks in many cases as large as $4 \times 6 \times 8$ feet—though in several places it seemed to be carved out of a solid, uneven bed rock of pre-Cambrian slate. The buildings were far from equal in size, there being innumerable honeycomb arrangements of enormous extent as well as smaller separate structures. The general shape of these things tended to be conical, pyramidal, or terraced; though there were many perfect cylinders, perfect cubes, clusters of cubes, and other rectangular forms, and a peculiar sprinkling of angled edifices whose five-pointed ground plan roughly suggested modern fortifications. The builders had made constant and expert use of the principle of the arch, and domes had probably existed in the city's heyday.

The whole tangle was monstrously weathered, and the glacial surface from which the towers projected was strewn with fallen blocks and immemorial debris. Where the glaciation was transparent we could see the lower parts of the gigantic piles, and we noticed the ice-preserved stone bridges which connected the different towers at varying distances above the ground. On the exposed walls we could detect the scarred places where other and higher bridges of the same sort had existed. Closer inspection revealed countless largish windows; some of which were closed with shutters of a petrified material originally wood, though most gaped open in a sinister and menacing fashion. Many of the ruins, of course, were roofless, and with uneven though wind-rounded upper edges; whilst others, of a more sharply conical or pyramidal model or else protected by higher surrounding structures, preserved intact outlines despite the omnipresent crumbling and pitting. With the field glass we could barely make out what seemed to be sculptural decorations in horizontal bands—decorations including those curious groups of dots whose presence on the ancient soapstones now assumed a vastly larger significance.

In many places the buildings were totally ruined and the ice sheet deeply riven from various geologic causes. In other places the stonework was worn down to the very level of the glaciation. One broad swath, extending from the plateau's interior to a cleft in the foothills about a mile to the left of the pass we had traversed, was wholly free from buildings. It probably represented, we concluded, the course of some great river which in Tertiary times—millions of years ago—had poured through the city and into some prodigious subterranean abyss of the great barrier range. Certainly, this was above all a region of caves, gulfs, and underground secrets beyond human penetration.

Looking back to our sensations, and recalling our dazedness at viewing this monstrous survival from aeons we had thought prehuman, I can only wonder that we preserved the semblance of equilibrium, which we did. Of course, we knew that something—chronology, scientific theory, or our own consciousness—was woefully awry; yet we kept enough poise to guide the plane, observe many things quite minutely, and take a careful series of photographs which may yet serve both us and the world in good stead. In my case, ingrained scientific habit may have helped; for above all my bewilderment and sense of menace, there burned a dominant curiosity to fathom more of this age-old secret—to know what sort of beings had built and lived in this incalculably gigantic place, and what relation to the general world of its time or of other times so unique a concentration of life could have had.

For this place could be no ordinary city. It must have formed the primary nucleus and center of some archaic and unbelievable chapter of earth's history whose outward ramifications, recalled only dimly in the most obscure and distorted myths, had vanished utterly amidst the chaos of terrene convulsions long before any human race we know had shambled out of apedom.

Here sprawled a Palaeogaean megalopolis compared with which the fabled Atlantis and Lemuria, Commoriom and Uzuldaroum, and Olathoc in the land of Lomar, are recent things of today—not even of yesterday; a megalopolis ranking with such whispered prehuman blasphemies as Valusia, R'lyeh, Ib in the land of Mnar, and the Nameless city of Arabia Deserta. As we flew above that tangle of stark titan towers my imagination sometimes escaped all bounds and roved aimlessly in realms of fantastic associations—even weaving links betwixt this lost world and some of my own wildest dreams concerning the mad horror at the camp.

The plane's fuel tank, in the interest of greater lightness, had been only partly filled; hence we now had to exert caution in our explorations. Even so, however, we covered an enormous extent of ground—or, rather, air—after swooping down to a level where the wind became virtually negligible. There seemed to be no limit to the mountain range, or to the length of the frightful stone city which bordered its inner foothills. Fifty miles of flight in each direction showed no major change in the labyrinth of rock and masonry that clawed up corpselike through the eternal ice. There were, though, some highly absorbing diversifications; such as the carvings on the canyon where that broad river had once pierced the foothills and approached its sinking place in the great range. The headlands at the stream's entrance had been boldly carved into Cyclopean pylons; and something about the ridgy, barrel-shaped designs stirred up oddly vague, hateful, and confusing semi-remembrances in both Danforth and me.

We also came upon several star-shaped open spaces, evidently public squares, and noted various undulations in the terrain. Where a sharp hill rose, it was generally hollowed out into some sort of rambling-stone edifice; but there were at least two exceptions. Of these latter, one was too badly weathered to disclose what had been on the jutting eminence, while the other still bore a fantastic conical monument carved out of the solid rock and roughly resembling such things as the well-known Snake Tomb in the ancient valley of Petra.

Flying inland from the mountains, we discovered that the city was not of infinite width, even though its length along the foothills seemed endless. After about thirty miles the grotesque stone buildings began to thin out, and in ten more miles we came to an unbroken waste virtually without signs of sentient artifice. The course of the river beyond the city seemed marked by a broad, depressed line, while the land assumed a somewhat greater ruggedness, seeming to slope slightly upward as it receded in the mist-hazed west.

So far we had made no landing, yet to leave the plateau without an attempt at entering some of the monstrous structures would have been inconceivable. Accordingly, we decided to find a smooth place on the foothills near our navigable pass, there grounding the plane and preparing to do some exploration on foot. Though these gradual slopes were partly covered with a scattering of ruins, low flying soon disclosed an ampler number of possible landing places. Selecting that nearest to the pass, since our flight would be across the great range and back to camp, we succeeded about 12:30 P.M. in effecting a landing on a smooth, hard snow field wholly devoid of obstacles and well adapted to a swift and favorable take-off later on.

It did not seem necessary to protect the plane with a snow banking for so brief a time and in so comfortable an absence of high winds at this level; hence we merely saw that the landing skis were safely lodged, and that the vital parts of the mechanism were guarded against the cold. For our foot journey we discarded the heaviest of our flying furs, and took with us a small outfit consisting of pocket compass, hand camera, light provisions, voluminous notebooks and paper, geologist's hammer and chisel, specimen bags, coil of climbing rope, and powerful electric torches with extra batteries; this equipment having been carried in the plane on the chance that we might be able to effect a landing, take ground pictures, make drawings and topographical sketches, and obtain rock specimens from some bare slope, outcropping, or mountain cave. Fortunately we had a supply of extra paper to tear up, place in a spare specimen bag, and use on the ancient principle of hare and hounds for marking our course in any interior mazes we might be able to penetrate. This had been brought in case we found some cave system with air quiet enough to allow such a rapid and easy method in place of the usual rock-chipping method of trail blazing.

Walking cautiously downhill over the crusted snow toward the stupendous stone labyrinth that loomed against the opalescent west, we felt almost as keen a sense of imminent marvels as we had felt on approaching the unfathomed mountain pass four hours previously. True, we had become visually familiar with the incredible secret concealed by the barrier peaks; yet the prospect of actually entering primordial walls reared by conscious beings perhaps millions of years ago—before any known race of men could have existed—was none the less awesome and potentially terrible in its implications of cosmic abnormality. Though the thinness of the air at this prodigious altitude made exertion somewhat more difficult than usual, both Danforth and I found ourselves bearing up very well, and felt equal to almost any task which might fall to our lot. It took only a few steps to bring us to a shapeless ruin worn level with the snow, while ten or fifteen rods farther on there was a huge, roofless rampart still complete in its gigantic five-pointed outline and rising to an irregular height of ten or eleven feet. For this latter we headed; and when at last we were actually able to touch its weathered Cyclopean blocks, we felt that we had established an unprecedented and almost blasphemous link with forgotten aeons normally closed to our species.

This rampart, shaped like a star and perhaps three hundred feet from point to point, was built of Jurassic sandstone blocks of irregular size, averaging 6×8 feet in surface. There was a row of arched loopholes or windows about four feet wide and five feet high, spaced quite symmetrically along the points of the star and at its inner angles, and with the bottoms about four feet from the glaciated surface. Looking through these, we could see that the masonry was fully five feet thick, that there were no partitions remaining within, and that there were traces of banded carvings or bas-reliefs on the interior walls—facts we had indeed guessed before, when flying low over this rampart and others like it. Though lower parts must have originally existed, all traces of such things were now wholly obscured by the deep layer of ice and snow at this point.

We crawled through one of the windows and vainly tried to decipher the nearly effaced mural designs, but did not attempt to disturb the glaciated floor. Our orientation flights had indicated that many buildings in the city proper were less ice-choked, and that we might perhaps find wholly clear interiors leading down to the true ground level if we entered those structures still roofed at the top. Before we left the rampart we photographed it carefully, and studied its mortar-less Cyclopean masonry with complete bewilderment. We wished that Pabodie were present, for his engineering knowledge might have helped us guess how such titanic blocks could have been handled in that unbelievably remote age when the city and its outskirts were built up.

The half-mile walk downhill to the actual city, with the upper wind shrieking vainly and savagely through the skyward peaks in the background, was something of which the smallest details will always remain engraved on my mind. Only in fantastic nightmares could any human beings but Danforth and me conceive such optical effects. Between us and the churning vapors of the west lay that monstrous tangle of dark stone towers, its outré and incredible forms impressing us afresh at every new angle of vision. It was a mirage in solid stone, and were it not for the photographs, I would still doubt that such a thing could be. The general type of masonry was identical with that of the rampart we had examined; but the extravagant shapes which this masonry took in its urban manifestations were past all description.

Even the pictures illustrate only one or two phases of its endless variety, preternatural massiveness, and utterly alien exoticism. There were geometrical forms for which an Euclid would scarcely find a name—cones of all degrees of irregularity and truncation, terraces of every sort of provocative disproportion, shafts with odd bulbous enlargements, broken columns in curious groups, and five-pointed or five-ridged arrangements of mad grotesqueness. As we drew nearer we could see beneath certain transparent parts of the ice sheet, and detect some of the tubular stone bridges that connected the crazily sprinkled structures at various heights. Of orderly streets there seemed to be none, the only broad open swath being a mile to the left, where the ancient river had doubtless flowed through the town into the mountains.

Our field glasses showed the external, horizontal bands of nearly effaced sculptures and dot groups to be very prevalent, and we could half imagine what the city must once have looked like—even though most of the roofs and tower tops had necessarily perished. As a whole, it had been a complex tangle of twisted lanes and alleys, all of them deep canyons, and some little better than tunnels because of the overhanging masonry or overarching bridges. Now, outspread below us, it loomed like a dream fantasy against a westward mist through whose northern end the low, reddish antarctic sun of early afternoon was struggling to shine; and when, for a moment, that sun encountered a denser obstruction and plunged the scene into temporary shadow, the effect was subtly menacing in a way I can never hope to depict. Even the faint howling and piping of the unfelt wind in the great mountain passes behind us took on a wilder note of purposeful malignity. The last stage of our descent to the town was unusually steep and abrupt, and a rock outcropping at the edge where the grade changed led us to think that an artificial terrace had once existed there. Under the glaciation, we believed, there must be a flight of steps or its equivalent.

When at last we plunged into the town itself, clambering over fallen masonry and shrinking from the oppressive nearness and dwarfing height of omnipresent crumbling and pitted walls, our sensations again became such that I marvel at the amount of self-control we retained. Danforth was frankly jumpy, and began making some offensively irrelevant speculations about the horror at the camp—which I resented all the more because I could not help sharing certain conclusions forced upon us by many features of this morbid survival from nightmare antiquity. The speculations worked on his imagination, too; for in one place—where a debris-littered alley turned a sharp corner—he insisted that he saw faint traces of ground markings which he did not like; whilst elsewhere he stopped to listen to a subtle, imaginary sound from some undefined point—a muffled musical piping, he said, not unlike that of the wind in the mountain caves, yet somehow disturbingly different. The ceaseless five-pointedness of the surrounding architecture and of the few distinguishable mural arabesques had a dimly sinister suggestiveness we could not escape, and gave us a touch of terrible subconscious certainty concerning the primal entities which had reared and dwelt in this unhallowed place.

SEA DADDY, STORMALONG JOHN AND MINUTE MAID

from *Life at the Bottom* (1977)

John Langone

John Langone was a journalist who set out to capture the timbre of daily life in American Antarctica, just as the phase of military dominance was drawing to a close. Here, three U.S. Navy Seabees talk about overwintering at a South Pole which is no longer an abstract spot on the map, or a goal of heroic struggle, but a posting; one place, exceptionally cold and weird and sex-deprived, in the roster of places a guy might be sent to work construction and maintenance under difficult circumstances. Blue-collar Antarctica would remain, but it would never again be this purely male, or this much a pure extension of military culture.

"THE HARD PART ABOUT IT," says Minute Maid, sipping his coffee, "is when you come back to the world, to the hustle and bustle. Down here, once you get into the winter, you just get into a routine and it's nice and slow. Comin' off the ice can be rougher, to my way of thinkin', than goin' on. Like, you know what one of the first things that just hits at you is when you get off here? It's the shrill sounds that a woman makes. You get to really notice her voice, that shrill kind of piercing voice. And you notice the hardness of the pavements, and the different smells, and the traffic noises."

"Yeah," says Sea Daddy, "there's no red lights here, no traffic like that, no nothin', and you're on your own, that's why I like it, it's real fine."

"Different at Palmer than here, though," says Minute Maid. "This here place's a metropolis; down at Palmer we had ten, eleven of us."

"Whooee, beach party place, man," laughs Sea Daddy, showing white teeth through his black beard. "How'd you guys stand the fuckin' heat?"

"Buried myself in my work," says Minute Maid. "Mechanic and equipment operator, but you wouldn't know nothin' about that shit, Daddy; I do more accidentally than you do on purpose. Let me tell you, there was only but four Seabees in the whole place, so you sort of did just about everything that came up, like a lot of projects. We got a little bit of everything from building to utilities, which is plumbing. And I didn't volunteer for it, neither, not like that ass-hole, Daddy over there, and I just came up for orders. When they came in they had the words 'Deep Freeze' on there, you know? I called the detailer, which is the mechanic chief, Buddy Dew, and I talked to him and he says, well, he's been down there and he says he thinks it's a good thing for me, and I ought to go down one time anyway. I says, thanks a lot, you fat fuck, I could really care what you think of the place, I don't want to go, no way. But they shipped my ass down anyway."

"I had thirteen seasons down and three winters-over," Sea Daddy says proudly, thumping the table three times. "I own the fuckin' place. Pole, Byrd, McMurdo, name it, men. I was there when Crazy Charlie was down."

Stormalong John comes alive at Charlie's name, smiles wistfully, shakes his head. "Oh Jesus, Crazy Charlie. Used to get a little drunk down there, paint a face on the back of his head, didn't know whether he was comin' or goin'. Pasted a battleship on the top of his head and went to sleep like that at night. Crazy stuff like that to get morale goin', that's why they called him Crazy Charlie."

"Yeah, lots of laughs in those days," says Sea Daddy. "It's changed now. Place is becomin' Skirt City. Used to get lots of things in the old days. It's all over now, Navy's gettin' out in a couple years once we get Pole built, I'm gettin' out. Used to be like when you wintered over you got a promotion, sometimes got a first duty choice, sometimes got somethin' named after you. Now they're siftin' so many people in and out, it's like a Ford plant down here, not what they call . . . used to call, exotic duty, you know."

"I'll say one thing, though," says Minute Maid. "Everyone who asked for their first choice while down there this year got it. Palmer is good that way. Only thing is I didn't ask to go there in the first place, and I didn't think it was so great, duty or not. One year, right? Nine months before that I got Vietnam. Separated from my wife for a year then, right? I really didn't care that much for it. Now I'm down here. This time, all the guys were tellin' me that when I get the screenin', you know, all I got to do is run in there and kiss the first guy I see, and they'll let me out. Well, I didn't exactly do that, but when the shrink asks me how I felt about goin' onto the ice, I tell him anybody goes to the Antarctic and winters over is a nut. And you know what he says? He says, you're all right, get your gear together, buddy. Well, I got down there, and I didn't think too much about it until my ship, *Hero,* pulls out. Last one to leave, you know, pulls out in April and that's it, no more for a long time.

"Well, I hated it, but it really wasn't too bad, lowest it got was ten below, highest was fifty-five. But you still got all your Antarctic differences, changing winds that blow in circles, the isolation. They kept us busy, just busy. Only thing was I got tired of seeing the same guys all the time. Only two buildings at Palmer. To get away from the guys you had to go on a hike or some such. Like this one guy and I we went out onto a glacier with our sleeping bags and slept up there, just to break the monotony. But you know, I wouldn't winter over again. This is fine for a few months here at McMurdo, but I wouldn't winter over again. I've asked for summer support and I'll be goin' down a few more times, and there is an advantage of sorts, you know. You get to meet different people from other countries and you never find any hostility. Everybody is human down here. And there's somethin' about the Antarctic that you just can't describe. It has its own . . . environment or somethin'. I can't get the right words . . . But it's just . . . somethin' . . ."

Stormalong John says, "Let me say something about that. Last year I was out in the Dry Valleys at Vanda, and there was nobody around, nobody. I couldn't see a soul. The helo left and went over to somewhere else for a few hours and I couldn't see a soul. In my whole life, I never been in a situation like that, where you feel like it's the cleanest, nicest place in the world. But I think I've had enough, too; I've put in four times, given 'em three winters. And I put in for twenty choices of duty, too, and I got none of 'em. Last time, before this one, I got orders for Cape Hatteras, and I just had left one year of isolated duty at Pole, so they sent me to Cape Hatteras, must have figured I needed a rest. But that place. Nearest dime store is fifty miles away and that's in the so-called States. Base was fallin' apart, they wanted it repaired, and there I was with a ulcer tour starin' me in the face, and I just got done with one ulcer tour in the Antarctic. Finished up the Cape Hatteras tour, and come back here, and figured that was the least of two evils. I don't really mind it, though. The wintering over is really the best part of it. Once you get the summer support people and most of the scientists the hell out, then it's good duty. During the winter you don't have fifteen, twenty bosses trying to tell you how to run a job, from the skipper on down."

"Yeah," says Sea Daddy. "More damned tourists comin' down, plus scientists, tellin' you what to do, tryin' to do five hundred things when they should be doin' one, or maybe none, lookin' at some of these ding-dongs."

[. . .]

"It's better, like I say, in the winter. But, about halfway through, things start to happen, like you may be topside throwin' snow, and one time I turned out the lights and started lookin' out there, just standin' and lookin' and it's so pitch black, and I thought, why you dumb sonofabitch, what are you doin' here anyway? You got to be crazy. Periodically, you'll do that. Everybody'll sit there and all of a sudden someone will say, hey, what the hell am I doin' here? I must be fuckin' nuts. But all in all, I liked it. It ain't such a bad place. You're more or less your own boss, particularly at the outlying stations, and if you want to try something new they'll let you, build somethin' or try an experiment. If it works, great, if it doesn't, well, WTF. There's just no place in the service that you have that except on the ice. There's no place to go, but there's freedom.

"You either like it or you don't. And if you're winterin' in and you don't like it, well, man, you're in trouble.

"I was up at Pole when they locked up the first guy they ever locked up in the Antarctic. We built a brig and shoved his ass in it. He was with the weather people, an ex-Air Force guy, and he seemed like a real nice fella during the summer. Well, the day the last plane left he did a one-eight. He became a problem. Drinking, liar, a thief. Finally, one night they put him on medication, and he stayed on it and he seemed to do pretty good. Well, then he said the hell with it and he went bananas. When he first started up like that, the OIC says we're going to build a brig, and this guy got the message, he knew it was for him. They held that over him for a while, and he stuck straight. But, just when we got to where we had six to eight weeks left before we got relieved, he flipped again, and he got hold of some booze and some medicine, and he just went snaky. He decked the medical officer, and he decked me when I jumped in, and he run out and got a fire ax and started heavin' that around. Well, we finally got hold of him and quieted him down, and I told the watch that if anything happened don't mess around with this dude, just hit the fire bell. Well, I'm in the sack and this alarm goes off, and it was him, out on a rampage again. We finally found him hidin' out in the club, and there were twenty of us, and we gave him a choice, beat his ass off or he goes into the brig peacefully. Well, he says he's going to call his lawyer in New York, going to sue the OIC, me, the whole fuckin' Navy, goin' to contact his lawyer with a ham radio, he says.

"Well, we canned him for three days, and we sent a message out when we locked this dude up, and this was the first word, our message, that the admiral back in Christchurch had that we had a problem. He gets this message that we done locked up a civilian at Pole Station, and he went right up through the overhead. Well, he sent a message to McMurdo, told 'em to tell us to unlock this dude, we don't confine nobody, but nobody, much less a civilian, turn him loose or it's our asses. Well, the OIC, he starts sweatin', he's got twenty years in, and he says to me, hey, what do you make for base pay 'cause I think I'm goin' to be busted to chief after this is over.

"So, we let his ass out and put locks on all our doors, and he's out, roaming around the corridors night and mornin', and we stayed away from him as far as we could. We put up with that six weeks, scared shitless. Nothin' came of it that I know; they yanked six weeks off his pay. I think we would have killed him if he started in on us again during that six weeks. There was also this other civilian at Byrd Station, just started walkin' and just took off one night."

"That guy was so depressed," says Sea Daddy, "we never found a trace of him. Found the dog, but not him. We had some guys like that fella up at Pole, but after a while they straightened out. You do get a little downhearted when the planes go out, and I missed all the football.

[. . .]

"But the funny part of it was when the reliefs come in, and that's somethin' to see, when they come in. Scared, don't know where they're at, in a state of shock, they are. Well, you're so glad to see these guys you start a party. The admiral come out this one time on the first plane and stayed overnight, and we're in the club havin' our party. These new guys have altitude sickness, you know, and they're tired and not climatized and they're just sittin' around sort of stunned lookin' at these here animals. Well, the admiral is out there talkin' to the OIC, and our cook, Herbie, has got a beanie on. We built him this propeller on top, and he's really blowed out of his mind on them vodka freezes he used to whip up; he did that with snow and a eggbeater and they'll send you higher than helium. Well, Herbie walks over to the admiral and he says, hey admiral, have a drink. The admiral says, no thanks. And you know what that crazy-ass Herbie does? He pours one right over the admiral's head, right over the rear admiral's head, it's dripping right down his parka. Well, the admiral, he's a good shit, no one like him that I know, and he says to the OIC, ah, doctor, I uh think you ought to secure this here little party. And the OIC says, admiral, I ain't never secured one of their parties and I can't start now."

Stormalong John laughs hard, and Minute Maid asks him about the time they built this cage, for the benefit of the relief party, and they put this guy in it, and when the plane comes in there it is waiting for them.

"Oh Jesus, we used to do that shit all the time for the reliefs, scared the piss out of 'em. One time in sixty-nine our foreman came dressed as Mickey Mouse, made this plaster of paris Mickey Mouse suit, a beautiful thing. When the first plane drops down, there's the reliefs on board, eyes buggin' out, and there's ol' Myron standin' out there with this suit on, wavin' the plane to the fuel pit. Another year we ran up weather balloons, up about two hundred feet, strung a big sign between 'em, WATCH YOUR ASS. The best thing we did, though, was get out there with nothin' on, just your bunny boots, and you'd stand there, wavin' bottles of booze, and they'd just about shit, think we'd all gone Asiatic."

Stormalong John pulls out a card and waves it. It reads, "This is to certify that Stormalong John Wheeler, being of sufficient courage and questionable sanity, is a member of South Pole 200 Degree Club. Temperature: −108 F. Nutus Extremis."

"We started this club, and we built the first steam bath out there in sixty-four. The OIC, being a doctor, he's not familiar with Seabees, and they'll do anything. So we decided we're goin' to build a steam bath. Took a fifty-five-gallon drum, cut two-thirds of it off, stuck some holes in her, welded in some fittings, put in some electric hot water tank heating elements, and a float valve in there with a water line comin' into it so it maintained its own level, and we wired that up to a thermostat and took a piece of sheet metal and cut a hole out of it and set that on top of this, and we run a pipe out into this room we built under the ice at Pole. In July, we decided to commission her, and we called the OIC up and said, doc, we goin' to fire up the steam bath. By this time, he's ready for anything, but all he says—he's shakin' his head and stayin' in his quarters a lot—he says, it won't work. So we told him, oh yeah, that's what they told Orville, and we set the thermostat at one hundred eighteen, and we're in there in our skivvies, about twenty minutes. Somebody says, let's go out and roll in the snow. So we opened the door and run upstairs, and that night at Pole it was a record cold, one hundred twelve point five below. And that ain't includin' any wind-chill factor, either, like they do at McMurdo in them familygrams to let the folks back home figure it's a lot colder than it is. That's straight cold, I don't know what it is with wind. Well, we rolled around out there for a few minutes, and it wasn't too bad; you couldn't feel it at the beginning. When they took our pictures, you couldn't see nothin', only this blob of steam your body's throwin' off. Breaks the monotony. It's entertainment."

WHITE LANTERNS

from *The Moon by Whale Light* (1991)

Diane Ackerman

The New Yorker *writer Diane Ackerman travelled to the Antarctic Peninsula as a tourist, which had become possible for the first time in a few years earlier, and which provided a whole, new, shipboard social setting from which to experience the continent. She went to write about penguins, to explore the science of the zone of intense marine life which rings the dead interior of the continent. But she found herself reflecting, too, on the intensity with which a traveller to Antarctica feels alive, when cold air touches warm skin and the summer light glows on 24 hours a day.*

AT FIRST LIGHT, on calmer seas, I opened the two porthole covers like a second pair of eyes and looked out onto Antarctica for the first time. White upon white with white borders was all I expected to see; instead, colossal icebergs of palest blue and mint-green floated across the vista. Beyond them, long chalky cliffs stretched out of view. Throwing on a parka, I raced upstairs to the deck and looked all around. As far as I could see in any direction, icebergs meandered against a backdrop of tall, crumbly Antarctic glaciers, which were still pure and unexplored. Human feet had not touched the glaciers I saw; nor had many pairs of eyes beheld them. In many ways, the Antarctic is a world of suspended animation. Suspended between outer space and the fertile continents. Suspended in time—without a local civilization to make history. Civilization has been brought to it; it has never sustained any of its own. It sits suspended in a hanging nest of world politics. When things die in the Antarctic, they decay slowly. What has been is still there and will always be, unless we interfere. *Interfere* is such a simple word for what is happening: the ozone hole, the greenhouse effect, disputes over territory, pollution, mining. Who discovered Antarctica we may never know. We remember the Shackletons and the Scotts, but it was the whalers and sealers who opened up the Antarctic, not the explorers. Because the whalers and sealers didn't talk much about their good hunting grounds, they have sifted between the seams of history.

Soon we dropped anchor at Harmony Cove, Nelson Island, in the South Shetland Islands, whose ice cliffs are layered with volcanic ash from the Deception Island eruption of 1970. Piling into the Zodiacs, we dashed toward a cobble beach, where one of the crew, who had gone ashore early, had teased the other Zodiac drivers by spelling out LANDING in stones against the snow. A blue odalisque of ice floated offshore. Hundred-foot white glacial cliffs stood next to huge rooms of pure aquamarine ice. Ah-hah! A small welcoming party of gentoo penguins, ashore to feed their waiting chicks, waddled close to look us over. One penguin tilted its head one way, then the other, as it stared at me. This made the bird look like an art dealer, quietly thinking and appraising. Penguins don't have binocular vision as humans do, so they turn one eye to an object, then the other, to see it. Although they can see well underwater, they don't need long vision when they're on land. The last time I saw a look quite like that gentoo's was in the Penguin House at the Central Park Zoo. There, in a shower of artificial Antarctic light, in a display created by a theatrical-lighting designer, gentoos and chinstraps had eyed the crowd of people watching them—including some of the homeless of Manhattan, who used the Penguin House as a favorite warming-up spot.

According to one saying, "There are two kinds of penguins in the Antarctic, the white ones coming toward you, and the black ones going away from you." All penguins are essentially black and white on their bodies, a feature known as countershading. Their white bellies and chins blend in with the shimmery light filtering through the water, so they're less likely to be spotted from below when they're in the ocean. That makes hunting fish easier, as well as escaping leopard seals. Their black backs also make them less visible from above as they fly through the murky waters. To the krill, the white belly of the penguin looks like a pale orb, harmless as the sky. To the leopard seal, the black back of the penguin looks like a shadow on the ocean bottom, unpalatable. Researchers found that if they marked penguins with aluminum bands, the tags flashed and leopard seals could spot them too easily. A lot of their study birds were killed, and they switched to black tags. Another advantage of being black and white: If they're too hot, they can turn their white parts to the sun and reflect heat; if too cold, they can turn their black parts to the sun and absorb heat. Because most of a penguin's body is below water, it's the head that has developed so many interesting designs and colors. A field guide to penguins would only need to show you the heads. Adélie penguins (named after Adélie Land, a stretch of Antarctic coast below Australia that was itself named after Adélie Dumont d'Urville, the wife of the nineteenth-century French explorer Jules-Sébastian-César Dumont d'Urville, who first sighted it; among Captain d'Urville's other accomplishments was sending the Venus de Milo to the Louvre) have black heads, with chalk-white eye rings. They are the little men in the tuxedo suits we see in cartoons. Rock-hoppers have lively red eyes, long yellow and black head feathers resembling a crewcut that's been allowed to grow out, and thick yellow satanic eyebrows that slant up and away from their eyes, giving their face an expression that says, *I dare you!* Chinstraps get their name from the helmet of black feathers that seems to be attached by a thin black "strap" across their white throats. Their amber eyes, outlined in thick black, look Egyptian, like a hieroglyph for some as-yet-undecipherable verb. Emperors have black heads, a tawny stripe on the bill, and a bib of egg-yolk yellow around their neck and cheeks. The most flamboyant of all, king penguins display a large, velvety-orange comma on each cheek, as if always in the act of being quoted about something. A throbbing orange at their neck melts into radiant yellow. And, on either side of their bills, a comet of apricot or lavender flies toward their mouth. Fairy penguins are tiny and blue-headed. Each of the seventeen different species of penguins, though essentially black-and-white, differs from all others in head pattern.

On shore, of course, a mass of penguins with predominantly black-and-white bodies looks a bit like linoleum in a cheap diner. Human beings tend to be obsessed with black-and-white animals, like killer whales, giant pandas, and penguins. "We live in a world of grays, could be's, ambiguities," Frank Todd once observed. "Maybe it's just nice to see something that's cut-and-dried. It's black-and-white. It's there, and that's the way it is."

Gentoos feed their chicks every eighteen to twenty-four hours, and the adults that had just arrived were fat-bellied, crammed with fish and krill. Native to the Antarctic and sub-Antarctic, the gentoos were white-breasted and black-backed like all penguins, but they had a white bonnet on their heads. Though quiet and friendly, they drifted just out of reach. Along with Adélies and chinstraps, the gentoos belong to the genus *Pygoscelis* ("brush-tailed"), because of their short, paintbrush-shaped tail, and they are shy penguins, whose chicks grow slowly, staying close to a parent's warm body for weeks after hatching. The gentle gentoos are docile and may not have to pair-bond as vigorously as other penguins; otherwise they would need to declare their territory and mate more stridently and become more aggressive about intruders. The name gentoo is from the deceit of a British Museum man, who received a gentoo skin from an Antarctic explorer, thought it was a new species of bird, and decided to hide the information for a while. Later, he went off to Papua New Guinea, and when he returned, he described the bird as if it were one of the local species, naming it after the Gentoo, a religious sect on Papua New Guinea.

As we straggled along the shore toward granite outcroppings where penguins nest, two large brown birds began forays, dive-bombing. This was our first close encounter with the skua, nemesis of the baby penguin, and I held an arm above my head because, like lightning, skuas strike at the highest spot. They can pick an animal's eyes out before it knows what has happened. Hawklike, cunning, and bold, they are the ace predators of sick, young, or abandoned penguins. Some claim that skuas divide up a penguin colony into thousand-pair lots and that if you want a quick population estimate of a penguin colony, count the skuas and multiply by two thousand. A skua will carefully monitor a rookery, find a deserted chick, knock the bird senseless on the back of the head to kill it, then consume almost every scrap. When it devours an adult, it eats the viscera first, turning the carcass inside out like a sleeve, leaving only the head, skin, and bones. A big skua landed in front of us, spread its wings, and noisily proclaimed its territory. Then we saw why it was so anxious: A fluffy skua chick, head tucked into its shoulders, scurried away in the other direction and crouched. Another skua arrived, and both parents tried to draw our attention away. A little farther on, we found a small rookery of chinstraps, one with its flipper out straight, as if it were signaling a left turn, all looking like a gathering of crosswalk guards. Another was lying on its stomach and turned the soles of its feet up to cool off. Moving its flippers, it revealed an underside that had gone pink in the penguin equivalent of a blush. It was a warm day for them in Antarctica. A group of gentoo penguins ambled by, going anywhere, going nowhere. Penguins are born followers. If one begins to move with purpose, the others fall in behind it.

"Those poor penguins, living in this awful cold!" one woman lamented in a Southern drawl as she pulled her red parka tight around her neck and dragged a knitted cap down almost over her eyes. In fact, penguins rarely mind the cold. Quite the opposite. More often, they overheat. Like mammals, penguins are warm-blooded, which means that they're able to make their own heat and carry it with them wherever they go, instead of taking on the temperature of their environment. This allows them to migrate and to live in otherwise inhospitable regions of the earth. Of course, keeping warm can become something of a problem in the Antarctic. Penguins have evolved thick layers of blubber, which their bodies make from krill and planktonic oils, and because blubber conducts heat poorly, a layer of it below the skin acts as an excellent insulator. It is also a place to store fuel for the long, cold breeding season. The farther south you go, the bigger the penguins get, since big animals find it easier to stay warm. About one third of the weight of the emperor penguin, which lives in the coldest regions, is blubber. In addition, as anyone who skis or spends much time outdoors in the winter knows, air makes one of the best insulators. Travelers to the Antarctic are advised to dress in many layers of clothing with plenty of air in between them. Penguins do that with tightly overlapping feathers, which don't ruffle very easily and, as a result, trap a layer of warm air against the skin. Also, each feather develops a fluffy down at the base of its shaft, and that downy layer adds even more insulation. Penguins are watertight and airtight and thought to have more feathers than any other bird. The feathers are shiny, long, and curved, arranged like carefully laid roof shingles. Dipping into the oil gland at the base of the tail, a penguin spreads a layer of oil on the feathers to keep them slick and tight. Of course, feathers do get tattered after a year or so, and then the bird must molt, to slough off the old feathers and grow new ones. If it molted gradually, it wouldn't be waterproof any longer, so it goes through all the steps of molting at the same time, a process that takes about thirty days. New feathers grow in underneath the ones that are molting and push the old ones out. It makes the penguin look scruffy and slightly crazed, as if it were ripping its feathers out in some avian delirium. What is worse, since they're not waterproof while they're molting, they can't go hunting food in the ocean. *Fasting* is what it's called by scientists, although that word suggests choice on the part of the penguin, which loses about 30 percent of its body weight and is bound to be hungry and is not exactly a volunteer.

But heat is a problem. There are few things as ridiculous as a penguin suffering the equivalent of heat stroke in the middle of the coldest place on earth. All around the rookery, overheated penguins resort to what look like vaudeville moves: Ruffling their feathers, they release some of the hot insulating air next to the skin. They hold one arm out, as if parking a 747, then they pirouette and signal a turn in the other direction. They flush pink under the wings, where capillaries swell with blood. Baby penguins like to lie down on their bellies and stick their feet up behind them, so that they can lose heat through the soles of their feet. They radiate heat through the few featherless zones on their bodies (usually around the eyes, flippers, and feet). A large adult suddenly ruffles up all over and extends its flippers at the same time, as if someone had scraped a fingernail across a blackboard.

The cold, on the other hand, isn't really a problem. If the temperature drops too low (around 15° F. with a strong wind), thousands of birds will huddle together to stay warm in what the French researchers call *tortues* (turtles). Using one another for insulation, they don't burn up their fat stores quickly. Huddling birds lose only half as much weight as birds braving the winds solo, because only a small portion of their bodies are exposed to the wind. It is akin to the protection apartment dwellers get, surrounded by apartments on either side that act as insulation.

"Come and look at this krill poop," Harrison said, bending down to consider some guano. "It's not very fresh. See those black spots in it?" He held up a handful and smudged it between his fingers. "That's the eyes of krill, which are indigestible, like tomato seeds to us. When you see the ground stained red like this, it's probably a chinstrap or an Adélie rookery rather than a gentoo, because the chinstraps eat krill and poop red. White poop comes from a diet of fish or squid. And green poop means they're not eating at all; what you're seeing is bile." Across the hillside and around the large slabs of rock, the ground was stained pink. Even if the rookery had been deserted, we could tell chinstraps or Adélies lived there. Most of the zoologists I know are, by necessity, coprophiles. A living system leaves its imprint on what passes through it. So I'm no longer surprised to find a naturalist sifting through bat, alligator, or penguin excrement. Some even study petrified dinosaur excrement, or coprolites, as they're called.

Beyond the rookery, molting elephant seals snoozed on the shore like overgrown salamis. They rolled around the sand together and against each other, to rub off the old fur, which wears out and has to be replaced each year. Pieces of molted fur and skin littered the beach. One often finds elephant seals with penguins, lying on the beach like so many old cast-off horsehair couches. It takes seven or eight years for the long nose of the bulls to develop. These pug-nosed ones were young males, which would grow larger, although they were already around twelve feet long and weighed about three thousand pounds. A gang of penguins strolled among them, seemingly without care.

One in the center scratched his neck with a five-clawed flipper. Sluggish as they may look, elephant seals can dive to more than three thousand feet to feed on squid and fish.

On a rise, three fur seals sat up and stared aggressively as we passed. If they wanted to, they could gallop across the sand at great speed, tucking their pelvic girdles and undulating like fast worms. Fur seals will attack human beings. The previous year, a fur seal had grabbed a lecturer as he was getting into a Zodiac and punctured his lung. The man needed thirty-six stitches, and it took many months for the wounds to heal, since, to add to their armament, fur seals have an enzyme in their mucus that keeps their bites from healing properly. "I hate these," one of the guides said under his breath. "I have nothing whatsoever against fur-seal coats. I tell you, I'm sincere about this." As a territorial male started toward him, Stonehouse clapped his gloved hands, shouted, and kicked black volcanic sand up at its face. The seal stopped, huffed loudly, and sidled back to its original spot.

Between two rock knobs, a chinstrap-penguin colony sat on red krill-stained rocks. The gentoos choose a flat shelf area to nest and breed on; chinstraps prefer rocks, and gentoos a flatter terrain, so even though living in close quarters, they don't compete with one another for nesting sites. A baby gentoo put its head up and made a metallic gargling sound. The babies, forming little crèches with their flippers wrapped around each other, achieved a look of intense mateyness. (Other animals, like young flamingos, eider ducklings, and baby bats, form crèches, too.) While the parents are away hunting food, the babies are open to attack from skuas and other birds and are a lot safer in a nursery of chicks. Not only is there strength in numbers, but adults wandering through to feed their young can help ward off attacks by skuas. King penguins feed their chicks for nine or ten months, so their young spend a long time in crèches. Returning from the sea, adult gentoos easily recognize their babies by voice. The pattern of white dots, bar over the eyes, and other characteristics also varies slightly from one individual to another.

A chick flapped rubbery flippers. It takes time for the bones to set into the strong, hard flight-muscles of the adult. As immature birds, gentoos have a great tendency to wander and may migrate as much as two hundred miles. But in the second year they will return to the rookery, ready to breed. Because they're a mated pair, the gentoo couple doesn't have to go back to the same site each year to nest. Like all other penguins, they take two to three weeks to build a nest and copulate, but they're mobile and can change their nests. Because they don't split up when the breeding season is over, they probably remain together as mates year-round. Gentoos are the most passive penguins, and perhaps that has been their undoing. There are only about forty thousand gentoos left in the world, but at least their numbers are not declining.

In some areas, the ground was streaked with beautiful white star-bursts—squirted guano—so that it looked like a moonscape. And it was pungent! Sailors have been known to use the smell of a well-known rookery as a navigation aid, especially when fog is too thick for them to see any of the birds. A pink tinge of algae glowed from beneath the snow, which acted as a greenhouse. Frost polygons had turned the sod into a six-sided design. A chinstrap raised its bill into the air, its air sacs puffed up, and it worked the bellows of its chest. Just offshore, a row of giant petrels waited for the chicks. The range of light was so wide it was taxing for the eye to take everything in—the round, dark, wet, sullen rocks, the brilliant white snow reflecting against low clouds in a visual echo chamber of white.

On the ground were the remains of that morning's breakfast for a skua: a pair of orange penguin feet, a head, a skeleton. Stonehouse picked up the half-eaten chinstrap, showed me the flight muscles, the thick red ribbons that were the salt glands, and the concertina ribs. He handed me a small white feather, revealing the main shaft, and then, at its base, a second feather of silky down.

"Why would *this* one have died?" I said. "Why would a skua have singled it out?"

Turning the skeleton over, he discovered its eccentric bottom bill, bent at a ninety-degree angle. "You occasionally see penguins like this, with deformed bills. There's no way they can feed correctly, but even if they did manage to feed, they still wouldn't be able to preen themselves. So they would get heavily infested with lice. It's very sad, like seeing a deformed and neglected child. That's a simple, small thing to go wrong, when you think about it—just a misshapen bottom bill. But the chick would lead a difficult life for about nine months and then die of starvation. Before that happens, a skua usually identifies it as a weakling. Chicks running around without parents in this colony soon die, and they eventually form the debris on the floor of the colony that you see."

Looking more closely at the ground, I saw the long scatter of bones for the first time and was stunned. We were standing in an ancient cemetery. This penguin colony lived on top of a graveyard that may have been thousands of years old. Under the feet and nests of the birds lay all the frozen, partially mummified remains of their ancestors. The cold had preserved their carcasses for as long as three thousand years. Most adult penguins die in the ocean, but babies die right on land, and no one removes their skeletons. The bones gradually sink into the permafreeze like designs into some fantastic paperweight. The chicks are born astride a grave. A wind gust sent feathers blowing into the air as if in a pillow fight. Penguins molt each year, some even shedding parts of their bills. The ground litter included not just corpses, but also pungent guano, spilled krill, blood, feathers, molted elephant-seal skin and hair, and miscellaneous bits of animal too dismembered for an amateur to identify.

Seeing Stonehouse with the chinstrap skeleton in his hands, his wife, Sally, walked up and smiled. Her lovely English complexion had gone ruddy in the brisk Antarctic air and a few wisps of brown hair strayed from under her knitted hat. "We're so used to Daddy bringing home dead finnies," she said cheerfully. "When we were in New Zealand, if we could find a dead penguin, he was always so pleased."

Just then a lone male chinstrap tossed its helmeted head to the sky, arched its flippers back, and trumpeted an "ecstatic display" loud enough to stop a train. Its chest and throat rippled rhythmically as it called, as if with all its soul it hoped to lure a willing female by telling her that he was available and ready with a lovely little nesting site. Penguins are not profound thinkers, but their instincts guide them through all the demands of the landscape and of their hormones. An ecstatic display sounds both desperate and automatic. It may happen at any time, sometimes with good reason, sometimes by mistake, sometimes in a chorus of tens of thousands of voices screaming at the top of their lungs the equivalent of *Tell me you love me! I said, Tell me you love me!* It is a little like overhearing thousands of actors auditioning simultaneously for a Sam Shepard play. All summer, their frantic ecstasy fills the Antarctic air. Plighting their troth, an Adélie pair will do an ecstatic display, then the male will give her a precious and, to her eye, perfect stone. The actual copulation takes only seconds, and has been termed a cloacal kiss. Foreplay is everything—a complex drama of eye contact and body language. Courting males repeatedly bow to females, and the female has her own balletic gestures to use in reply.

"Well, he *is* eager, isn't he?" Sally said good-naturedly. We laughed. Life goes on, having nowhere else to go.

✳

Somewhere along the way, we had lost the nighttime. Where did we lose it? In the deserted whaling station, in whose smoky hall we ate a barbecue of reindeer meat and danced to Glenn Miller? At the Polish station, whose greenhouses grew snapdragons and tomatoes? In the volcanic ring of Deception Island? Watching rippling terraces of Adélie penguins go about their lapidary business, obsessed with nesting stones? At the small British base, Signy, whose young men had not seen strangers, or women, for nearly two years? (Visiting with us in the lounge, some of them were trembling; and we sent them away with handshakes, good wishes, and sacks of potatoes, onions, and other fresh food.) At the Valentine's Day dance, on seas so rough that dancers held on to the ceiling? On Elephant Island, a forbidding snag of mountain but a thriving chinstrap rookery, where Shackleton and his men landed after their trials on the pack ice? At the eerie, deserted penguin rookery, where watching a lone penguin chick face the death machine of a rampaging skua, which played out its instincts blow by blow, tortured our hearts? Among the guano-thick beaches, where waves of hungry penguins bobbed in the sea and babies clamored to be fed? Listening to the assistant cruise director, a fine pianist, give recitals of Debussy, Haydn, Bach, and Beethoven against a backdrop of sunstruck glaciers?

Now we lived only in a late summer twilight. Icebergs clustered around us like statuary as the ship sailed through the Gerlache Strait, which separates the Antarctic Peninsula from Brabant and Anvers islands. Each narrow waterway seemed to lead into another one, until finally we sailed through the Lemaire Channel, which narrows to a mere sixteen hundred feet wide at its southern end. This too was the penguins' world. On either side of the ship, glaciers spilled into the sea, jagged mountain peaks rose into the clouds, and icebergs roamed freely. In the channel, the water was like lucid tar, with icebergs of all sizes drifting through it, their white tops a thin reflection of mortality—their blue bases pale and inscrutable. The blazing white of an iceberg lay on a thick wide base of blue ancient as the earth, older than all of the people who had ever seen it or who had ever visited the Antarctic combined. The icebergs took all sorts of shapes, and some had fissures through which a searing blue light shone. In the wake of an iceberg the water looked like oiled silk because the surface of the water had been smoothed by the ice's palm. On both sides of the boat, black, jagged, ice-drizzled mountains reflected in the mirror surface. On a small berg, five gentoo penguins sat, their white bonnets sparkling in the sun. On another small berg sat their death—a leopard seal, sprawling on one side, idly scratching its flank with a five-fingered paw.

"*Seehund,*" a woman from Frankfurt said solemnly. We were all on the side of the penguins, though nature should have no partisans, no sides, no center, except the center that is forever shifting, as Emerson said, a center that moves within circles and circles that move.

Great tongues of ice stretch out from the continent and speak in a language like music, with no words but with undeniable meaning. And like music, the vista is a language we don't have to learn to be profoundly moved—we who do not just use our environment but also appreciate, admire, even worship it. True, we kill other lifeforms to survive, but we feel a kinship to them, we apologize for stealing their life from them. We are the most vital creatures ever to inhabit the earth, and the one truth we live by is that life loves life. Still, nature proceeds "red in tooth and claw," as Tennyson said. This becomes simpler to see in a simpler environment. When you walk through a penguin rookery, where the underweight chicks stand doomed and the skuas maneuver like custodians of death, pages of Darwin's *Origin of Species* spring to life in front of you. All the cozy denials we use as shields fall aside.

The sky that day was clear, and the air as astringent as ammonia vapor. The sun poured down but had no heat, and the ice mountains occasionally revealed weavings of blue and green. After the darkness of winter, the five months of summer sun did not warm things up much. Because the sun rode so low on the horizon, it seemed to have little warmth. The ice reflected the heat back into the sky. Most people on board had greatly dilated pupils by then, a side effect of the scopolamine patch they wore behind one ear to ward off seasickness. It made them look a little like zombies, but it also allowed their eyes to take in big gulps of light. In Zodiacs, we drifted along the peninsula, through an ice-sculpture garden. Heraclitus said you never step in the same stream twice. The Antarctic version of that is that you never see the same iceberg twice. Because each iceberg is always changing, one sees a personal and unique iceberg that no one else has ever seen or will ever see. They are not always smooth. Many had textures, waffle patterns, pockmarks, and some looked pounded by Persian metalsmiths. A newly calved iceberg lay like a chunk of glass honeycomb, spongy from being underwater. (At some point it was other-side-up.) Another had beautiful blue ridges like muscles running along one side. So many icelets thickened the water, each one quivering with sparkle, that the sea looked like aluminum foil shaken in the sun. There were baths of ice with blue lotion, ice grottos, ice curved round the fleecy pelt of a lamb, razor-backed ice, sixteen ice swans on an ice merry-go-round, ice pedestals, ice combs, ice dragons with wings spread, an ice garden where icebergs grew and died, ice tongs with blue ice between their claws, an ice egret stretching its wings and a long rippling neck out of the water. Apricot light spilled over the distant snow-tipped mountains. Chunky wedges of peppermint-blue ice drifted past us. Behind us, the Zodiac left a frothy white petticoat. And farther beyond, shapes arched out of the water—penguins feeding, oblivious to what we call beauty.

We paused at Paradise Bay, where blue-eyed shags nested along the cliffs, a whale maneuvered at a distance, penguins porpoised to feed, and crabeater seals lazed on small icebergs, red krill juice dripping down their chins. Through pale green water, clear and calm, a gray rocky bottom was shining, along with red and brown seaweeds and patches of yellow-blooming phytoplankton. A loud explosion startled us and, turning, we saw ice breaking off a glacier to become an iceberg, which would float for four years or so before it succumbed to the sea. I looked down through the fathoms of crystal water to the smooth rocks on the bottom. Suddenly an eight-foot leopard seal swiveled below the boat, surfaced to breathe, cut a fast turn, and began circling the Zodiac, around and around, underneath it and alongside. Each time it spun underwater, large blue air bubbles rose to the surface like jellyfish. Mouth open, baring its sharp yellow teeth, it lunged up through the water and bit a pontoon on the Zodiac. "Back away from the edge!" David Kaplan, our driver, said with contained urgency, and the twelve passengers leaned inward, away from the attacking seal, which could leap out of the water and seize an arm, pulling a person under. Circling, fast, handsome, wild, ferocious, it spun below again, dove, and leapt to the surface. It was attacking us as it would penguins on an ice floe. We who live at the top of our food chain rarely get the chance to feel like prey, to watch a predator maneuver around us with a deftness that's instinctive, cunning, and persistent—and live to tell about it.

"Just an average day in Paradise," Kaplan said, brightening the motor and heading for shore. We climbed out at an abandoned Argentinian base. As the clouds drifted behind the peninsula, the continent itself appeared to be moving, as of course it was. On a rock ledge, an Antarctic tern—a small white bird with black cap and startling red beak and matching red feet—thrilled the sky with song. A teal vein of copper ore cascaded down the rocks among patches of fiery orange lichen and green moss. As the rest of our party climbed up a steep slope of glacier to the top of a mountain where a wooden cross had been planted, I stood like a sentinel, still as a penguin, watching my kind struggle up the hill from the sea. Across the bay, the snow mountains were glazed in a dusky pink light and the water was cerulean blue. Gray clouds hung in front of and below the powdered tops of the mountains. The air was so pure that the clouds looked cut-out and solid, suspended by a sleight of hand, a magician's trick. Mirrors lay scattered on the surface of the water, where there was no ice to disrupt the flowing light. A blue iceberg shaped like open jaws a hundred feet high floated near shore. Corrugated-metal buildings shot off hot orange. A long hem of brash ice undulated across the south end of the bay. Somewhere the leopard seal sat looking for less elusive quarry, and would find it.

The birders were up early as we approached Coronation Island, in the South Orkneys; they were desperate to spot an Antarctic petrel, a bird that resembles but is slightly larger than the many pintados, or painted petrels, swirling in small tornadoes behind the boat. Most Antarctic petrels are in their rookeries as much as a hundred miles inland from the ice shelf. When one finally winged across the water, the birders went berserk.

"Oh! There it is! Beautiful!" a woman cried.

"Wow! The nape is almost buff!" said an enraptured man.

All the Antarctic petrels we saw that day were pale. They were in molt, a wonderful coffee-tan color, and flapped stiffly because their new feathers weren't in yet. To see one or two at this time of year so far out was a bonus. However, many of us were not looking for petrels but for a rarer sight: the emperor penguin. Largest of all penguins, emperors can dive to nearly nine hundred feet to feed on squid and stay submerged for nearly twenty minutes or more. When they stand in the snow like vigilant UFO watchers, they have the usual black-and-white coloring, but also a spill of honey at their throats and cheeks. Emperor penguins are such altruistic parents—or fanatical, depending on your point of view—that they will even pick up a frozen or ruined egg and try to incubate it, or try to incubate stones or an old dead chick. An abandoned or wayward chick will immediately be adopted. Sometimes adults even squabble so much over a chick that the chick gets hurt or killed in the process. Emperors rarely, if ever, touch bare ground. They live out their whole lives standing sentry on shelf ice or swimming in the ocean. Unfortunately, their rookeries were too far south for us to see them.

A small flock of Antarctic prions hydroplaned over the water, plowing a furrow through it, using their tongues to sift krill into a feeding pouch under the jaw, feeding the way baleen whales do.

"Bird alert! Bird alert!" sounded over the intercom, waking passengers from their slumber and early diners from drowsy breakfasts. "EP alert! EP alert! A juvenile emperor penguin has been spotted off the stern of the boat!" I ran to the stern, colliding with people frantically running up the stairs from the cabins below. HOLD THESE RAILS it said on brass plates at each stairway on every deck, as if in rebuke to excited birders who had been turning the ship into an aviary. The stampede ended with a clash of bodies on the stern deck. And there it was: porpoising out of the water, looking like part of an inner tube with a flash of yellow showing every now and again. Then it vanished, and we were left standing quietly with our amazement. To glimpse an emperor penguin in the wild, feeding this far from its home, was a benediction.

Stowing my binoculars, I went downstairs to breakfast, which I barely touched. Despite the elaborate meals, I'd been losing weight at a reckless pace. It was as if I was being nourished so thoroughly through my senses that I felt too full to eat. Before coming to the Antarctic, I had thought that penguins lived in a world of extreme sensory deprivation. But I had found just the opposite—a landscape of the greatest sensuality. For one thing, there was so much life, great herds of animal life to rival those in East Africa. Many people have compared Antarctica to a wasteland; instead, it is robust with life. For another, the range of colors was breathtaking; though subtle, it had changing depths and illuminations, like flesh tones. The many colors were in the ever-bluing sky, in the cloud formations, the muted light, the midnight sun, the auroras dancing over still waters with icebergs and crash ice, and in areas that dazzled like small hand mirrors, through which black-and-white penguins dove. Who would have imagined the depth of blue in the icebergs, appearing as sugar-frosted cakes with muted sunlight bursting off them?

Or the scale, the massiveness—sitting alongside an iceberg, you couldn't see around it in either direction. One day the water was so smooth that you could use it as a mirror, and four hours later the wind was howling at ninety knots. And was as beautiful at ninety knots as when crystal-calm. Huge ice caverns formed arches of pastel ice. *Glare* had so many moods that it seemed another pure color. The mountains, glaciers, and fiords bulged and rolled through endless displays of inter-flowing shapes. The continent kept turning its shimmery hips, and jutting up hard pinnacles of ice, in a sensuality of rolling, sifting, cascading landscapes. There was such a liquefaction to its limbs. And yet it could also be blindingly abstract, harrowing and remote, the closest thing to being on another planet, so far from human life that its desolation and iciness made you want to do impetuous, life-affirming things: commit acts of love, skip Zodiacs at reckless speeds over the bays, touch voices with a loved one by way of satellite, work out in the gym on thrones of steel until your muscles quit, drink all night, be passionate and daring, renew the outlines of your humanity.

PARTICLES

from *Water, Ice and Stone* (1995)

Bill Green

Here a specialist in the science of lakes meditates on the connection between the small, slow, subtle mystery he is trying to solve in the Dry Valleys of Antarctica, and the globe's biggest processes. For Bill Green, scientific procedures are forms of questioning which naturally belong alongside poetry and mythology; and as one of its scientific inheritors, he takes the continent of Antarctica to be the richest possible playground, somewhere whose layers of system and meaning deserve lifetimes of attention.

When I tell people there are lakes in Antarctica, they think surely I am joking. "Lakes there?" they ask. "How can that be? It's all ice and snow. Penguins running around." Then, when I assure them that it's true, they ask, in a more assertive tone, "But they're frozen, of course?" And I say, "Well, yes, there's ice on the surface, but below there's liquid water, sometimes as deep as two hundred feet." Then they ask—and this is inevitable—"Are there fish?" I say, "No, not a single one." "Hmmm," they respond, incredulous, "a lake without fish. Does anything live in them, at all?" And they emphasize "at all." "Only algae and bacteria," I say. "Nothing you can actually see with your eyes. Except for the mats of algae, which are tiny columns and pinnacles on the bottom, far below the ice."

But then it is precisely what is not there, what has never been there, that makes the lakes—indeed, the whole continent on which they lie—so strange and so important.

For me these absences, and the simplicity to which they give rise, were the key. The lakes are the most isolated inland waters in the world. Landlocked, they are without spillage or outflow; each has only a few streams, and these hold water for only a few weeks out of the year. They are ice-covered, so that very little in the form of dust or snow enters them from the air. And, of course, there is never rain. That in itself makes them magic. How can you have a lake without rain? A lake without fish, maybe, but a lake without rain? A land without rain. A whole continent. Such living things as there are are mostly microscopic—algae, bacteria, yeast, a minimalist's tableau. And into this setting, stark and largely inorganic, Martian almost, the elements come—nitrogen, phosphorus, the metals—unheralded, but replete with possibilities, with lives to be lived.

It would be no exaggeration to say that I was obsessed with the lakes, and especially with the metals that coursed through them like bits and pieces of an invisible wind. In this seemingly fantastical concern, I was not unlike Borges, who once wrote of a silver coin he had dropped into the sea. The coin had become, in consequence, a kind of persona in the drama of the world, its destiny unfolding alongside that of the poet Borges himself. I had my coins, too, by the countless billions.

I knew, for example, that the Onyx River in Wright Valley had brought tons of cobalt and lead and copper into Lake Vanda over its long history. Yet there were virtually no metals in the lake. I knew this. But where had they gone? What was removing them? What thin veil of purity had caught them in its mesh? And whatever veil it was, did it fall elsewhere across the Earth and its seas, purifying as it went? Did the Earth, or this tiny piece of it, regenerate itself? At what speeds? By what agencies? Last year I had set particle "traps" in the lakes, had left them there for a whole year. They were nothing more than clear plastic tubes, capped at one end and suspended below the ice. But in time, if all went well, I would get them back and I would know the answers.

We took two sleds. Into each we threw a trap catcher and a bunch of plastic caps so that we could stopper the traps if we were lucky enough to retrieve them. Dr. Yu and I headed off toward the deep hole; Mike and Tim headed east toward the center of the lake. We had the sled tethered to a long rope that opened up into a skinny isosceles triangle—the sled at one corner, Dr. Yu and I at the other two. The sled pulled easily and I held the rope near my shoulder with just one hand. We were moving toward a huge glacial erratic that lay among much finer debris on the lower slopes of the Olympus Range. The boulder was a landmark. The line between it and our Scott tent ran through the point on the ice where we had placed our most important trap.

This was the same boulder that Canfield had mentioned in his note, the stone behind which he had secured the bamboo pole. "Thought you might need this, guys," he had written in that casual voice. It was pure Canfield.

As we approached the site, I began to feel apprehensive. The traps had been suspended for twelve months. In that much time, anything can go wrong. Even here, where we knew no one had been. We were certain, and yet there was the question: What if something had been here? What if the rope had been severed, been eaten through? In the distance I was beginning to make out the rock cairn we had left as a marker. As I drew closer, I could see that a curious separation had occurred. The dark basalts, warmed by the sun, had melted deep into the lake ice. They were looking up at me as if from a crystal sarcophagus. The white granites had hardly melted through at all. They lay in smooth hollows on the surface of the ice. I was not expecting this. But I should have been.

Dr. Yu stood there with his hands on his hips. He was turning in a slow circle, looking over the valley from beneath his parka hood. "It is like Qaidam Basin," he said in his clipped English, which was a kind of poetry. "In Qaidam," he said, "No birds fly in sky. No green grass on land. Stones run before wind. Mountains are knives. I must climb icy mountain, go down fiery sea. People say this of Qaidam, of salt lake. What say of this place?" He was facing down valley, looking over the length of the lake, down toward the dry Onyx. Tim and Mike were still moving across the ice, although I could no longer hear their sled. But I was not thinking of the valley. I was not thinking of China. I only wanted the traps back.

We had laid the traps where we had to, up and down the lake: near the Onyx, about half a mile from the debouchment; by the peninsula, where the lake constricts before it opens into its major basin; and a mile or so west of the peninsula. In the deep lake, we suspended three sets of traps from a single line. The first hung at forty-eight meters, just at the top of the calcium chloride brine layer; the second was at sixty meters, where the oxygen-rich region met the oxygen-starved region of the lake; the third was at sixty-five meters, deep within the sulfide brine. We placed the traps where the lake's long history foreordained.

It was twelve hundred years ago, near the close of the first Christian millennium, while the young Charlemagne was uniting Europe, that an epochal event occurred in the Wright Valley. At that time, in the years of the Crusades, Vanda had evaporated to become little more than a salt flat, a glistening whiteness in the center of which lay a shallow circle of water. Then the change came, and it came rapidly. The climate warmed and the Lower Wright Glacier began to discharge water, not the springtime trickle that had gone before, but whole sheets of water. The lake level rose. The fresh waters from the glacier overlaid the dense brine. There was little or no mixing, only diffusion, the slow transfer of matter. Now it was as though there were two lakes: one large and made of fresh water that had rolled suddenly down the Onyx like a deluge; the other a dense brine, the sweating evaporite remains of the ancient lake.

It was this single medieval event that had set the structure of Lake Vanda, that had ensured that the upper waters would be light and clear and oxygen-filled, and that the depths would be heavy and dark and rife with decay. Knowing these things, we had set the traps.

As I lowered the new trap catcher hand over hand, the bottom folded at the hinge, twelve feet of bamboo slowly disappeared through the opaque slurry that clogged the hole. I was holding my breath. Then I felt the horizontal arm fall open when it cleared the bottom of the ice. I lay on my stomach, spread my legs, wrapped my gloves around the pole, and slowly turned it, first clockwise, then counterclockwise, in small, gentle arcs, fishing for the line that I knew was hanging parallel, just a foot away. After a few minutes I thought I touched it. I began to twist the pole. It was resisting, as though something were being wound on it, as though something were coming in. I could hear the bamboo creak below the ice and I thought it might splinter. I bent my head around to look at Dr. Yu. "I think I have something," I said. "But it feels like it's going to snap. It's very heavy."

He lay on the ice across from me, his head only a few inches from mine. He took hold of the pole just above where my hands were placed. We both began to turn the bamboo. The sounds were becoming louder, like the creaking of a wooden ship at the docks. I could hear nothing else. We were looking at each other, talking with our eyes: I imagined Dr. Yu thinking, *Maybe a little more, just a little, it's okay; bamboo, bamboo strong.* Then we were looking at the slurry and at the pole turning through it. There was a cloud of fog over the hole from my breath and his. I could feel the weight of the line in my fingers. It seemed too much. I wanted to just hold it there in place, for a few seconds, not risk anything. We were touching the traps. The particles were down there at the end of the line. I wanted to pretend we already had them.

*

Sometimes the geochemists made it sound like a riddle. "Why isn't the sea a copper blue?" they would ask. Not the blue of the sky, but the deep blue of a copper solution. After all, the rivers have been bringing copper to the sea from the beginning of time. And yet the seas are not full. Why?

To convince yourself of this truth, you could do the calculation on the back of an envelope. Each year the rivers pour into the oceans about seven thousand tons of copper. Over the last sixty million years, just since the demise of dinosaurs, the rivers have excavated and deposited whole mountain ranges of copper. But there is virtually none in the sea—a few nanomoles per liter, a few million tons in all.

Where has it gone? Why are the seas not absolutely deadly with copper? In just the last two decades, the geochemists have found the answer: particles.

It was one of those things you knew and didn't know. Science was full of them. Democritus knew there were atoms, little flicks of cold that pricked your skin, that made your nostrils flare. And Lucretius knew this too. You can look at *De Rerum Natura* and hear the wind as it shakes the mountains and rolls the seas and wracks the vessels that go there. And in Lucretius you think you "see" the wind, not as something insubstantial but as a maelstrom of tiny bodies, of corporeal beings, of atoms. And so on up through history, through Boyle and Newton and Descartes. But did they really know, know in the same keen predictive sense that Dalton knew, or that Rutherford and Bohr and Schroedinger knew? Or was *atom* just a convenient name you gave to things, to causes you could not fathom, to whatever invisible mystery it was that could hollow a rock or float a tree or sweep the sky clean of clouds? The Greeks knew that light was particulate, but did they really *know?* Did anyone really know about light until Newton and Young, or perhaps until Einstein?

So it is sometimes said that the Danish oceanographer, Forschhammer, in one of the most eloquent and condensed passages about the chemistry of the sea, knew about particles. For what Forschammer saw was that the amounts of the elements found in the seas were not so much dependent on the rivers that poured them in, but on what happened to them once they arrived. Applied to the problem of copper, this seems to suggest that despite the burnished mountains of metal that come, dissolved, to the sea each year, we have no reason to expect the oceans to be awash in it. What is really important is what the sea is doing. And the sea is not passive. The sea, here, is Borges's sea, "violent and ancient, who gnaws the foundations of earth." It is the site of chemical and organochemical action, the place in which elements are rendered, in various ways and at various rates, insoluble. Where ions are transformed into particles. Where particles sink and are carried away. Where things, for a time, are lost.

In the oceanography of Karl Turekian and Wallace Broecker and Edward Goldberg, the shadowy particles of Forschhammer have been given flesh and identity. So abundant are they in the rivers and oceans of the world, and so varied and pervasive in their influence—these clays, these oxides of iron and manganese, these cool flakes and platelets of calcite and cells of sinking plankton, and these circling desert dusts—that Turekian has referred to them as agents in a "great particle conspiracy." It is this "conspiracy," this great passage of grain and spore and flocculi, this pulverant earth-wide storm, that has in great measure removed metals from the sea. It is this simple process that has cleansed the sea of its most toxic bodies, leaving behind only traces of copper and zinc and cadmium and lead.

What particles exactly? There were thousands of them. Which ones were the most effective "scavengers"—the cells of sinking organisms, the surfaces of calcite, the microscopic umber of the manganese and iron oxides? And did what happened in the sea happen also in lakes? And to what extent, and how fast? As Keith had said, "the proof is in the particles." I couldn't wait to see what particles our traps held.

The sounds coming from below had gotten sharper. The cold bamboo, as it strained, made noises that rang out like shots. We were still winding the rope around the pole, inching the trap line up. Dr. Yu and I got to our feet. We bent over the bamboo that stood a foot or so above the surface of the ice. I was breathing hard, knowing that it was time to bring it up, but not wanting to, afraid that the leg of the L would break off. We were standing in a cloud of condensate from our own breath, delaying what we had to do.

I began to raise the pole. I felt the leg touch the bottom of the ice sheet. I could almost sense the texture of the ice, its smoothness, twelve feet below. It was as though the organic fibers of the bamboo were feeding into the circuitry of my own arm. It was as though my fingers were tracing along the underside of the sheet.

The hole was only ten inches in diameter—the width of the drill bit. The leg of the trap catcher was two feet long, long enough to allow us to snag the line. Tim had designed a hinge that would allow the leg of the L to straighten when it hit the ice—on the way up it would become an I. That way it could be pulled through the hole. If everything worked.

It resisted. Maybe we had wound the rope around the hinge. I pulled a little harder, but it still wouldn't budge. The leg was down. It would not come through the hole. I got on my knees again. I removed my gloves and began to scoop out the netting of ice crystals and slush that lay in the hole, that obscured what was happening below. As I pulled the ice from the surface, new rime from the drilling floated up. Five minutes, ten minutes, I don't know how long it took. It seemed I would never clear the hole. Then the last crystal bobbed to the surface. I lifted it with my thumb and forefinger and put it on the ice.

There was blue light coming off the hole. Its sides were gently corrugated. You could see how the drill had cut. The water was so clear it seemed invisible. I could see to the bottom of the ice sheet and below. Far below.

The rope had knotted around the hinge, but through the sparkling shaft of water it appeared that there was room for some play. We began to lift, slowly, so as not to ripple the surface. My head was down in the hole, as close to the waterline as I could get. I was almost breathing the water of the lake. The leg of the trap catcher began to bend. It was creaking. I could count the degrees as the angle opened up. Dr. Yu was above me. He was whispering as he pulled, as though he were repeating a mantra: "Bamboo strong. No worry. Bamboo very strong. Very, very strong." He repeated it over and over. "I know this," he said. "I know this."

It had opened to 150 or maybe 160 degrees when it cracked. I saw the leg detach from the long pole. I saw it wobble a bit and begin to sink under the weight of the traps. The rope was uncoiling, falling away. I felt my heart stop. Dr. Yu said "Oooooooh," his voice sinking with the traps.

Then I felt a tug on the line. The rope had thrown a coil over the catcher. Suddenly nothing was moving. There was no sound. The bamboo leg hung limp in the water. The traps were attached.

We were both on our feet now. We were bringing the pole up through the water. It was high above our heads, a line of bamboo against the mountains and the sky. As it rose, there was a length of rope and then the leg and then the hook and the coil wound around it. I grabbed the rope with both hands, held it tightly. I could feel the weight of the traps. For the first time in a year I knew we would get them back. Dr. Yu was standing there grinning, saying "Goooood, goooood, very goooooooood." It sounded to me as if he was singing.

"You will never believe me," Pablo Neruda said, "but it sings, the salt sings, the hide of the salt plains, it sings through a mouth smothered by earth." It sings, but we cannot hear. A voice beneath things, but earthwise, and everywhere, not just in the hide of the salt plains or in the rounded grapeskin of water. The comings and goings, the small visitations, the nanosecond or millennial lingerings, the departures—there is music in all of these. Not the music of a few distant spheres, but the music of a million cycles, like hooped bracelets, fine-spun silver twirling and whispering.

In this valley and in this lake and in the sediments beneath, these cycles turned in miniature, in a space that was comprehensible, but that was tied nonetheless into the larger space of the world. The river came in the springtime. It was sound and it was light, but it was also the head-over-heels tumbling of each water molecule, the combined energies of those molecules, their separated charges like torch fires, burning at the tips. What sound did the loosening of cobalt make, the adsorbed ion wavering a little like a minnow at the surface of a rock, then heading off downstream? How long did it stay in the lake after it had glided there on the current, after it had moved faceup, eyeing the blue Antarctic sky through the prism of ice? Maybe a year, maybe five. Not as long as sodium or chloride; longer certainly than iron. Then what? Perhaps an encounter with the surface of clay, glazed with a few atom-thicknesses of manganese oxide. Then capture. The cobalt transferred from water to stone, perhaps oxidized even, an electron transferred in the wink of an eye—now you see it, now it's gone, over there!—from the cobalt to the manganese. The stone sinks, first swiftly through fresh water, then more slowly through salt, the cobalt all the while clinging, being basketed and woven in like Moses by the manganese.

And this is the way it goes. A downward journey of a few weeks. Transit out of light. Transit into darkness. Transit into the deep. And in the oxygen-poor waters the manganese is reduced, falls away, unravels like thread. The atom of cobalt is free again, waterbound. The dazzling little motions of the water molecules, like tiny boomerangs whizzing about it, coming close to its charge, then retreating, are familiar and welcome. So it stays. Perhaps a year. Then another encounter: Something that was once living, a few cells still clinging together drift by. To the cobalt it is as though the roots and branches of a great elm were being dragged by in a flood. The branches reach out, enfold it: chelation. It is on its way to the sediments. Possibly to a small eternity there. Until the next ice sheet comes. But even buried you can hear it, you can hear the cobalt. Like the salt plains, you can hear it sing.

This was only a hypothesis, of course. Possibly it was extravagantly wishful thinking. Who knew whether the cobalt's fate was linked to manganese? I wanted it to be, but the proof was in the particles. I wanted to imagine it drawn from the water to the oxide surface and then released. I wanted to imagine the sound, the susurrus of that exchange, repeated over and over and over in every lake and ocean and river on earth, and for every element, modulated and toned and hallowed and joined in a single lifting chant. I wanted to speak the name of every element, every zinc and copper and mercury and lead, every iron and manganese and every compound of these and of hydrogen and oxygen and carbon and millions and millions more, intertwined and twirling on the wrists of the world.

The hypothesis was not without foundation, however. One year, Canfield and I had sampled for manganese and the trace metals. We did the manganese right there on the lake, on our knees on the ice, as soon as the samples came up. As soon as we got the points off the spectrophotometer, we plotted them on a sheet of graph paper. At the same time we plotted a set of oxygen values. It was remarkable. The data sets were mirror images. Where the oxygen was high, the manganese was low. Where the oxygen was low, the manganese was high. The little decrease in oxygen below fifty meters was matched by an increase in manganese at the same depth. And when the oxygen disappeared at sixty meters, the manganese rose to its highest value.

When I saw the graph, I thought of Keith's party and of what he had said about manganese and oxygen, how they were like kids on a seesaw, how when one went up the other went down, how in nature they seemed poised like yin and yang. I thought of oxidation and reduction, of the oxygen minimum zone above the crusts of the Hawaiian Ridge; about the hypolimnion of Acton Lake.

At fifty-five meters, manganese was being reduced, it was gaining electrons from all of the decaying carbon down there; you could see that in the profile. It was as clear as anything. I began to construct a little story about the manganese. I didn't know whether it was true or not, but I knew eventually we could test it and find out. That was the way science worked. You wrote a story. It was pure imagination bounded by a few, usually weak, constraints. Then you tested it, saw whether the world out there could really abide your notions of what was so. Usually it could not. So you tried again and again until you got it. Until you had something that might actually be so.

It appeared for all the world that manganese—probably in the form of solid manganese oxide, probably clinging to the flat surfaces of clay particles that had been weathered by the Onyx—was sinking into the deep lake and was dissolving away there into manganese ions. The solid, with a sigh and an exhalation, was becoming mere charge in the reducing, electron-rich waters below fifty meters. But if that was so, shouldn't there be a great release of other things as the oxide fell apart and crumbled in upon itself? Shouldn't there be a great release of cobalt, for example, or nickel or other metals that might be riding upon the oxide surface? Shouldn't the profiles for these tell a similar story?

I remember how I had looked forward to the metal analyses. No sooner had the samples arrived back in Ohio, in their sturdy wooden boxes, than I was looking at them. When you study metals, you become obsessed with purity. You can't escape. I worried about every stray breeze, about whether it carried with it a scintilla of lead or cobalt. I worried overtime about the reagents. Every working hour—which is to say every hour I was awake. Was the Freon TF as clean as it could be? What about the DDDC and the APDC and the nitric acid? And the bottles in which the samples had been collected, and the separatory funnels of expensive Teflon? I began to dream of metals. They were everywhere, truckloads of them, and I was counting every atom. Just another of my counting dreams!

After treating them and going through all of the preparatory steps, I had only a tiny extract from each sample. I took the extracts and lined them along the bench in Boyd Hall. The windows let in the winter light from outside; the trees were bare, snow had fallen in the woods. I switched on the instrument, dialed in the wavelength for cobalt, turned on the argon tank and the cooling water, and programmed in the temperatures for the graphite furnace. Into the cups of the automatic sampler I put a few milliliters of extract from each depth, in exactly the same order in which we had collected them. For a second I imagined I was in the field, facing the lake.

Then I set the instrument going. The little arm of the autosampler began to twist. Then it came down and took a few microliters from the cup, barely a drop. It halted a second before it rose again and moved toward the graphite tube. It looked like the pitching machine my father had used for throwing batting practice. Just like that. When it reached the tube, it hung there, deposited its precious droplet, then came back and rested. The furnace kicked in. The temperature of the tube climbed to 110, stopped; climbed to 250, stopped; then shot instantly to 2,300 degrees. You could hear the controller work. There was a surge of current through the tube. Then a burst of white steam, as though a tiny volcano had just erupted right there in the lab. A trace of light shot across the face of the instrument. A number appeared on the screen. Then it appeared on the printer. There were little clicks as the paper moved into position for the next reading.

It was working, but I couldn't watch. I was too nervous. I went back to my office and let the instrument tack downward into the depths of the lake: five meters, ten meters, fifteen meters; it was moving slowly, doing triplicates on each sample. I had the instrument interfaced with the computer. After each analysis the point was placed on a graph whose vertical axis represented depth and whose horizontal axis was the concentration of cobalt. I waited. I thought about what Keith had said at the party. About how manganese might be controlling everything in the oxygen minimum zone of the Pacific. How oxygen and manganese were linked, "like this," as Keith had said. And how the metals might be linked to manganese.

When I finally summoned the courage to go into the lab, the analysis was complete. On the blue computer screen there was a curve connecting the points, from the ice surface down through the water column to the sediments. The whole journey, a whole year, was laid out there on the screen in a single trace. It was frightening to look at. I wanted to shield my eyes.

But there it was. Point for point, the curve for cobalt analysis matched the curve for manganese. Matched it to a T. Where the dissolved maganese was low, so too was the cobalt. Where the manganese rose, in response to the disappearance of oxygen, so too did the cobalt. Even the fine structure between fifty and sixty meters matched up. It was perfect. The story was beginning to write itself.

It was night when I left for home. The woods behind Boyd Hall were filling up with snow. The drifts came nearly to my knees. At times as I walked I thought I was floating.

But as Keith had said that evening, overlooking the flat darkness of the Pacific, the proof is in the particles. If the manganese oxides were really transporting cobalt and lead and the other metals, as the overlapping profiles in Vanda had suggested, then we should see it in the particles, in their composition, in the way the metals were distributed among the complex tangle of plankton cells and minerals and bits of clay and organic ooze. Each particle was, indeed, a universe, a world more extraordinary than even Blake had imagined. On every particle there was generation and decay, the comings and goings of ions from a million knotted surfaces; breathing and exhalation conjoined on a sphere no larger than a pinhead. How many angels could dance on the skin of a sinking particle? Maybe a million. Maybe more. And how many manganese atoms and how much cobalt?

So we had put the traps in. We had left them there suspended. They had seen the coming of winter. They had seen the sun extinguished. They had seen the moon rise over Linneaus Terrace and cast shadows on the mountains. They had heard the fierce Antarctic wind, the wind of the early explorers, the wind of Scott's death. They had seen the lake scoured with sand. They had heard the ice crack like a bullwhip.

✳

Before I saw anything else, I saw the rope. It had gone in white. But in the lake it had turned brown and dark green. Filaments of algae clung to it and wove about it. It had the hoary look of things brought up from the sea.

I had not been expecting this. Vanda was known for its ultra-oligotrophy: biologically, it was among the least productive lakes in the world. In all the water I had drawn from it, all the thousands of liters, I had not seen a single rotifer. Nothing. Even under the microscope, I might as well have been looking at a drop of liquid mercury. Thoreau had said that Walden was "not fertile in fish." Vanda was fertile in nothing. It would have made the waters of Concord look fecund and teeming by comparison.

I had done some rough calculations, before we put them in, on just how much sediment we might expect to find in the traps. There was very little to go on—a single estimate of sedimentation rate that Alex Wilson had made several decades ago. Taking into account the diameter of the traps, I calculated that there should be about a hundred milligrams. Mostly sand, I thought. Sand that had worked its way through the ice cover. But my estimate could be off by a factor of ten either way. Maybe out here, so far from the river, we wouldn't find anything. Maybe like the rotifers, there wouldn't be a single particle.

The first trap, the one we had suspended at forty-eight meters, brimmed with crystalline water. Nothing broke its transparent perfection. I raised it like a dry martini to the afternoon sun. There was nothing to scatter the light. It moved in unbroken rays through the trap chambers. Had we come all this way, I wondered, waited so long, thought so much about the particles, about the rain of manganese oxides cleansing the lake, only to find this?

"Empty," Dr. Yu was saying. "Empty. Only water." He had removed his balaclava and was scratching his head, wiping his forehead.

I put the traps on the ice, stood the chambers upright. I reached into my parka and took out some orange plastic caps. I placed these over the top of the cylinders. Then I lifted the whole assembly over my head and began to shake it. I don't know what I was expecting. Maybe that the water would somehow transform itself into earth, the way the Greeks thought it did at the mouth of the Hellespont. Maybe that it would become air. Maybe that it would burn. Water could do anything. You just had to believe.

Dr. Yu was saying "Ohhhhhh, careful. No break." I shook the tubes for a full minute. A dry snow had begun to fall. The flakes were large. They clung to my skin without melting. I put the tubes down on the ice, lay on my stomach, and held them in front of my face. I couldn't tell what I was seeing. Something was drifting down inside. Maybe it was just the snow. Maybe the water *was* turning to earth.

Gradually I began to discern what was happening. There were tiny bits of clay suspended in the water. It was faint, but you could see it. It was settling, but very, very slowly, the way small particles settle. Stoke's beautiful law. It was the dust of the lake, the dust of the river, the dust of the valley. We had waited a year for it. I let out a whooop.

We were on the lake all evening and well into the morning. It continued to snow out of a gray sky. The whole valley was enveloped in cloud. The mountains folded in around us. A wan light spread up toward the sound. The glaciers seemed to be floating again. The parkas whispered against themselves as we moved.

We repaired the trap catcher, and each time it held. It folded tight against itself and went down the hole. It opened and groped its way through the lake. It touched the line, gathered it in, wound the traps to itself. Then it released at the ice edge and straightened again, just the way Tim and Dr. Yu had designed it. Bamboo could do anything.

We covered the whole length of the lake, all the way down valley to where the Onyx would soon begin to flow in. We turned the sleds to shore and pulled them along the smooth annual ice, five miles back to camp. You could hear the water sloshing in the traps as we moved.

Just where the ice meets the shore, you could see a thin band of open water. It was no more than a few inches wide. Dr. Yu was pointing toward it and saying "Ahhhhh, water! In Qaidam Basin, we say maybe spring. Maybe not long."

CABIN 532

from *Skating to Antarctica* (1997)

Jenny Diski

Not everybody comes to the Antarctic wide-eyed, or with their psyche neatly tuned for enthusiasm. The British novelist Jenny Diski took herself south in a state of wisecracking, deadpan melancholy, looking for a landscape that would soothe her the same way the sight of an all-white bedroom did. She found one, and in the process created a small classic of Antarctic iconoclasm. This is Antarctica, the neurotic version; very nearly the indoor *version.*

THE NEXT MORNING around 6.15, I lay shivery in my bunk, with a sore throat. I had a streaming, screaming head cold and felt ghastly, but it was OK since it was a travelling day and I didn't have to do anything if I didn't want to. I had woken earlier, at four, to the unrelenting daylight with a small damp weight inside of me. A tiny blueness. Perhaps it was a moment of panic at being out of reach of anywhere, hundreds of miles out of sight of land. A kind of bleakness. I closed the beige curtains across the bed, making a soft half-light, enclosing myself, and dozed on and off.

Later, I got my achey self up to look out of my window. Birds: pintados, petrels, black-browed albatross (I was getting good at this) were wheeling around, dipping and diving on the wind to feast off Melville's "effulgent Antarctic sea." The surface of the water was dark, hardly lively, but active enough to create a gentle rocking motion in the boat that sent me back to bed to enjoy it. I carried on with *Moby Dick* and Ahab's search for the great white whale. Such a pleasure. Grand, huge and free. Taking all the freedom a novelist needs. I couldn't think of a circumstance that would improve my life—apart from not having a cold. Though soon I would have to get up to have breakfast—which meant other people, greeting, smiling and talking. That felt like a bit of a trial. My wish was to stay where I was, in my cabin, in bed. As I put my book down for a moment, I looked through the window and realized that it was snowing. The sky was dove grey with heavy snow-weighted cloud, and the snow was falling softly, making the windswept wilderness of the Antarctic sea as silent as any suburban winter garden. The horizon was a very long way away.

I skipped breakfast and both of the lectures designed to help fill our at-sea time, devoting the day to my cold, my bunk, my book and the view from my window. I was perfectly contented. Sleeping, reading and staring out at the snow and sea, I could have done this forever. No Sister Winniki to nag. Just left alone to take all the pleasure I wanted in indolence. What more could anyone want?

Then in the evening the first iceberg floated by.

The iceberg emerged before my lazy gaze at the window, like a mirage, a dream appearance, a matt white edifice ghostly in the misty grey light and falling snow. A sudden, smoothly gliding event in the great empty sea under the great empty sky. I blinked at it. There was none of the disappointing familiarity of something seen too often on TV or in picture books. This startled with its brand-new reality, with its quality of not-like-anything-else. Even the birds seemed to have hushed for our entrance into the land of ice. The tannoy squawked into life, and Butch announced:

"Ladies and gentlemen, we have icebergs."

Time to get up. I pulled on a tracksuit and headed up to the bridge where I discovered that like a momentous theatrical production we were proceeding into real Antarctica through a corridor of icebergs. For as far as the eye could see, to either side of us, icebergs lined our route. We were journeying along iceberg alley. It was absurdly symmetrical, like a boulevard in space. The bergs were tabular; as their name suggests they are flat as a table on top, as if someone has planed away their peaks, too smooth to be real, too real to be true. It took about an hour before we had sailed the length of iceberg alley, and all the time I and many of the other passengers stood and watched in silence, broken only by gasped oohs and ahhs and the inevitable camera clicks and camcorder whirr. For a while after that we seemed to travel through debris, odd-shaped bergy bits, and the larger growlers—some just the dimension of buoys bobbing and rocking like corks in the waves, others the size of a dinghy or a small cottage. These were the remnants of big bergs, worn down by wind and water, breaking off and melting away eventually to nothing. Finally we sailed through cracked pieces of ice, chips and fragments, as if we were making our way through a bowl of granita.

Later, in the early hours of the morning, more big bergs appeared, not in formation any more, but dotted about on the sea in gatherings of two and three, some of them huge, the size of our ship and bigger. I watched one go past that, we were told the next morning, was four miles long, an island of ice that took an age to sail along.

I didn't get much sleep. Irma may not have been so pathological in her attitude towards the steak. I was feeling decidedly nauseous and having stomach cramps. The light-soaked night was spent divided between hovering over my toilet without quite throwing up (there is very little I hate more than vomiting), and leaning out of the open window with my chin on my folded arms watching the superlunary white mountains float by. Sometimes, as I watched one huge berg sailing towards us, it would shift shape, playing parallax games, and getting nearer, became two or three separate bergs at what I finally saw were a considerable distance apart. Then when we had left it behind, it returned again to its singular form.

As a matter of fact, in keeping with the interesting but not fatal disparity between my fantasies and rest of the trip so far, the icebergs close up, even quite far away, were not daydream white at all. Blue. Icebergs are blue. At their bluest, they are the colour of David Hockney swimming pools, Californian blue, neon blue, Daz blue-whiteness blue, sometimes even indigo. Blue is an odd colour: the signifier of good things to come ("blue skies, nothing but blue skies") and of dark thoughts ("Mood Indigo"). Different shades, different promises, same colour. It can be bright, clean, even cold; and it can be mysterious, deep, the colour of night and dreams. Like Melville's notion of whiteness, it makes clear and it obscures; it is purity and complexity. The colour blue does no violence to my hankering for white. It belonged with and in the ice, making it seem colder, emotionally empty, and yet more dense, layered beyond what could be observed. The bergs were deepest blue at sea level, and where cracks and crevices gave a view of the depths of the berg where the ice was the oldest and so compacted that all the air had been forced out. Why this should cause them to be aching blue is not a question that someone who spent physics lessons turning notes into poetry can answer. The blue tinge higher up was, I was told, because ice absorbs all wavelengths of light except the shorter blue ones. I told this proudly to Marjorie who pointed out that this account would cover the reason why anything looks any colour. Melville again: there is no colour, only an appearance of colour that conceals the universal negation of white light. The explanation didn't after all explain anything much, and once again I wished I hadn't dicked around during physics and deprived myself of answers to most of the questions I now found myself asking.

These floating mountains of blue ice shaded with white, white ice shaded blue, were not slick and shiny like ice-rink ice. They were dense, matt islands of compressed icing sugar. Confections that lit up the lead-coloured sea and sky. All night they floated along, carved by wind and water into ancient shapes—ships, castles, monuments, mythic creatures. None of the explorers who described them managed to avoid these descriptions. They are as unavoidable as a sigh at the sight of the elephant seals having sex. I passed great craggy faces in profile, ominous fingers pointing, lions crouching, birds leaping. The mind couldn't help it. But sometimes, and most breathtakingly, they were simply vast walls of ice, passing so near to my window that they cut out the rest of the world. A great blank wall of ancient compacted snow that had travelled from the blank centre of the Antarctic continent for centuries—the deepest ice is 10,000 years old—to its edges to become a tongue of glacier or the very periphery of the ice cap, and finally broken off—calved is the correct term—to sail away on its own. They head north and west at a rate of five miles a day until they reach the convergence where the Antarctic Ocean meets warmer seas. Some of them last for ten years, until the sea erodes and melts them away.

Antarctica has seventy per cent of the world's fresh water locked up in the ice. In places the ice sheet is two and a half miles deep. Over the centuries people have planned to utilize the bigger icebergs as a source of water in drought-stricken regions. The only problem has been how to get the bergs from where they are to where they're needed. Grandiose plans to tow them have come to nothing: once they are into warmer water the chances of landing more than enough ice to cool a gin and tonic are unreasonably small. Icebergs are one resource that the human race has failed to find a way to make use of—apart from Daniel and my Manhattans. While the sea ice, pack-ice, frazil, nilas, pancake-ice are salt—though not as salt as sea water—the bergs that break off from the continent are pure fresh water—as pure and fresh as anything can be on the planet. Certainly, for all the pollution, purer than the stuff that comes out of our taps. And these mountains of frozen water I watched floating on the surface of the sea are, as everyone knows, just the tip of the iceberg. Nine-tenths of each berg is indeed below the waterline, and melts faster than the bit you can see, so that eventually they become top-heavy and turn turtle into the sea. The smaller, most rounded bergs are actually upside down, tip-turned and unstable. Growlers, they are called, and not loved by sailors; they are the ones that creep up and scuttle ships in the dark, being harder to see and often invisible to radar.

The scene from my cabin window was otherworldly, I am afraid there is no other really apposite description. In the early hours of the morning, the light was pale silver, slightly misty. Half-close my eyes and there was nothing but a spectrum of grey, blue and off-white. It could have been bluey grey, greyish blue, hues of blue. There were huge bergs coming now. I suppose they were related. They came in waves. Three, four, but no, they were two big bergs. They must have broken off at the same time, or it could have been one great berg that split. On the horizon the cloud seemed to have settled down on to the sea and turned into another iceberg. It was impossible to tell what was cloud and what was berg in the distance. It looked fluffy, the same colour as the clouds higher up, only distinct, and squat, flat on the sea, but still cloudlike for all that. But it wasn't a cloud. Soon it turned into one . . . two separate bergs. It was all clouds and bergs, bergs and clouds. And there, close by the ship, was a single penguin on an iceberg that looked like a lion. The bird was riding on the lion's mane, standing stark still, looking ahead in the direction in which it travelled. I'd never seen a penguin look heroic before. It was 3.20 A.M. and luminous, misty daylight.

We stopped off at Deception Island, making the approach after breakfast, in silvery light. The name alone was worth getting up for. Who would want to lie abed when they were sailing towards Deception? The sight from the bridge was immortal. Deception Island is a caldera, a volcano summit that has collapsed to form a crater. One section sank far enough to allow the sea water to flood into the interior of the caldera. As we approached "Neptune's Bellows" at the entrance to the crater, fleets of penguins came out to see what we were and swam, ducking and diving, alongside the ship. The sea was veneered with a pattern of sea ice made up of perfectly sharp-edged rectangles that looked as thin as wafers and pure white, floating side by side with narrow channels of dark water in between. It was a chessboard of flat floating squares of ice, as strange and orderly as anything I'd ever seen. It was shocking when the ship broke through one of the bigger rectangles on its way to the narrow entrance to Deception. The bow of the ship split the ice floe into two pieces which grew jaggedly ever wider apart as we pressed on through. There was something terrible about this, about breaking up the pattern just because we wanted to get beyond it. No real damage was done, but something artful in nature was dislocated. The design was spoiled. And yet the further into the checkerboard we got the more we seemed to be part of it. Soon it was surrounding us, an abstract picture made with great but delicate four-cornered slabs of ice, floating in the hazy sunlight, with a small white ship slipping through. The Zodiacs went ashore but I remained on board, on the deck at the back of the boat, and stared all morning at the chequered ocean. Checkmate.

The following morning I woke, still feeling rotten. Everything ached, my head was thick and I still felt sick and crampy. I couldn't move, not even to take a pill, not even to check out the state of the icebergs. I'd been up and down all night, at the window, then dozing, until I finally went back to sleep at 6.30 and slept until late. It was 11 A.M. and as light as it had been in the early hours. It was always light now, we were so far south. The lack of alternating dark and light made it very easy for me to stay where I was. The call to breakfast, lunch and the time-passing lectures failed to intrude on the structureless nature of continual daylight and feeling ill.

What was going to happen to my Antarctic adventure if I sank for days into my bunk, I wondered? Unless, of course, this *was* my Antarctic adventure. I was quite suspicious of my malaise. Through the aches and the nausea, pleasure shone through, at escaping the timetable, the events, the socializing. I guessed it was probably my psyche, rather than my immune system, that had revolted against being snuggled in with so many people. Still, the psyche had effectively persuaded the body to produce some impressive symptoms—good enough to justify hermiting without having to accuse myself of being anti-social. Though I did, in fact, accuse myself.

We were coming into Admiralty Bay, where humpback whales were supposed to hang out. We hadn't sighted any whales so far. Well, I could look out of the window if there were any sightings. Though only if they happened to be on my side of the ship. But I didn't want to see whales with a whole bunch of people on the bridge or the deck. I didn't want to be in a crowd watching whales, even if it meant missing them. I wanted to see my own whale out of my own window, all by myself. Or not. I imagined explaining to friends on my return to England how I didn't manage to see a whale. Wrong side of the ship. Didn't get up. Oh dear.

The following morning we were due to land at Paulet Island. Outside it looked very weird and alien. I had to keep telling myself I was on the same planet. Not on the moon. Maybe this sickness or not-sickness (I'd managed to throw up some yellow bile that night, which made me feel a little more authentic) was panic at not escaping, at never escaping. Here we were at the end of the earth and still it's the same planet. I would have liked to get somewhere else. Further off. One morning I was alarmed at being out of sight of the world, the next I was blue about not being able to escape the world altogether. Not very consistent, I told myself. Myself shrugged. I wondered, at last, if I was going to prevent myself from landing on Antarctica. What an odd thing, to have come all this way and then not land on the peninsula. There was a small but unmistakable internal smile at the thought. I located a tight subterranean knot of unwillingness to set foot on the last continent just because I happened to be there. Though of course I didn't just happen to be there. This was a place where no one just happens to be. Which made it all the more pleasing and/or distressing (the two emotions were inextricable) to consider the possibility of not stepping on to the land. There was considerable satisfaction at the thought that I might not set foot on the continent that I had taken a good deal of trouble to get to. I considered further ventures: how I might fail to let the sand run through my fingers in the Gobi desert; how I could turn back twenty-five feet from the summit of Anapurna; how, gaining disguised entrance to a Masonic initiation ceremony, I would shrug my shoulders at the door and wander away. I could easily go to Agra and fail to clap eyes on the Taj Mahal—too easily. What about a trip to Brazil spent entirely in the air-conditioned confines of the Brasilia Sheraton? Or a visit to every airport in Africa without setting foot on the continent itself? I could keep myself busy as the resistant traveller until my last gasp and then, at the gates of Heaven, make my excuses and turn away. It's not the arriving but the not-arriving . . . it's not the seeing of the whales, but the possibility of choosing not to see them. This was an aspect of me that I recognized from every period in my life. The Fuck-it Factor. *I don't have to if I don't want to.* Sometimes it has looked like a lack of persistence. "What you lack, my girl, is stickability" my oh-so-resolute father would say to me. I didn't finish pictures, knitting, stamp collections. Lost interest two-thirds, or even nine-tenths of the way through. I got thrown out of school. I left school two weeks before my A-levels. I wandered away from relationships. Sometimes, even now, I turn off movies on TV minutes before the end. Lots of people have nodded knowingly to each other, "She just can't finish things." But no one ever mentions the exhilaration of not finishing things. The rush of pleasure at not doing what is expected of you, of not doing what you expect of yourself. Of not doing. If it was originally about disappointing other people, it has become refined into a matter of pleasing myself. Of making choice less inevitable. But it's not a policy. Only something I notice happening from time to time, and the genuine satisfaction that goes with it. This time I'd caught myself in the early stages. Unless, of course, I was just not well. You can't say, and I may not choose to say.

I became gripped by the idea of willingness. As I gazed out of the window, I imagined myself into the future: the trip over, me back in London. Did I or didn't I get to Antarctica? At that delicious moment I really didn't know what the answer would be. It wouldn't make an iota of difference to the world, or in reality to me, if I didn't actually stand on the Antarctic landmass. Been there, haven't done that. I liked the absurdity of it, and the privacy. It's a matter entirely between me and myself. Indeed, I could say, back home, that I did, when I didn't. What difference would it make? Or, come to think of it, I could say that I didn't, when I did. And once I'd had this thought, it didn't matter whether I actually did or I didn't. The quality of my life wouldn't alter one bit, and either way only I, and a handful of fellow travellers, would know, and none of us would care. The decision became entirely academic. I could toss a coin: heads I land, tails I stay in my bunk—and then not go if it comes up heads. Or contrariwise. I might not have been feeling well, but my spirit was soaring. There was no longer a choice that had to be made, or an effort of will (I should, I ought, I must), no moral quandary, but something quite arbitrary. A great sense of freedom settled gently over me like a pure white goose-down quilt, and freedom, from that angle, looked very like uncertainty, as Antarctica slipped into Schrödinger's box and closed the lid quietly on itself. I had no idea whether I would get up and land the next day or not. And no one else, save a few scattered U.S. citizens, would know for sure what I eventually did. This, for reasons I don't choose to examine too closely, was a huge comfort. What was all this getting to Antarctica thing anyway?

The book I was going to write about the trip, and my mother, was to be my first full-length non-fiction, it had been agreed. I took this to mean that I would not be writing a novel about it. Obviously. Who would agree to write a novel about a real trip they hadn't yet been on? Novels aren't like that. But beyond that I reserved my judgement on what non-fiction is. There are infinite ways of telling the truth, including fiction, and infinite ways of evading the truth, including non-fiction. The truth or otherwise of a book about

Antarctica and my mother, I saw from my swaying bunk in Cabin 532, didn't depend on arriving at a destination. Nor in failing to arrive. I found myself beginning to get a taste for non-fiction.

Another day, and the tannoy announced that we were as far south as we were going to get: 63.42S 55.57W. We had arrived in the early hours at the Antarctic Sound. My landing quandary was resolved for this morning at least. The pack-ice was too thick to get through to Paulet Island, and things weren't looking good for the landing on the Peninsula at Hope Bay, scheduled for the afternoon. It looked possible that no one would be landing on Antarctica, and my decision—as so often—would be no decision at all in the event. However, Butch was doing his best for us and had arranged, instead of a walkabout on Paulet, a Zodiac cruise around the great field of icebergs surrounding us. The sun was blazing beyond my window and the sea was glacially still. There wasn't a ripple to be seen on the turquoise surface, nothing but the sharp reflections of the bergs. Suddenly, I thought I might be feeling a bit better. Time to get up. I very much wanted to get close up to the icing sugar mountains, and down near the surface of the brilliant sea.

After my sullen, anti-social couple of days in my cabin, I found myself greeted in the corridors, in the mud-room and on deck by everyone I saw, welcoming me back and asking how I was. They had noticed I was not around and asked Marjorie and Phoenix John if I was OK. I felt a little ashamed at the concern people were showing for me. Part of my sense of shame was that I slightly resented their attention. I don't like the idea of being thought or wondered about if I have gone to ground. This was so ungenerous compared to people I hardly knew telling me they were glad to see me up and about, that I was not pleased with myself.

The only noise was the buzz of the Zodiac engine. There was no wind, no screaming penguins, and when the Zodiac stopped at the foot of an iceberg, no sound at all, except for water lapping against the wall of ice. The sun shone and the sea was a deep green with sparkling lights, twinkling like sequins. The surface was calm, just slightly wrinkled by the gentle breeze, almost syrupy. A Crimplene ocean. Between the bergs, so common now but never boring, were flat ice floes, slabs several feet thick, on which the odd crab-eater seal lolled soaking up the sun. So long as we didn't get too near they refused to be bothered by our pottering about. If we went too close they slithered unhurriedly into the sea and disappeared.

To be at the base of an iceberg, rocking on the sea, is a remarkable feeling. The cold radiated off the wall of the berg and I peered into secret crevices that went to the deepest blue heart of the ice. The world was flat and still except for the bergs ranged above us as we wove in and out between them. The ship at anchor, as white and still as another berg, belonged there, another mythic shape in the landscape and didn't seem to impose itself. It was uncanny and peaceful, a near oblivion, but deceptive. This was not a place, though it was a position on a navigation chart. Nothing about this region would be quite the same again, as the floes and bergs floated and melted, winds whipped up the presently calm sea, seals made temporary lodging, and flotillas of penguins porpoised around the ship in the distance like flying fish. Everything about this seascape would change, but it would also remain essentially the same, its elements merely rejigged. It was so untroubled by itself that the heart ached. Other landscapes fidget—rainforests full of plants and creatures clamouring for a living, moors troubled and ruffled by scathing, distorting winds, mountains trembling with the weight of snow—but this was truly a dream place where melting and movement seemed only to increase immutability. Nothing there stays the same, but nothing changes.

But what, I wondered, was the point of witnessing this sublime empty landscape and then passing on? That question was one reason, I suppose, for the rate at which the cameras clicked away. The photograph was evidence for oneself, not others really, that you'd been there. The only proof that anything had once happened beyond an attack of imagination and fallible memory. It also caused there to be an event during the moment of experiencing, as if the moment of experiencing doesn't feel like enough all by itself. If you merely looked and left, what, when you returned home, was the point of having been? It was not hard to imagine such a landscape, to build one in your head in the comfort of your own home, and spend unrestricted time there all alone. In real life, you look, you pass through, you leave—you take a photo to make the activity less absurd. It provides something to do with your hands while you are trying to experience yourself experiencing this experience. But how do you become, as I wanted to be, part of this landscape, to be of it, not making a quick tour through it? What I was doing was having a taster of something, watching the trailer of a movie I would never see. I would take this memory of a place in motionless flux back with me, and add it to the Antarctic in my mind. I wondered if it would be a useful addition once the experience was in the past.

But I had forgotten about Cabin 532. That *had* been a new experience, something I hadn't already dreamed up or dreamed of, somewhere I couldn't have visited through pictures taken by someone else. Cabin 532 was something really new to carry back to London and play with.

HEY, WOO

from *Terra Incognita* (1996)

Sara Wheeler

Sara Wheeler went to Antarctica to do for it, for the first time, what a good travel writer does for a place: to cast a beady, evocative eye on the manners and behaviour of the inhabitants, and to knit that sociable human reality together with the sublime ice in the background. In bases run by the British Antarctic Survey, the perpetually boyish scientists put spoonfuls of sherry trifle in her snow boots. At McMurdo, the National Science Foundation put her on their Writers and Artists Program, and assigned her office W–002: hence the nickname "Woo" that follows her, here, out to a seismology field camp in the continent's big, empty center.

RESUMING THE QUEST for Seismic Man and his group, I wheedled my way on to a fuel flight to Central West Antarctica, and after a series of false starts I was transported to the skiway with four members of a science project staging at CWA en route to Ice Stream B. The West Antarctic ice streams—fast-flowing currents of ice up to 50 miles wide and 310 miles long—are cited as evidence of possible glacial retreat and the much-touted imminent rise in global sea levels. The project leader pulled out *The Road to Oxiana*, the greatest travel book ever written and one which lies so close to my heart that it gave me a shock to see it there, as if the paraphernalia of home had followed me. He was a beatific man in his mid-fifties with a round, mottled face like a moon, and his name was Hermann. Ten years previously, he had climbed out of a crashed plane in Antarctica.

Later, when we were airborne, the scientists retreated into the hoods of their parkas, jamming unwieldily booted feet among the trellis of rollers, survival bags and naked machinery. I loitered on the flight deck for a while, but I couldn't see much. It grew colder.

The previous evening, in the galley at McMurdo, I had run into a mountaineer from a science group which had recently pulled out of CWA.

"Hey!" he had said when I told him I was on my way there. "You can sublease the igloo I built just outside camp. It's the coolest igloo on the West Antarctic ice sheet."

When we landed at eighty-two degrees south, the back flap lifted and light flooded into the plane. Tornadoes of powder snow were careering over the blanched wasteland like spectral spinning tops. There were no topographical features, just an ice sheet, boundless and burnished. Lesser (or West) Antarctica is a hypothesised rift system—a jumble of unstable plates—separated from the stable shield of Greater (or East) Antarctica by the Transantarctic Mountains. On top, most of Lesser Antarctica consists of the world's only marine-based ice sheet. This means that the bottom of the ice is far below sea level, and if it all melted, the western half of Antarctica would consist of a group of islands. The assemblage of plates which make up Lesser Antarctica have been moving both relative to one another and to the east for something like 230 million years, whereas Greater Antarctica, home of the polar plateau, has existed relatively intact for many hundreds of millions of years. In Gondwanaland, the prehistoric supercontinent, what we now know as South America and the Antipodes were glued to Antarctica. Gondwanaland started to break up early in the Jurassic Period—say 175 million years ago—and geologists like to speculate on the relationship of Antarctica to still earlier supercontinents. Most exciting of all, Antarctica once had its own dinosaurs.

The crewmen began rolling pallets off the back of the plane. We walked down after them, and the wind stung our faces. The engines roared behind us as we struggled to pull our balaclavas down around our goggles.

In the sepulchral light ahead I could see a scattering of Jamesways, a row of sledges, half a dozen tents, and Lars, the shaggy-haired Norwegian-American from Survival School. He was looking even shaggier, and proffering a mug of cocoa. We hugged one another. Lars led the way into the first Jamesway, where half a dozen weatherbeaten individuals were slumped around folding formica tables.

"Welcome, Woo!" somebody shouted. I had brought them cookies and a stack of magazines, and as I handed these over we all talked at once; a lot seemed to have happened in two months.

"Guess what?" said Lars. "We saw a bird."

The CWA field camp was probably the largest on the continent. Fifty people were based here for most of the summer season, working on four separate geological projects. Often small groups temporarily left camp, travelled over the ice sheet on snowmobiles or tracked vehicles, pitched their tents for a few days and tried to find out what the earth looked like under that particular bit of ice. They were creating a relief map of Antarctica without its white blanket.

Seeing Seismic Man's lightweight parka hanging on a hook in the Jamesway, I suspected he was away working at one of these small satellite camps. I was thinking about this, just as Lars produced another round of cocoa, when a familiar figure flew through the door of the Jamesway and clattered to a standstill beside me. It was José, the diminutive Mexican-American biker who grinned like a satyr and with whom I had failed to get to CWA on my first attempt. He had made it here a week before me. In one long exhalation of breath he said that he had heard I'd come, that he and two others were about to set off to strike a satellite camp thirty miles away, that it would take about twenty-four hours and they wouldn't be sleeping, that I could go too if I wanted . . . and then he trailed off, like his bike running out of fuel.

Having trekked halfway across the continent to find Seismic Man, I left immediately without seeing him at all. It was the idea of the quest that had appealed to me. Feeling vaguely irritated about this, as if the whole expedition had been someone else's idea, I climbed into the back of a tracked vehicle and shook hands with a tall loose-limbed Alaskan in the driver's seat.

"They call me Too-Tall Dave," he said as he pumped my hand, crushing a few unimportant bones. "Pleased to meet you."

The man next to him—a medical corpsman on loan from the Navy—looked as if he had just got up. His name was Chuck, and apparently he had forgotten the American president's name one day and asked Too-Tall Dave to remind him. José and I spread out over the two bench seats in the back of the vehicle. It was a temperamental Tucker which only liked travelling between eight and ten miles an hour, and we were towing a flat, open trailer and a sledge loaded with survival gear. As we were following a flagged route to the small camp you couldn't really call what Too Tall was doing driving: it was more a question of stabilising the steering wheel with his elbow and looking at the dash every so often to make sure he was maintaining the correct rpm to keep the water and oil at a stable temperature. It was very warm in the Tucker. The ice was dappled with watery sunlight, and the sky pale, streaky blue.

"This," said José, "is what travelling in a covered wagon across the United States must have been like."

After five hours, we reached a weatherhaven and a Scott tent.

"Is this it?" I asked.

"Yep," said Too Tall, swinging nimbly out of the Tucker.

When I saw the tent, its flap still open, sunlit against the white prairies, an image flashed across my mind, and after a moment I recognised it as J. C. Dollman's painting of Captain Oates staggering off to die, arms outstretched and wearing a blue bobble hat. The lone tent in the background of the picture was identical to the one I was looking at, except that Dollman hadn't painted something which looked like a Land Rover parked outside.

I packed up the contents of the weatherhaven while José and Too Tall set about dismantling it from the outside. Fortunately the wind had dropped, but it was bitterly cold.

"Why didn't the beakers do this themselves?" asked Too Tall irritably. "Next time they'll be asking us to wipe their butts."

By the time we had finished loading the gear on to the trailer it was six o'clock in the morning. We squatted in a banana sledge we had forgotten to pack up, and opened three cartons of orange juice and a large bag of trail-mix.

As we rearranged our own gear in the back of the Tucker afterwards I noticed that fuel had leaked all over my sleeping bag, not for the first time or the last. I wasn't the only one in Antarctica who smelt like an oil rig.

I drove for the first three hours on the way back to camp. It was a mesmerising occupation, and as I wandered into a reverie or stared blankly out at the ice sheet, the needle crept up on the rev. counter dial.

"Less gas!" Too Tall would then say, delivering a karate chop on my shoulder from the bench in the back. The mongtony was broken by the appearance of a bottle of bourbon. José set up a Walkman with a pair of speakers.

"We need tortured blues," said Too Tall. He was right. It was the perfect accompaniment to the inescapable monotony of the landscape and the hypnotic rhythm of the Tucker.

I got accused of picking all the cashews out of the trail-mix, a crime of which I was indeed guilty. Everyone started talking.

"Are you married?" José asked me.

"No," I said. "Are you?"

"No." There was a pause, which something was waiting to fill.

"Go on José, tell her!" said Chuck.

José cleared his throat.

"Actually, I married my Harley Davidson," he said.

I choked on the last cashew.

"Oh, really?" I said, in an English kind of way. "Who performed the, er—ceremony?"

"Owner of my local bike shop. He does it a lot."

This information was almost more than the human spirit could bear. Fortunately an empty fuel drum chose that moment to fall off our trailer and roll over the ice sheet, and after we had dealt with that, the topic was forgotten.

A fresh one, however, was looming.

"You know that Captain Scott," said Too Tall in my direction as the bourbon went round again. "Was he a bit of a dude, or what?"

I had just begun to grapple with a reply to this weighty question when Chuck, his face puckered in concentration, chipped in with "Hey, is that the guy they named Scott's hut after?"

"No," I said, quickly grasping the opportunity to divert the conversation away from the dude issue. "That was Mr Hut."

It took us eight hours to get back, and then I had to put up my tent. It was snowing lightly, and I was too tired to dig out the igloo. I chose a place at the back of camp, facing the horizon. My metal tent pegs weren't deep enough, so I hijacked a bunch of bamboo flagpoles, and after the bottle-green and maroon tent was up I collapsed into a deep sleep.

When I woke up, a face was hovering a foot above mine.

"Hi Woo," it growled. "Didn't want to wake you."

"This is a funny way to go about not waking me," I said as the face drew closer.

*

They were using explosives to find out what the ground was like under 6,000 feet of ice. "We're not particularly interested in ice," someone commented breezily. Because of the inconvenient ice cover, most Antarctic geology can only be studied by remote-sensing methods like seismology. This involves setting off explosions, bouncing the soundwaves down through the ice to the earth's crust, and recording them on their way back up.

Before they could be detonated, the explosives had to be buried, and twelve itinerant drillers had been travelling around the ice sheet within a 200-mile radius of CWA boring a series of ninety-foot holes. They began each hole using a self-contained unit which heated water and sprinkled it on the ice like a shower head. This unit fulfilled a secondary function as a hot tub, and we got in four at a time, draping our clothes carefully over the pipes to prevent them from turning to deep-frozen sandpaper. This was a task requiring consummate skill. A square inch of fabric inadvertendy exposed to the air could have excruciating consequences.

Five members of the drill team were women, and in the hot tub one day I found myself next to Diane, a lead driller. She was tall and willowy with long hair the colour of cornflakes. I asked her how long she had been away.

"Thirty-five days," she said. "And my feet were never dry."

"What did you do out there?" I asked. "I mean, when you weren't drilling?"

"Well, just living took all our time. We worked twelve-hour shifts on the drill, and then we'd have to set up the cook tent and all that. We had to plan what we were going to eat carefully, as even if it was going to be a can of peaches it had to be hung up in the sleep tent overnight to thaw."

"Was it your, er, ambition to do this kind of work?" I asked, struggling to grasp the concept that a woman could enjoy spending weeks in sub-zero conditions manipulating a drill for twelve hours a day.

"I do love it," she said. "I think this is the most magical place in the world. People say—'But all you can see is white!' That's true, but I could never, ever get bored on the drill when I can watch the dancing ice crystals, and the haloes twinkling round the sun. It's another world."

The evening before they flew back to McMurdo, the drillers brought in ice from a deep core and hacked it up on the chopping board in the galley. It was over 300 years old, and packed with oxygen bubbles. It fizzed like Alka-Seltzer in our drinks. Diane was baking cinnamon rolls. When she opened the oven door a rich, spicy aroma filled the Jamesway. It was like a souk.

Diane inhaled deeply. "Heaven!" she said.

The next day I moved into the igloo. It was at the back of what they called Tent City, and it took me two hours to dig out the trench leading down to the entrance. Like all good igloos, the sleeping area was higher than the entrance, thereby creating a cold sink. Inside, there was a carpet of rubber mats, and a ledge ran all the way round about six inches off the floor. I spent a further two hours clearing away the pyramids of snow that had accumulated through the cracks. When my new home was ready, I spread out my sleep kit and sat on it. The bricks spiralled to a tapering cork, filtering a blue fluorescent light which threw everything inside into muted focus. I was filled with the same sense of peace that I get in church. Yes, that was it—it was as if I had entered a temple.

The previous inhabitant had suspended a string across the ceiling like a washing line, so after hanging up my goggles, glacier glasses, damp socks and thermometer, I fished out the beaten-up postcards that I always carry around. These could be conveniently propped on the ledge. The blue light falling on the "Birth of Venus" highlighted her knee-length auburn hair with an emerald sheen, and the flying angels had never looked more at home. I felt that Botticelli would have approved.

In the mornings I sat underneath rows of cuphooks at one of the formica tables in the galley Jamesway, watching the beakers making sandwiches and filling waterbottles before setting out to explode their bombs. The cooks were the fixed point of camp. Bob and Mary were a great team. Every morning they dragged banana sledges over to what they called their shop, a storage chamber seventeen feet under the ice from which they winched up filmy cardboard boxes on a kind of Antarctic dumb waiter. Mary was relentlessly cheerful, and she loped rather than walked. Bob had an Assyrian beard, a penguin tattoo on his thigh and a reputation as the best cook on the ice. He was hyperenergetic, very popular, and seven seasons in Antarctica, including two winters, had left him with a healthy disrespect for beakerdom.

"What's going on out there?" someone asked one day after an explosion of historic volume.

"They're just trying to melt the West Antarctic ice sheet," Bob said, scrubbing a frozen leg of lamb. He could seem abrasive, but really he was as soft as a marshmallow.

Seismic Man had spent so long in the field over the past six weeks that he said "Over" as he reached the end of whatever he was saying. When he had to set off an explosion we rode far out from camp on the back of Trigger, his snowmobile. The ice was mottled and ridged like a relief map, and a hint of wind blew a fine layer of white powder over the surface. You could almost absorb the psychic energy out there.

"You know what?" I said to him one day as I unpacked orange sausages of nitroglycerine. "People call this a sterile landscape, because nothing grows or lives. But I think it's *pulsating* with energy—as if it's about to explode, like one of these bombs."

"Hell, yes," he drawled. "I've often felt as if it's alive out here. Hey, look at that," he said, pointing to where the china-blue sky grew pale.

"It looks like a bunch of fuel drums," I said.

"Ha!" he replied. "It's the distorted image of camp, thrown up by refraction of the light. It's caused by temperature inversion in the atmosphere."

Every few minutes a sharp tirade would issue forth alarmingly from somewhere within the folds of Seismic Man's parka. The beakers were forever gabbling to each other over the radio. They had developed their own language, and entire conversations took place between Lars and Seismic Man consisting of acronyms, nicknames and long-running, impenetrable jokes.

I had never met anyone who found life as effortless as Seismic Man. He approached everything with a positive attitude, and saw something to laugh about in every situation. As a result, everyone loved him. In addition, he was disarmingly perceptive. He seemed to have got me taped, anyway. He exemplified the easy-going languor I associate with Texas, without any of the cowboy-hat brashness.

"Can you tape the explosives into bundles of three?" he said, handing me a roll of tape. "I have to set up the shotbox."

The drillers had already made a hole, and after attaching the first two orange bundles to an electric line, we lowered them both into it. Then we tossed down the other 400 pounds of explosives. When the time came to initiate the detonator, I pressed the button on the shotbox and a black plume shot up like a geyser. A sound that could have come from Cape Canaveral followed in a second.

"Wow," I said.

"That's it, Woo!" said Seismic Man throwing an empty tube of explosive into the air and heading it like a football.

"How are we measuring the soundwaves, then?" I asked. "When they bounce back up from the earth's crust?"

"Well," he said, packing up the shotbox, "what we're trying to do here is image the geology under 6,000 feet of ice. Seismology is the tool we use, and it operates either by refraction or reflection, the difference between the two being largely a function of scale in that reflection facilitates the imaging of a smaller area in greater detail. With me so far?"

I nodded.

"What you've just been doing is refracting. The soundwaves we send down are refracted back to the surface from the earth's sediment and recorded by a line of Ref Teks, the soundwave equivalent of the tape recorder. The Ref

Teks contain computers hooked up to geophones, and we have 90 Ref Teks 200 yards apart on a line right now, recording away. So all you and I have left to do is pack up!"

On the way back we stopped about ten miles from camp to eat our sandwiches (tinned ham and mustard). A narrow strip of incandescent purply blue light lay on the horizon between ice and sky, looking for all the world like the sea. It seemed to me that it would be almost impossible, in this landscape, not to reflect on forces beyond the human plane. Here, palpably, was something better than the realm of abandoned dreams and narrowing choices that loomed outside the rain-splattered windows of home.

"You're right," said Seismic Man when I mentioned this. "It's like plugging yourself in to the spiritual equivalent of the National Grid out here. Wasn't it Barry Lopez who wrote that Antarctica 'reflects the mystery that we call God'?"

I called what I sensed there God too; but you could give it many names. It was more straightforward for me than it had been for some, as I brought faith with me. I can't say where the faith came from, because I don't know; it certainly wasn't from my upbringing, since neither of my parents have ever had it. I remember first being aware of it when I was about fourteen, the same time that a lot of other things were happening to me. At first, it embarrassed me, like a virulent pimple on the end of my nose. I have no problems of that kind with it now, though I have persistently abused the giver by following the siren voices of the opposition, also dwelling in the rocky terrain of my interior life and determined to fight to the death.

Despite a good deal of high-mindedness and a sprightly ongoing dialogue with God, in the day-to-day hustle I constantly failed to do what I knew to be the right thing. A sense of spirituality all too often stopped short of influencing action. I was a hopeless case. But I believed that what mattered to God was the direction I was facing, not how far away I was. Sin, it seemed to me, was the refusal to let God be God. I admit that it was a handy credo to espouse—but I did it from the heart. The inner journey, like my route on the ice, was not a linear one. It was an uncharted meandering descent through layers and layers of consciousness, and I was intermittently tossed backwards or sideways like a diver in a current.

On Friday 13 January a Hercules appeared in the sky. It was going to take the drill team and a few others back to McMurdo. When it landed, incoming mail was borne inside in a metal turquoise-and-red striped crate like a crown before a coronation. Everyone leapt up, plunging their arms into the crate and calling out names as packages were passed eagerly from hand to hand.

My own mail was supposed to be waiting back at base. It could have been worse: Shackleton and the crew of the *Endurance* missed their mail by two hours when the ship sailed out of South Georgia, and they got it eighteen months later. Then I heard my name being called. Someone in the post room must have known where I was and slung my bundle into the metal crate. It was like a minor Old Testament miracle. The bundle included eight Christmas cards, three of them featuring polar bears, two pairs of knickers from my friend Alison and a bill from the taxman, the bastard. I took the cards to the igloo later and put them up to block the cracks between the ice bricks, and blue light shone through the polar bears.

We went out to wave goodbye to the drillers. The plane attempted to take off four times. It was too light at the back, so the drillers, we heard over the radio, had to stand in the tail.

That night I found a hillock of snow on my sleeping bag and was obliged to reseal the igloo bricks from the outside. It was perishing cold in there all the time and getting to sleep was an unmeetable challenge. I tried to listen to my Walkman to take my mind off the pain but the earphones got twisted under my balaclava and the batteries died in minutes. All my clothes froze in the night. Besides the waterbottle, I was obliged to stow my VHF radio and various spare batteries in between the bag liner and the sleeping bag to prevent them from freezing.

"It's like sleeping in a cutlery drawer," said Seismic Man, who had made valiant efforts to stay in the igloo. "Why are you putting yourself through it, when there are warm Jamesways a few hundred yards away?"

It was the romance of it, if I was honest. I liked the idea of living in my own igloo, slightly apart from camp, on the West Antarctic ice sheet. Besides that, during the periods when I didn't have to devote every ounce of energy to maintaining my core temperature, I did love the blue haze very much. I had noticed that when the sun was in a certain position it was faintly tinted with a deep, translucent claret. The surface of the bricks gleamed like white silver all around me. When I crawled out in the mornings (this had to be accomplished backwards) and twisted round on my sunken front path, I looked up and blinked at a pair of pale sundogs glimmering on either side of the sun, joined by a circular rainbow.

Each night, however, produced a new torment. That evening my knees got wet (this was caused by a rogue patch of ice on the bag liner), so I moved the windpants doing service as a pillow down under them. This meant that the mummy-style hood of the bag flopped down over my head, raising the problem (they were queuing up for recognition now) of imminent suffocation. The digital display on my watch faded. Out of the corner of my eye I spied a fresh cone of snow on the floor near the entrance. Forced out of the bag to plug the hole with a sock, I brushed my head against the ceiling and precipitated a rush of ice crystals down the back of the neck. I began nurturing uncharitable thoughts about Eskimos.

The morning after the departure of the drillers I went straight to the galley to thaw out, noting that I had forgotten to stand the shovel upright with the result that it was now lost in accumulated snow. Camp had shrunk from forty-five to twenty-two overnight. Patsy Cline was blaring out of the speakers and Bob and Mary were playing frisbee with a piece of French toast.

In the end, the igloo defeated me. As I walked back to it the next night I eyed the drums nestling in cradles outside the Jamesways, pumping diesel into the Preways. Sneaking guiltily into one of the two berthing Jamesways, I lay on the floor behind a curtain. It was so easy.

Ice streams A, B, C, D and E were located on the West Antarctic ice sheet. There was also a little F, but no one ever talked about that. Hermann, the moon-faced *Road to Oxiana* scientist, was investigating Ice Stream B. He wanted me to go out there with him and his team—they were staying for a week—but I knew I wouldn't be able to get back easily, and I couldn't risk being stranded anywhere at that point in the season. I was sad.

"Just come for the put-in," he said. I looked at him. It was an extraordinarily kind gesture: the put-in involved two Otter flights, and taking me along would seriously complicate logistics. "You must see it," he said. "*I must support you as a writer.*"

Hermann was buzzing around his pallets like a wasp as the Otter arrived. When we took off, I sat in the back of the hold with him. The Whitmore Mountains appeared in the distance. Hermann's eyes lit up.

"Look!" he said, pointing to a hollow above a deeply crevassed area. "The beginning of Ice Stream B!" The ice there looked like a holey old sheet. Hermann pressed a hand-held Global Positioning System unit against the pebble window and said solemnly, "We are entering the chromosome zone." It sounded like the opening sequence of a science-fiction movie. "The crevasses change direction as the glacier moves," he said, "and they turn into thousands of Y chromosomes."

After that we entered the transition zone between the moving ice and the stable ice. It was called the Dragon, a highly deformed, heavily crevassed area streaked with slots. Hermann tapped his propelling pencil against the thick glass of the porthole and held forth. "The ice streams are not well understood. The boreholes we have drilled to the bottom of this stream reveal that the base of the stream is at melting point. So they move"—tap, tap, tap—"these motions provide a process for rapid dispersal and disintegration of this vast quantity of ice. I mean that most of the drainage of this unstable western ice sheet occurs through the ice streams. The mechanics"—tapping—"of ice streaming play a role in the response of the ice sheet to climactic change. In other words the ice streams are telling us about the interactive role of the ice sheet in global change."

So it seemed that if the ice melted, resulting in the fabled Great Flood of the popular press, water would pour out of the continent, via the ice streams, on to the Siple Coast, virtually the only part of Antarctica not bounded by mountains.

Hermann settled back in his seat. "The aim of investigating ice-stream dynamics," he concluded, "is to establish whether the ice sheet is stable."

"Geology," Lars had told me, "is an art as well as a science."

Hermann stowed the pencil in his top pocket, my ears popped and we landed at a few dozen ragged flags on a relatively stable island in the middle of Ice Stream B. This island, shaped like a teardrop, was called the Unicorn, and Hermann's eye glittered like the mariner's at three or four flags flapping on bamboo poles in the distance. "The flags mark our boreholes," he said, "and we have left equipment down in these holes, gathering data. Those two boreholes"—he pointed to a pair of ragged red flags—"are called Lost Love and Mount Chaos." It was like entering a private kingdom. The Dragon, which resembled a slender windblown channel of ice you could walk over in five minutes, was really a two-mile-wide band of chaotic crevassing running for forty miles down one side of the Unicorn. It was a dramatic landscape, its appeal sharpened by the fact that fewer than twenty people had ever seen it.

Hermann's long-standing field assistant, who had travelled with us, was a gazelle-like woman called Keri. When the plane took off and the sound of the engines faded she began spooling out the antenna.

"You be my deputy field assistant," Hermann said to me. We crunched off to a flag where he dug around until he found a plywood board encrusted with crystals. Fishing out a skein of wires from underneath it, and attaching them to a small measuring device, he began sucking up data. After a few minutes he beamed, an expression he retained until I left the camp, and possibly much longer. He started inscribing a neat column of figures in pencil in a yellow waterproof notebook.

"These bits of data," he said, "are all little clues to the big puzzle."

Back at CWA they were detonating the last blasts of the season. Everyone went outside one morning to watch 750 pounds of explosives go up half a mile away. The blaster was close to the site. A black and grey mushroom cloud surged 500 feet into the air, followed, seconds later, by a prolonged muffled boom.

"One less for lunch, Bob," said José.

I skied out to see the crater. It was forty-five feet in diameter with a conical mound in the middle, and a delicate film of black soot had settled over the ice. The blaster was admiring his work. "My hundredth of the season," he said proudly.

He was taciturn, as cold as the ice in which he buried his explosives, but once I showed an interest in his bombs his face mobilised and he began opening boxes to show me different kinds of powder and expounding on the apparently limitless virtues of nitroglycerine.

"Largest charge I've used this season," he intoned with the treacly vowels of Mississippi, "was 9,000 pounds," and I tut-tutted admiringly as he ran his fingers through baby-pink balls of explosive which looked like candy and smelt of diesel.

[. . .]

As the days were slipping away from me, I decided I ought to catch a lift back to McMurdo in the Otter, rather than wait for a Herc which might not come. I whipped up a bread-and-butter pudding as a farewell gift. While they were eating it, Jen, a feisty individual working her second season as a field assistant, filled a tin bowl with hot water, rolled up her long-johns and perched on a chair in the galley shaving her legs.

"But who's gonna see those legs, Jen?" someone yelled. This was followed by a ripple of laughter.

"Get outta here," she called. "I wanna be a girlie for once."

I went over to the igloo and lay down one last time, looking up at the spiral of bricks and the blue haze.

Everyone came over to the Otter to say goodbye. "See you in Mactown," said Seismic Man, squinting into the sunlight.

I watched them get smaller and smaller until they disappeared into the ice.

[. . .]

The pilot wanted to play cards with the air mechanic, so I moved into the cockpit. A stack of cassettes were jammed between the front seats—most of it was 1970s stuff I hadn't heard since school, and it was perfect cruising music. *Crime of the Century*, *The Best of the Eagles*, early Bowie, that Fleetwood Mac album we all had.

[. . .]

Looking down at the earth from 12,000 feet, I felt then that my life was in perfect perspective. It was a sense of oneness with the universe—I belonged to it, just like the crystals forming on the wing tip. At that moment I knew that all my anxieties and failures and pain were shadows on the wasteland.

BLUE COLLAR

from *Big Dead Place* (2005)

Nicholas Johnson

Far beneath the scientists and the administrators, down at the bottom of the status pyramid, contemporary Antarctica also contains a little world of contract workers, for whom the continent can be less a wide-open wilderness than a screw-top jar full of managerial idiocies you just can't escape. Nicholas Johnson's scabrous memoir of working through the winter at McMurdo today is a kind of Dilbert *on ice.*

McMurdo lies in the shadow of Mount Erebus, a smoldering volcano encrusted with thick slabs of ice. To make room for McMurdo, a ripple of frozen hills on the edge of Ross Island have been hacked away to form an alcove sloped like the back of a shovel, and then affixed with green and brown cartridges with doors and windows. Silver fuel tanks sparkle on the hillside like giant watch batteries. As if unloosed from a specimen jar, a colony of machines scours the dirt roads among the simple buildings, digesting snow and cargo dumped by the wind and the planes, rattling like cracked armor and beeping loudly in reverse.

[. . .]

In the distance, framed by ratty utility poles and twisted electrical lines, the gleaming mountains of the Royal Society Range spill glaciers that glow like molten gold onto the far rim of the frozen white sea. Near Castle Rock, skiing toward Mount Erebus, in the middle of nowhere, you can stop at the bright red emergency shelter that looks like a giant red larva and call your bank to dispute your credit card fees.

[. . .]

Most of the population work for NSF's prime support contractor, which employs everyone from dishwashers and mechanics, to hairdressers and explosives-handlers. While the National Science Foundation is known as a proud sponsor of public television programming, Raytheon is known for making the Exoatmospheric Kill Vehicle and other top-shelf weapons systems.

[. . .]

In Antarctic parlance, all of the United States divides into "Washington," referring to NSF's sphere of influence, and "Denver," referring to a vague suburban belt of Sheratons and brewpubs on the outskirts of Denver, where the support contractor has long been headquartered. Toward Denver is the most immediate over-the-shoulder check for the Antarctic lackey. "I'm going to have to 'okay' that with Denver first," they say, or "I'm not the one who made the decision—if you have a problem with it, talk to Denver." Denver is where most of the managers and full-time employees work, and where strategies for improving morale are formulated. Some of the clocks in McMurdo and at South Pole are set to Denver time.

*

One Saturday in mid-May we had an All-Hands Meeting, our first since the medevac plane had left in late April. We crowded into the Galley in our work clothes. The first speaker was the technician who maintains lab equipment. There was not a single scientist in McMurdo this winter, but if anyone could be labeled a "researcher" to make it easy on the newspapers in the U.S., it would be he. He showed us some video footage of the volcanic activity in Mt. Erebus accompanied by the Pink Floyd song "One of These Days (I'm Going to Cut You Into Little Pieces)." Afterwards he told us that the green laser in the Crary Lab had been fixed, and that if we wanted to see it we could call him and he would turn it on for us. He reminded us not to climb on the roof to look at it from above, because it would permanently blind us.

In the winter the green laser shoots from the top of Crary into the sky while everyone goes about their business. To see this green laser while I hauled garbage was the central reason I had decided to winter. In a strange world hardened by routine, the rub between the fantastic and the mundane creates a spellbinding itch.

The next speaker was the winter Galley Manager, with whom I had worked my first year down here, when he was the baker. I would take breaks from washing pots to hear his stories about being a chef in the Playboy Mansion, where he spent much of the time making hot dogs and PBJs. Now he told us that his new oven had been stolen. Parts of it had been sent to him with ransom notes, threatening its destruction unless he sang "I'm a Little Teapot" at the All-Hands Meeting. He sang one verse and sat down, but everyone cheered for more, so he did it again, and we cheered even louder.

Then the Operations Manager took the floor and reminded us to check the hours on our machines, and to remember to bring in our machines for PMs, or Preventive Maintenance. "Also," he said, "if you have an accident, you need to report it."

Then Franz, the new Station Manager, stood up to talk. He had replaced the previous station manager, who went out on the medevac flight. Before his new appointment he had been working as a supervisor in Materials. Though he was a fingee, Denver liked him for his management experience in a hotel in suburban Denver.

I had first met Franz in the summer, when he was Butch's roommate. One day after work Butch and I were hanging out in his room shooting the shit. When Franz came in from work, he busied himself on his side of the room for a while before saying to Butch:

"So, I'm going to take a shower now. I'll be five or ten minutes. Then the place is yours."

Butch looked at him without understanding.

"I'll just be five or ten minutes," Franz repeated, suggesting that we leave the room so he could take a shower, in the bathroom, behind a closed door that had a lock.

Franz read some statistics from a study he had scavenged on the Internet concerning the psychological effects of wintering in Antarctica. "Many of you at this time of year will have sleeping problems," he said, "and may become depressed, irritable, or bored. Five percent of you," he said, "will suffer effects that will clinically categorize you as in need of psychological treatment."

I was excited to see what personal dementias I would face, and realized that if my disrupted circadian rhythms or thyroid activity were to show any symptoms, since it was May now, they should be kicking in any day. I wondered if I would collect pictures of animals, or draw eyes on all my belongings, or come to despise asymmetrical shadows.

In the dark ages a withered priest might have warned us that the Devil was on the loose and that we had to purge ourselves of sin. Now we had scientific evidence to remind us, via a former hotel manager, that the individual's predetermined behavior and aberrations are the product of devilishly powerful external forces, such as the planet's tilt. There was little practical reason for any manager to warn us of winter psychological effects, since they are disregarded in daily affairs. You would still be written up for tardiness, regardless of "sleeping problems." A "support program" for the "depressed" would not be authorized until the end of winter. And "irritability" would still be met with dorm room inspections in your absence. Whatever the initial intent of these academic psychological studies, their field application is as an orientation to employee culpability.

After the meeting I stopped at the store to see if *Rosemary's Baby* was in the video collection and to buy some Skoal. The McMurdo store is miraculously well-stocked. Though someone is always likely to complain that this or that item has run out, the store has Pringles and Rolos and jars of hot salsa. There are a half dozen kinds of beer, most common types of liquor, and a considerable selection of red and white wine. (Every fourth bottle will be rancid by the end of the winter, because the wine is stored upright and the arid air will have ruined the corks.) There are hundreds of videos for free checkout just by giving the liquor clerk the last four digits of your Social Security number. There are windscreen facemasks for sale. There are aerial posters of Ross Island, a few kinds of soap and shampoo, nail clippers, and anti-fart medication. For years there have been hundreds of unsold postcards of a velvet Elvis painting that someone photographed at the Pole. When the Navy first opened the store, they stocked mosquito repellent that no one bought because there are no mosquitoes in Antarctica.

Much of the souvenir merchandise in the store is contracted for manufacture to a company in Denver. The souvenir t-shirt selection is large, but with two basic varieties. The first and most common variety centers around the penguin. These shirts may also include icicles or the sun, and their style staggers toward stark romance. They might say "Wild Antarctica" or "The Last Frontier" on them. The second variety may also include the penguin, but the styles imitate tired surf or snowboard designs. These designs include men in parkas with surfboards, or "Antarctica" written in the style of the Ford logo. They refer to Antarctic "powder" and say things like "eternal sun" and "chill out."

Some of the goods at the store are depressing, like the bumper sticker that says, "Antarctica: Been There, Done That," and some are confusing, like a cap embroidered with a colorful bass biting a fishhook and the text "Bite Me—Antarctica."

At the counter I browsed the Antarctica pins that I never buy, and scrutinized one that depicted the Antarctic continent flanked by American flags. The clerk told me that last summer NSF, which usually has little to do with the running of the store, instructed her to remove the "Made in Taiwan" sticker from the backs of the pins before displaying them.

The next day, in the middle of a Sunday afternoon, on a dark road coming down from T-site, Bighand flipped a truck. He came down too fast and went over the embankment by the Cosmic Ray Lab, launching the acetylene and oxygen tanks into the air from the back of the pickup. The truck rolled but landed on its wheels, and Bighand drove it back to town after gathering the pressurized canisters of volatile gases. He hid the truck behind one of the orange fish huts where scientists in summer stay warm while fishing for specimens (there are no Antarctic bass) through the sea ice. Then he found his boss at the bar and reported the incident.

The Heavy Shop assessed the truck's extensive damage. HR summoned Bighand for questions as to why the cab of the truck smelled like beer, and he said that the mechanics must have poured beer in the truck to frame him.

The grapevine lurched into action. At dinner the next day someone said, "Shit, I didn't even hear about it until after lunch." Bighand was in and out of the HR Office all day, and at break we speculated on the company's strategy. We determined that since Bighand was a foreman he wouldn't be fired straight away. He would stay on to work and would probably be sent out at Winfly in August. Had he been in some menial position, he might have been made an example of and put on minimum wage until he could be flown out, but since he was a foreman and necessary for construction of JSOC, he would be kept on until someone new could be brought in. HR would give him the impression that he had been forgiven, but he would be fired just before the first flight out.

The next few days were marked with investigations by HR and Safety. Both of those departments this winter totaled two people. They brought in everyone who was at T-site on Sunday and pumped them for information with which to convict Bighand.

T-site is the hub of all radio transmission in The Program. The road up to it is long and windy, with signs along the way warning of exposure to hazardous doses of radiation should you stray into the garden of transmitters. Because of the importance of uninterrupted communication around the continent, a couple of people, one of whom must always remain on-site, live at T-site in swank and roomy quarters. They have a pool table, a band room stocked with instruments, supplies for brewing beer, a well-stocked pantry and kitchen, and comfortable couches. Just off the ordinary living room lie corridors lined with banks of transmitter components: some of it state-of-the-art, some of it antiquated but reliable gear from the early Cold War era. One can get up from the couch in the warm and carpeted living room, pad in one's socks down the corridor full of vigilant technology sprouting bundles of wires and silently ricocheting voices or strands of data around the continent, and seat oneself on another couch by the pool table in the equally comfortable rec room. Looking out the window in the summer, one's view weaves through the dozen or so enormous spidery transmitters nearby for an otherwise clean view of the Transantarctics and of White Island and Black Island, where another transmission outpost stretches the range of communication from T-site. Going to visit the comm techs at T-site brings a change of scenery, where the relentless sound of loader back-up beepers in town is faint.

Bighand had been driving down from a band rehearsal held that afternoon in the T-site rec room. Franz, the new Station Manager, and the HR Guy called everyone who had been at T-site that day into the HR Office one by one to sign a "warning" acknowledging that the signer had violated an NSF policy by using government vehicles to enter a restricted area; presumably this aimed to fill a hole somewhere in the documentation of T-site's restricted status. Franz told Nero that HR was "just going to shred them up at Winfly anyway," but Nero didn't see why he should sign a "warning." Many people did sign it, but many refused.

By the time he flipped the truck, Bighand was already notorious around town. His drink of choice was a tall glass of Jim Beam topped with Wild Turkey and a splash of Sprite. Then several more. When I wore a skua'd priest shirt to the bar one night, he got down on one knee before me with his eyes rolling back in his head and began babbling incoherently, so I blessed him and howled "Demons be gone!" When he tried to leave, he walked into the door and almost fell over before going outside. I didn't think he would make it home, so I followed him outside, where he was just getting up from a fall on the ice, and walked him to his dorm. One time Bighand filled a truck with diesel instead of mogas. That's a pain in the ass for the Heavy Shop, who must then drain the lines. Now that he'd also flipped one of the new red trucks, he was a bona fide public buffoon. Trying to blame the Heavy Shop for the beer in the cab had also created enemies, as well as a potential rift between Ironworkers and Mechanics. Perhaps this was why someone crept into the JSOC job shack one night and took a shit in Bighand's hardhat, wiping their ass with a piece of the project's blueprint.

Aside from "Been there, done that" and "We need to touch base," managers are particularly fond of the phrase "It's a harsh continent," which has two uses. The first meaning is that of the manager speaking of some hassle or burden on himself. In this case, the manager says, "But hey, it's a harsh continent," expressing a noble resignation. In the other case, the manager, awed by the big decisions coming down from someone more powerful than herself and fantasizing about making such decisions herself, says, "Well, it's a harsh continent," which translates as "Tough shit for all of you." Though these uses may seem opposed, they really express different shades of the same sentiment: submission is survival. To work hard and increase one's competence is nearly irrelevant. The most important things are to occasionally seek decisive assistance, to mimic the mannerisms of the immediate superior, and to occasionally let out a squeak or yelp of fear or pain.

Managers also like to joke about "putting out fires." Fire is the direst threat at an Antarctic station, where the dry air makes the buildings tinderboxes that can crumple minutes after the first flame. In manager parlance, though, "fires" are problems of any kind, and the manager knows there is no end to the fires, so he usually follows the reference to extinguishing them with a fatigued sigh. Once a fire is put out, he moves onto the next fire. Each new flame is addressed as a unique problem, unrelated to anything that came before it. He rushes around the room extinguishing isolated flames, emphatically smothering anything in the vicinity of the smallest wisp of smoke, lest the snoozing overhead detector be aroused and its shrill scream betray his failure to control his sector.

[. . .]

By June our routine was hopelessly solid. Each of us in Waste could distinguish the sounds of different loaders even from afar. Every Saturday we checked the glass and aluminum bins at the bars to make sure they were empty. Every Monday we checked the same bins to see if they were full. The Galley pumped out a daily stream of Burnables and Cardboard and a medium stream of Plastic and Light Metal. FEMC produced a lot of Wood and Light Metal and Construction Debris. The Firehouse hardly put out anything at all, but when they did, they separated their trash poorly. The Heavy Shop made a lot of Construction Debris, and we had to make sure to pick up their Heavy Metal when it was only half-full; otherwise we might have trouble dumping it, because one of our loaders had some hydraulic problems at max capacity. The Carp Shop could fill a Wood dumpster in a day or two. The dorms were steady with Burnables and sporadic with everything else. The Coffeeshop Glass bin only filled with wine bottles, and we appreciated the bartenders' separating by glass color even though it wasn't their job. The power plant dumpsters had been requiring attention this winter, because the engineers were cleaning house, and they called us to pick up their cardboard frequently, but that's because they didn't break down their boxes. Crary Lab took forever to fill anything but Haz Waste or Plastic.

Passing conversation ever more often involved Christchurch, an Antarctican's Heaven, where the year's grinding work would be rewarded with sushi and botanical gardens, Thai food and titty bars. There would be rain on windows and the sound of wet tires on pavement. Fresh off the plane, we would seep into Christchurch like diesel into snow. We would be full of money. We would scatter about the hostels and hotels, then clump again into smaller groups at restaurants throughout the city. Ice people would be everywhere, stopping on the sidewalks to ask each other what they ate for lunch, because now lunch would not be the same for everyone. To avoid tables for ten with confusing bar tabs, one would avoid the Monkeybar Thai restaurant. Bailey's, a bar at the edge of Cathedral Square, draws so much business from the USAP that they have sent kegs of Guinness down to special parties on the ice. Bailey's would replace Southern Exposure, but without parkas by the door, and the work stories would be full of nostalgia instead of details. The talk would concern beautiful future beaches and bloodless Antarctica.

In Christchurch we look pale, weird, and menacing, but soft as adult-sized newborns. People who were attractive in thick brown Carhartts and all manner of accessories to cover necks, faces, and hands appear in the Christchurch summer as a mass of elbows, kneecaps, and toes. People wear shorts and sandals, exposing pasty flesh and propensities for camping. We are no longer Carps or Fuelies or Plumbers. Our cold-weather clothes are taken away, our intertwining community vines pruned; we suddenly have separate destinations.

By June, work was sometimes a wearisome prospect. I was sleeping longer on weekends and tired on weekday mornings, even when I went to bed early. When I did laundry, the clean clothes got put away just in time to do laundry again. Shaving was a chore. My room was getting messier. My memory seemed weaker.

One day while I was welding a dumpster I had problems explaining something to Jane.

"In this case it'd be better to hold the steel plate than to . . ." I hawed. Jane waited patiently. I couldn't think of it. I pointed to the clamp lying on the floor. "What's that called?" I asked her.

"Clamp," she said.

". . . than to clamp it," I finished. Jane said that pretty soon we would be speaking in grunts, merely pointing at things to name them, and staring into space.

IN THE FOOTSTEPS OF AMUNDSEN, 2016

from *Antarctica* (1997)

Kim Stanley Robinson

The best nature-writer in the U.S. at the moment happens also to write science fiction: this is a strand from the novel he wrote after a polar sojourn courtesy of the National Science Foundation's Artists & Writers Program. A wilderness adventure in the near future is going wrong for a group of wealthy clients and their burned-out guide, just as the party reaches the top of the Axel Heiberg Glacier, but as it does so, it generates a kind of tough-minded Californian elegy for the Heroic Age, wise about the difference between then and now, and about the space we ought to allot in our lives for Antarctic beauty.

THE OTHERS HAD ALL GONE around the ice block to the higher belay and got off the rope, and Val and Jack had clipped their harnesses onto the belay rope and were just beginning to pull the sledge into line, when the ice block above them leaned over with a groan and fell. Val leapt into the crevasse to the left, her only escape from being crushed under falling ice. She hit the inner wall of the crevasse with her forearms up to protect her. The rope finally caught her fall and yanked up by her harness; then she was pulled down again hard as Jack was arrested by the same rope below her. For a second or two she was yanked all over the place, up and down like a puppet, slammed hard into the wall. The rope was stretching almost like a bungee cord, as designed—it was very necessary to decelerate with some give—but it was a violent ride, totally out of her control.

But the belay above held, and the belayers too. As soon as she stopped bouncing, however, she twisted and kicked into the ice wall with both front-point crampons, then grabbed her ice axe and smacked the ice above her with the sharp end to place another tool.

A moment's stillness. Nothing hurt too badly. She was well down in the crevasse, the blue wall right in front of her nose. Ice axe in to the second notch, but she wanted more. Below her Jack was hanging freely from the same rope she was, holding onto it above his head with one hand. No sign of the sledge.

Voices from above. "We're okay!" she shouted up. "Hold the belay! Don't move it!" Don't do a thing! she wanted to add.

"Jack!" she called down. "Are you okay?"

"Mostly."

"Can you swing into the wall and get your tools in?"

"Trying."

He appeared to be below a slight overhang, and she above it. A very pure crevasse hang, in fact, with the rest of the group on the surface several metres above them, belaying them, hopefully listening to her shouts and tying off, rather than trying to haul them up by main strength; it couldn't be done, and might very well end in disaster. Val didn't even want to shout up to tell them to tie off: who knew what they would do? Not wanting to trust them, she took another ice-screw from her gear-rack, chipped out a hole in the wall in front of her, set the screw in it, then gave it little twists to screw it into the ice. Its ice coring shaved out of the aluminium cylinder. A long time passed while she did this, and it became obvious that they were underdressed for the situation; it was probably twenty below down here, and no sun or exertion now to warm them; and sweaty from adrenaline. They were chilling fast, and it added urgency to her operation. She had to perform a variation of the operation called Escaping the Situation—a standard crevasse technique, in fact, but one of those ingenious mountaineering manoeuvres that worked better in theory than in practice, and better in practice than in a true emergency.

The screw was in, and she clipped onto it with a carabiner and sling attached to her harness, then eased back down a bit. Now both she and Jack had the insurance belay of the screw.

From the surface came more shouts.

"We're okay!" she shouted up. "Are the belay-screws holding well?"

Jim shouted down that they were. "What should we do?"

"Just hold the belay!" she shouted up anxiously. "Tie it off as tight as you can!" More times than she cared to remember she had found herself in the hole with the clients up top, and they had often proved more dangerous than the crevasse.

She tied another prussik loop to the belay rope, then reached up and put her right boot into it. Then she stood in that loop, jack-knifing to the side, and unclipped her harness from the ice screw, then straightened out slowly as she slid the prussik attached to her harness as high on the rope as it would go. When she was standing straight in the lower loop, she tightened the upper prussik, then hung by the waist from it, and reached down and pulled the lower one up the rope, keeping her boot in the loop all the while. There was the temptation to pull the lower loop almost all the way up to the higher one, but that resulted in a really awkward jack-knife, and made it hard to put weight on her foot so she could move the higher one up. So it was a little bit up on each loop, over and over; tedious hard work, but not so hard if you had had a lot of practice, as Val had, and didn't get greedy for height.

"Jack, can you prussik up?"

"Just waiting for you to get off-rope," he said tightly.

"Go ahead and start!" she said sharply. "A little flopping around isn't going to hurt me now."

Soon enough she reached the edge of the crevasse, and the others on top helped haul her over the edge, where she was blinded by the harsh sunlight. She unclipped from the rope and went over to check the belay. It was holding as if nothing had ever even tugged on it. Bombproof indeed.

Then it was Jack's turn to huff and puff. Prussiking was both hard and meticulous, accomplished in awkward acrobatic positions while swinging in space all the while, unless you managed to balance against the ice wall of the crevasse. Jack appeared to be making the classic mistake of trying for too much height with each move of the loops, and he wasn't propping himself against the wall either. It took him a long, long time to get up the rope, and when he finally pulled up to the point where the others could haul him over, he was steaming and looked grim.

"Good," she said when he was sitting safely on the ramp. "Are you all right?"

"I will be when I catch my breath. I've cut my hand somehow." He showed them the bloody back of his right glove, a shocking red. The blood was flowing pretty heavily.

"Shit," she said, and hacked some firn off the ramp to give to him. "Pack this onto it for a while until the bleeding slows."

"A sledge runner caught me on its way down."

"Wow. That was close!"

"Very close."

"Where is the sledge?" Jorge said.

"Down there!" said Jack, pointing into the crevasse. "But it got knocked in and past us, rather than crushed outright by the block. I gave it a last big tug when I jumped in."

"Good work." Val looked around. "I'll go back down and have a look for it."

"I'll come along," said Jack, and Jim, and Jorge.

"You can all help, but I'll go down and check it out first."

So she took from her gear-rack a metal descending device known as an Air Traffic Controller, and attached it to the rope, then to her harness using a big locking carabiner. She leaned back to take the slack out of the rope between her and the anchor, then started feeding rope through the Air Traffic Controller as she walked backward toward the crevasse, putting her weight hard on the rope. Getting over the edge was the tricky part; she had to lean back right at the edge and hop over it and get her crampon bottoms flat against the wall, legs straight out from it and body at a forty-five degree angle. But it was a move she had done many times before, and in the heat of the moment she did it almost without thinking. After that she paid the rope slowly up through the descender, one hand above it and one running the rope behind her back for some extra friction. Down down down in recliner position, past the ice-screw she had placed, down and down into the blue cold. She was keeping her focus on the immediate situation, of course, but her pulse was hammering harder than her exertion justified, and she found herself distracted by an inventory that part of her mind was taking of the emergency contents of everyone's clothing. This was no help at the moment, and as she got deeper in the crevasse she banished all distracting thoughts.

Just past the tilt in the crevasse that blocked the view from above, there was a kind of floor. Her rope was almost entirely paid out, and she had not tied a figure of eight stopper-knot in the end of the line, which was stupid, a sign that she wasn't thinking. But it got her down to a floor, and it was possible to walk on this floor, she saw, still going down fairly steeply; and as she saw no sight of the sledge, but a lot of chunks of the broken ice block leading still onward, she called up that she was going off-rope, then unclipped, and moved cautiously over the drifted snow and ice filling the intersection of the walls underfoot—a floor by no means flat, but rather a matter of Vs and Us and Ws, the tilts all partly covered by drift. There was also no assurance at all that it was not a false floor, a kind of snowbridge in a narrow section, with more open crevasse below it; she would have stayed on-rope if there had been enough of it. As it was she crabbed along smack against the crevasse wall, hooking the pick of her axe into it as she went, testing each step as thoroughly as she could and hoping the bottom didn't drop out from under her.

She moved under the snowbridge she had noted from above, and the crevasse therefore became a tall blue tunnel. She moved farther down into it. Sometimes ice roofed the tunnel, at other times snowbridges, their white undersides great cauliflowers of ice crystal, glowing with white light. The view from below made it clear why snowbridges over crevasses were such dangerous things, so tenuous were they and so fatally deep the pits below them. But that was why people roped up.

The tunnel turned at an angle, and then opened downward into a much larger chamber. Val kept going.

This new space within the ice was really big, and a much deeper blue than what she had come through so far, the Rayleigh scattering of sunlight so far advanced that only the very bluest light made it down here, glowing from out of the ice in an intense creamy-translucent turquoise, or actually an unnamed blue unlike any other she had seen. The interior of the space was a magnificent shambles. Entire columns of pale-blue ice had peeled off the walls and fallen across the chamber intact, like broken pillars in a shattered temple. The walls were fractured in immense translucent planes, everything elongated and spacious—as if God had looked into Carlsbad Caverns and the other limestone caverns of the world and said *No no, too dark, too squat, too bulbous, I want something lighter in every way,* and so had tapped His fingernail against the great glacier and got these airy bubbles in the ice, which made limestone caves look oafish and troglodytic. Of course ice chambers like these were short-lived by comparison to regular caves, but this one appeared to have been here for a while, perhaps years, it was hard to tell. Certainly all the glassy, broken edges had long since sublimed away in the hyperarid air, so that the shatter was rounded and polished like blue driftglass, so polished that it gleamed as though melting, though it was far below freezing.

Val moved farther into the room, enchanted. A shattered cathedral, made of titanic columns of driftglass; a room of a thousand shapes; and all of it a blue that could not be described and could scarcely be apprehended, as it seemed to flood and then to overflood the eye. Val stared at it, rapt, trying to take it all in, realizing that it was likely to be one of the loveliest sights she would ever see in her life—unearthly, surreal—her breath caught, her cheeks burned, her spine tingled, and all just from seeing such a sight.

But no sledge. And back at the entrance to the blue chamber, there was a narrow crack running the other way, not much wider than the sledge itself; and looking down it, into an ever darkening blue, Val saw a smear of pale snow and ice shards, and below that, what appeared to be the sledge, wedged between the ice walls a hundred feet or more below; it was hard to judge, because the crevasse continued far down into the midnight blue depths below. There was no way she could get down there and get back up again; and even if she could have, the sledge was corked, as they said. Stuck and irretrievable. In this case crushed between the walls, it looked like, and broken open so that its contents were spilled even farther down. A very thorough corking. No—the sledge was gone.

✳

At first they walked over firn, which took their weight like a pavement. There were sastrugi of course, but they could easily step over them, and the different angles of hardpack that their boots landed on actually gave their feet and ankles and legs some variety in their work, so that no one set of muscles and ligaments got tired, as when pounding the pavement in cities or in the endless corridors of a museum. So in these sections it was good walking.

They spread out in little clumps: Jack and Jim up with Val, Jack going very strong, even pushing the pace a bit; Jorge and Elspeth behind them; Ta Shu back farther still, rubbernecking just as he had before their accident. Val set the pace and did not allow Jack to rush them. "Save it," she said to him once a little sharply, when she felt him right in her tracks. "Pace yourself for the long haul."

"I am."

But he dropped back a little, and on they walked. They were doing fine. As she always did on long hikes, Val stopped the group to rest for about fifteen minutes after every ninety of walking, in a system somewhat similar to Shackleton's. In ninety minutes their arm-flasks had melted the snow and ice chips stuffed into them, and so everybody had two big cups of water to drink. They could also eat a few inches of their belts, as Elspeth put it; their suits' emergency food supplies were sewn into an inner pocket wrapped all the way around the waist. The food was something like a triathlete's power bar, flattened and stretched into something very like a wide belt, in fact. It was good food for their situation. At some stops they chewed ravenously;

during others their appetites seemed to Val suppressed, by altitude or exertion no doubt. She made sure they didn't force it. In truth it was water that was crucial to this walk; they were breathing away gallons of it in the frigid, hyperarid air, and they were sweating off a little bit as well. Two flasks every ninety minutes was by no means enough, but it certainly staved off the worst of the dehydration effects, which could devastate a person faster even than the cold, and made one more susceptible to the cold as well.

So between the walks and the breaks they made steady progress. But after several of these had passed, they came upon swales of softer snow, which had been pushed by the winds into sastrugi like cross-hatched dunefields. These snowdrifts were new, the result of unusually heavy snowfalls on the edges of the polar cap in recent years, generally assumed to be an effect of the global warming generally, and of the shorter sea ice season in particular. Climatologists were still arguing what caused all the different kinds of superstorms, aside from the overall increase in the atmosphere's thermal energy. In any case the snow was here, one more manifestation of the changes in weather.

Val stopped for a meeting. "Follow me and step right in my footsteps, folks, and it will be a lot easier."

"We should trade the lead, so everyone saves the same amount of energy," Jack said.

"No no, I'll lead."

"Come on. I know we've got a long walk, but there's no reason to get macho on us."

Val looked at him for a while, counting on her ski-mask and shades to keep her expression hidden. When her teeth had unclenched she said, "I've got the crevasse detector."

"We could all carry that when it was our turn."

"I want to be the one using it, thanks. I know all its little quirks. It wouldn't do to have any falls now."

"You having the radar didn't keep it from happening last time."

They stood there under the low, dark sky.

"Go second, and make her steps better for us," Ta Shu suggested to Jack.

"We shouldn't have anyone lose more energy than anyone else."

"I have more energy than anyone else to start with," Val said. "Everyone except maybe you, but you're hurt. You cut your hand. You hit the wall of the crevasse. Let's not waste any more energy arguing about it. It'll work out."

It was very hard to be civil to him. She couldn't think of anything else to say, and so took off before she said something unpleasant.

He stayed right on her heels, like some kind of stalker. She could hear his breathing, and the dry squeak of his boots on the snow. Untrustworthy, disloyal, unhelpful, unfriendly, discourteous, unkind . . .

They followed her through the soft snow dunes in single file. She kept the pace easy, resisting the pressure from Jack. Never was the snow as soft as Rockies powder, of course, but it was extremely dry, and already had been tumbled by the winds until it was on its way to firn. It was more like loose sand than any snow back in the world, loose sand that gave underfoot, thus much more work than the firn. Then in the areas where it adhered, she had to pull her boots out of their holes after every step, and lift higher for the next one, which was also hard work. But she had put in her trail time—a lifetime's worth—and it would take many many hours of such walking to tire her. No, she would be fine; she could walk for ever. It was the clients she was worried about. She was responsible for them, and she had got them into trouble, as Jack had pointed out; but she couldn't carry them, they had to walk on their own. So it had to be made as easy for them as possible.

So she did what she could. But as they walked on, and hour after hour passed, under the sun that wheeled around them in a perpetual mid-afternoon slant, they began to lose speed and trail behind. Jack no longer trod in her bootprints the second she left them unoccupied, nor during the breaks did he again mention leading the way. In fact he spent the rest periods in silence now, a mute figure under his parka hood, behind his ski-mask and shades. He wasn't eating much of his belt, either. That worried Val, and she tried to inquire about it by asking the group generally how they were feeling, and getting a status report from everyone; Elspeth was developing blisters on her heels, she thought; Jorge's bad knee was tweaking; Jim and Ta Shu reported no problems in particular, but like everyone said they were tired, their quads in particular getting a little rubbery with all the loose, soft snow. Jack, however, only said, "Doing fine. 'Pacing myself for the long haul.' "

So, okay. End of that break. On they walked.

Her GPS was still out of commission, but occasionally when she turned it on it flickered and gave a reading, then blinked out again. The last one that had come through indicated that they were averaging about three kilometres an hour, which was normal on the plateau; a bit slower in the soft snow no doubt, hopefully a bit faster on the hardpack.

Then they came to a patch of blue ice, and Val groaned to herself. They had to stop and put on crampons, then scritch cautiously across the ice, which here was pocked and dimpled by big, polished suncups. The nobbly surface gave their ankles a hard workout indeed, as their crampon points forced them to step flush on the terrain underfoot no matter its angle. It was best to step right on the cusps and ridges between the little hollows, crampons sticking into the slopes on both sides and keeping the foot level; but that took a lot of attention and precise footwork. So *scritch, scritch, scritch,* they stepped along, making perhaps two kilometres an hour at best. Val headed directly for the nearest stretch of snow in the distance, so that as soon as possible they reached the far side of the blue ice, groaning with relief, and could sit down and take the crampons back off, and drink what water had melted at the bottom of their flasks, then restuff the flasks with hacked chips of the blue ice, which would yield more water when melted than snow. Then they were up and off again, on what felt like land, after a precarious crossing over water.

Jorge and Elspeth were clearly tiring now, though they did not complain. Jim too was getting tired, and Jack stuck with him, arms crossed over his chest. Jack still wasn't eating very much compared to the others, but he still wasn't responding to her questions about it, either.

"Aren't you hungry?"

"I'm fine."

"We're probably burning three or four hundred calories an hour doing this."

"I'm fine. Don't bother me."

So she shrugged and took off again. They were back on good firn again, and could make decent time with minimum effort. Just walking, a great relief after what had preceded it.

But now when she looked back, she saw that Jim and Jack were behind Ta Shu, bringing up the rear, and losing a couple of hundred yards per hour on all the rest of them. It didn't seem like much, but it added up. And it worried her. But there was nothing to do but carry on, and ratchet down the pace a bit so that no one pushed too hard, especially those bringing up the rear.

They had been hiking for ten hours when Val got another GPS fix. They had come some thirty kilometres, a good pace; but she had aimed them out to the south to avoid the crevasses at the top of the Hump Passage, at the head of the Liv Glacier. So they still had at least seventy kilometres to go, she reckoned, depending on how far south they would have to detour to get around the ice ridge extending southward from Last Cache Nunatak. Beyond that ice ridge lay the head of the Zaneveld Glacier, which was heavily crevassed; they would have to stay south of that; and then on the far side of the Zaneveld was Roberts Massif. All those features lay below the horizon, of course; they could see only about ten kilometres in all directions, which meant that they could see nothing but the ice plain, except for occasional glimpses of the peaks of the Queen Maud Range, poking over the horizon to their right.

Into her rhythm, taking it slow. So far of all the clients Ta Shu seemed the least affected by their long march. He spent all his rest time contemplating the distant peaks of the Queen Maud Range. While walking he stumped along steadily, and at times caught up with her and walked by her side. "We are doing well!"

"Yes."

[. . .]

But this group had only one Ta Shu in it, and the rest were slowly losing steam. In fact Jack and Jim were falling behind faster than ever. Puzzled, Val stopped and watched them closely for a while. It was not Jim who was slowing them down: it looked as if Jack had hit the wall. "Fuck," she said. He had gone out too fast, perhaps, and burnt out. Or was feeling the loss of blood from the cut in his hand. Or both. Anyway he was slowing down markedly.

Val called an early break, and waited for the two men to catch up, cursing to herself. They joined the group twenty minutes after Jorge and Elspeth came in, and during that time the others had eaten and drunk their flasks and refilled them, and were beginning to freeze. This was a serious problem, and she couldn't help thinking that it was Jack's fault. So often it happened that men like him took off too fast, on an adrenaline rush, thinking their emergency energy would be inexhaustible, and then they were the first to hit the wall. Pacing took a lot of self-discipline. And big, muscular men were generally not so good in ultralong-distance events: they had too many muscles to feed, and when they ran out of the day's carbo-load, they had too little body fat to throw on the fire.

So when Jim and Jack clumped into the group, Val suggested that they have a bite of their belts to give themselves more energy. Jim nodded, and pulled out some of his belt and tore it off and stuffed it under his ski-mask into his mouth before it froze.

Jack just shook his head irritably. "I'm just pacing myself," he snapped. "Like you said to do. Don't get neurotic about it, that's the last thing we need. Let people go what pace they want."

"Sure, sure. Try eating some food, though. We need the group to stay more or less together, or the people in front will freeze waiting for the ones behind."

"Don't wait then!"

She stared at him. "You should eat," she said finally. "And drink your arm-flasks and refill them, for God's sake."

And after a little while more she had taken off again, and was soon leading the way. No beeps, thank God; they were out on the big ice-cube itself now, a solid mass with very little cracking. Just a matter of walking. Pacing oneself, yes, and walking. Hour after hour. She shifted them to a ten-minute break every hour, which was exactly Shackleton's pattern. Frequently she glanced back over her shoulder, Jack was still falling behind, perhaps even more rapidly than before; and Jim was sticking with him.

At some time when she was not looking the sun was touched by a thin film of cloud, which had appeared out of nowhere. A white film, but heavily polarized by her sunglasses, so that it was banded prismatically.

As usual, it only took the slightest cloud cover for the day to go from blinding and hot to ominous and chill. Already they were pulling their ski-masks down over their faces, and zipping up their parkas; and as they did so the cloud thickened further, into a thin rippled patch thrown right over the sun, as if someone had tried to place it there. So often it happened that way; the cloud could have appeared anywhere in the sky, but ended up right between Val and the sun. It happened so frequently that she reckoned it must be some trick of perspective rather than a real phenomenon. In any case, there it was again.

Which was bad, bad news. The immediate effects were that their suits wouldn't be as warm, and worse, their arm-flasks would be much less efficient at melting snow and ice. It would take twice as long to melt snow now, maybe three times. So they were going to get thirsty.

The mental effect of the cloud was also bad. What had been a blazing plain was now shadowed and malign. Underfoot the beautifully elaborate crosshatching in the snow was revealed better than ever, a granulated fractal infinity of sharply cut micro-terracing. This complex world underfoot was as prismatic as any cloud whenever it flattened enough, and now when she looked in the direction of the sun Val saw diaphanous icebows, curving both in the cloud and across the snow itself. They walked forward into a geometry of rainbows. Val looked back at Ta Shu, and he raised a ski pole briefly, to let her know he had noticed the phenomena, and appreciated her thinking to bring it to his attention.

A beautiful sight; and yet still the world seemed dim and malignant. Clouds of any kind on the polar cap often presaged even worse weather, of course, which perhaps was part of the mood it cast. With luck her clients did not know that and so wouldn't be affected as much. They were still many hours' walk out from Roberts, and a lot could happen to Antarctic weather in that amount of time.

Nothing to do but forge on, of course, into a landscape turned alien; the awesome become awful, and all in the few minutes it had taken for a thin cloud to form. After which they were mere specks on a high plateau on Ice Planet, a place in which humans could not live except in spacesuits. And they could feel that palpably, in the penetrating cold.

At the next rest-stop they drank and ate in silence. There was no point, Val judged, in trying to cheer them on. She could have pointed out to them again that they were having an adventure at last, after trying so many times and paying so much money. But she doubted that would go over very well now. One of the distinguishing marks of a true adventure, she had found, is that they were often not fun at all while they were actually happening. And in one of their camp conversations Jim had quoted Amundsen to the effect that adventure was just bad planning. So that if she called it that, they might blame her for it. Jack was certainly ready to.

And she blamed herself. It had been a mistake to take the righthand route, as it turned out. Although still—as she walked on thinking about it, trying to cheer herself up—it seemed that what had happened showed that Amundsen was wrong, and that adventures could also be a matter of bad luck as well as bad planning. You could plan everything adequately, and still get struck down by sheer bad luck. It happened all the time. Chance could strike you down; that was what made these kinds of activities dangerous. That was what made all life dangerous. You couldn't plan your way out of some things. You had to walk your way out, if you could.

In any case, while there was no obvious way to cheer them up during the rests, there was also no great need to urge or cajole them along. The situation was plain; they either walked on or died. The intense cold they were living in reminded them of that at every moment.

She tried her GPS and it gave her a reading, showing them on the 172nd longitude. About the halfway mark of their hike. Not bad at all, except that they were getting very tired. They had hiked around thirty miles, after all, and were beginning to run out of steam; she could see it in the way they moved. Jorge was limping slightly. Elspeth was letting her ski poles drag from time to time, no doubt to give her arms a rest. Jack was doing the same, and moving like a pall-bearer. Jim was trying to keep to his friend's slow pace, though often he pulled ahead and then stopped and waited, not a good technique. Only Ta Shu still had the contained efficiency of someone with some strength left in his legs, placing each step precisely into her bootprints, using his ski poles in easy short strokes. He looked as if he could stump along for a long time.

Val herself was feeling the work, but was well into her long-distance rhythm, a feeling of perpetual motion that was not exactly effortless but a kind of contained low-level effort, one that she could sustain for ever; or so it felt. Obviously there would come a time when that feeling would wear away. But she had seldom reached it, especially when guiding clients, and right now it was still a long way off.

Her endurance, however, was not the point. They could only go at the speed of the weakest members of the team, and there was nothing she could do for them. Well; she could give them her meltwater. And so at the next break she did that, giving one cup to Jack and another to Elspeth, over their objections. "Drink it," she ordered, her tone peremptory in a way she had not let it be until now. "I'm not thirsty."

[. . .]

The little cloud was thickening, a white blanket thrown right over the sun, holding its position with maddening fixity. You could laugh at the Victorians for talking about a battle with Nature, but when you saw a cloud hold its position like that, in the freshening wind now striking them, it was hard not to feel there was some malignant perversity at work there, a Puckish delight in tormenting humans. It might be the pathetic fallacy, but when you were as thirsty as Val was it felt tragic.

She slowed down to hike with each of the others in turn, inquiring after them. Except for Ta Shu, and perhaps Jim, they were hurting; nearly on their last legs, it appeared, with more than thirty kilometres to go. Well, she would shepherd them there. Bring them on home. Give them her water, give them her mental energy. There was something about taking care of clients in such a way that felt so good. Others before self. Being a shepherd, or a sheepdog. Husbanding them along.

At the next stop, however, she tried again to give Jack her water, and suggested that he eat, and he refused the water and yanked the power bar out of his belt and tore off a piece savagely, muttering "Lay off, for Christ's sake. We're doing the best we can!"

Jim and Elspeth and Jorge all nodded. "This is hard for us," Elspeth said to Val wearily.

"Of course," Val said. "I know. Hard for anyone. You're doing great. We're making a very long walk, in excellent time. No problem. Let's just keep taking it easy, we'll get there."

And as soon as possible she had them moving again, despite Elspeth's suggestion that they take a longer break. That would only allow muscles to stiffen up; besides, the sheer impact of the cold made it impossible. They had to move to stay warm.

So she took off, trying to tread the fine line between going too fast and tiring them or going too slow and freezing them. She lost the glow she had felt during the previous march about the ethic of service and all that; in fact another part of her was taking over, and getting angry at these people for getting so tired so fast. Sure, she should have kept anything like this from happening. But they had no business coming down here to trek if they were not in shape. Even these so-called outdoorspeople were still very little more than brains in bottles—weekend warriors at best, exercising nothing but their fingertips in their work hours, the rest of their bodies turning as soft as sofa cushions. Watching computer screens, sitting in cars, watching TV, it was all the same thing—watching. Big-eyed brains in bottles. These clients of hers were actually among the fittest of the lot, they were the best the world had to offer! The best of the affluent Western world, anyway. And even they were falling apart after walking a mere seventy kilometres. And thinking they were doing something really hard.

But in their spacesuit gear the level of raw suffering was not that great, if they could just learn to thermostat properly. Indeed the whole idea of Antarctic travel as terrible suffering which required tremendous courage to attempt struck Val as bullshit, now more than ever. It was all wrapped up with this Footsteps phenomenon—people going out ill-prepared to repeat the earlier expeditions of people who had gone out ill-prepared, and thinking therefore that you were doing something difficult and courageous, when it was simply stupid, that was all. Dangerous, yes; courageous, no. Because there was no correlation between doing something dangerous and being courageous, just as there was no correlation between suffering and virtue. Of course if you went at it with Boy Scout equipment like Scott had, then you suffered. But that wasn't virtue, nor was it courage.

In fact, Val decided as she stamped along, most of the people who came to Antarctica to seek adventure and do something hard, came precisely because it was so much *easier* than staying at home and facing whatever they had to face there. Compared to life in the world it took no courage at all to walk across the polar cap; it was simple, it was safe, it was exhilarating. No, what took courage was staying at home and facing things, things like talking your grandma out of a tree, or reading the wanted ads when you know nothing is there, or running around the corner of the house when you hear the crash. Or waiting for test results to come back from the hospital. Or taking a dog to the vet to have it put down. Or taking a group of leukemia kids to a game. Or waiting to see if your partner will come home drunk that night or not. Or helping a fallen parent off the bathroom floor at four in the morning. Or telling a couple that their child has been killed. Or just sitting on the floor and playing a board game through the whole of a long afternoon. No, on the list could go, endlessly: the world was stuffed with things harder than walking in Antarctica. And compared to those kinds of things, walking for your life's sake across the polar ice cap was *nothing.* It was *fun.* It could kill you and it would *still be fun,* it would be a *fun death.* There were scores of ways to die that were immeasurably worse than getting killed by exposure to cold; in fact, freezing was one of the easiest ways to go. No, the whole game of adventure travel was essentially an escape from the hard things. Not necessarily bad because of that; a coping mechanism that Val herself had used heavily all her life; but not something that should ever be mistaken for being hard or heroic. It was daily life that was hard, and sticking it out that was heroic.

Val shuddered at this dark train of thought, stopped in her tracks. She looked back; she had been going too fast, and the people she was caring for had fallen far behind. "Come on, goddamn it!" she said at them. "You are so fucking slow. This is fun! This is your adventure! Are we having fun yet?" Almost shouting at them. But they were so far back there was no chance they would hear her.

A BRIEF CHRONOLOGY OF ANTARCTIC EXPLORATION

1839–43 British naval expedition in HMS *Erebus* and HMS *Terror*, led by James Clark Ross, discovers the Antarctic mainland

1882–83 First International Polar Year

1897–99 Adrien de Gerlache's *Belgica* expedition overwinters in the Antarctic pack ice

1898–1900 Australian/Norwegian/British *Southern Cross* expedition, led by Carsten Borchgrevink, spends first winter on the continent

1901–03 German expedition under Erich von Drygalski maps Kaiser Wilhelm II Land. First Antarctic balloon flight

1901–04 British *Discovery* expedition, led by Robert Falcon Scott and based on Ross Island. Southern journey reaches 82°17′S

1903–05, 1908–10 French expeditions led by Jean-Baptiste Charcot aboard the *Francais* and the *Pourquoi-Pas?* map the coast of the Antarctic Peninsula

1907–9 Ernest Shackleton's British *Nimrod* expedition at Cape Royds, Ross Island. Polar journey reaches 88°23′S

1910–12 Norwegian expedition led by Roald Amundsen reaches South Pole (90°S) on December 14, 1911

1912	Japanese expedition led by Nobu Shirase
1910–13	R. F. Scott's *Terra Nova* expedition at Cape Evans, Ross Island. Scott's party die on the return journey from the Pole, 1912
1911–12	German expedition in the Weddell Sea discovers the Filchner Ice Shelf
1911–14	Australian/New Zealand expedition, led byDouglas Mawson, in Adélie Land
1914–17	Shackleton's "Imperial Transantarctic Expedition," failed attempt to cross the continent from Weddell Sea to Ross Sea
1922	Shackleton dies in South Georgia
1928–30	American expedition, led by Richard Byrd, establishes Little America base on Ross Ice Shelf. First overflight of South Pole, November 28, 1929
1932–33	Second International Polar Year
1933–35	Byrd's second expedition. Byrd overwinters solo
1935	Lincoln Ellsworth makes first air crossing of Antarctica
1938–39	German expedition to Queen Maud Land, renamed "Neuschwabenland"
1943–45	British Royal Navy's "Operation Tabarin" establishes bases on the Antarctic Peninsula
1946–48	U.S. Navy's "Operation Highjump" and "Operation Wind-mill"
1955–58	Commmonwealth Trans-Antarctic Expedition, led by Vivien Fuchs, makes first land crossing of the continent

1955	U.S. Navy's "Operation Deep Freeze" begins
1956	Construction of South Pole base (U.S.) and McMurdo Station (U.S.)
1957	Construction of Vostok Station (USSR)
1957–58	Third International Polar Year
1961	Antarctic Treaty enters into force
1966	First ascent of Vinson Massif (16,067 ft), Antarctica's tallest mountain
1966	Regular tourist cruises begin
1969–70	First women included in the United States Antarctic Program
1983	Lowest temperature ever recorded on earth (–89.2° C) at Vostok
1985	Discovery of hole in ozone layer announced by British scientists at Halley and Rothera bases
1987	Signing of Montreal Protocol banning ozone-depleting chemicals. Patriot Hills runway first used for private flights
1995	Collapse of 2000 km^2 Larsen A Ice Shelf, Antarctic Peninsula
1996	Ice core completed at Vostok, showing 420,000 years of earth's atmospheric history. Discovery of "Lake Vostok," freshwater lake as big as Lake Ontario, sealed for 25 million years beneath nearly 4 km of ice
1997	Kyoto Protocol negotiated, aimed at reducing emission of greenhouse gases
1998	Antarctic Treaty's Protocol on Environmental Protection enters into force

2002	Collapse of 3250 km^2 Larsen B Ice Shelf, Antarctic Peninsula. Records over 50 years show a 2.5°C rise in the Peninsula's average temperature
2007–9	International Polar Year

ACKNOWLEDGMENTS

Various publishers and estates have generously given permission to use extracts from the following works:

"Lt. Shirase's Calling Card" by Nobu Shirase from *Shirase,* translated by Hilary Shibata and Lara Dagnell and published by Bluntisham Books. Used by permission of Hilary Shibata. "Scott Dies" by Francis Spufford from *I May Be Some Time,* copyright © 1997 by Francis Spufford. Used by permission of the author and Faber and Faber Ltd. "The Blow" by Richard Byrd from *Alone,* copyright © 1938 by Richard E. Byrd, renewed 1966 by Marie A. Byrd. "The Blasphemous City" by H. P. Lovecraft from *At the Mountains of Madness,* copyright © 1936 by Arkham House Publishers, Inc. First published in *Astounding Stories.* The definitive edition previously appeared in *At the Mountains of Madness,* published by Arkham House Publishers, Inc., in 1964. Copyright 1964 by August Derleth. Copyright renewed 1992. "White Lanterns" by Diane Ackerman from *The Moon by Whale Light,* copyright © 1991 by Diane Ackerman. Used by permission of Random House, Inc., "Particles" by Bill Green from *Water, Ice and Stone,* copyright © 1995 by Bill Green. Used by permission of the author. "Cabin 532" by Jenny Diski from *Skating to Antarctica,* copyright © 1997. Used by permission of Granta Books. "Hey, Woo" by Sara Wheeler from *Terra Incognita: Travels in Antarctica,* copyright © 1996 by Sara Wheeler. Used by permission of Random House, Inc., and by permission of the Random House Group Ltd. "Blue Collar" by Nicholas Johnson from *Big Dead Place,* copyright © 2005 by Nicholas Johnson. Used by permission of Feral House and the author. "In Amundsen's Footsteps, 2016" by Kim Stanley Robinson from *Antarctica,* copyright © 1998 by Kim Stanley Robinson. Used by permission of Bantam Books, a division of Random House, Inc. Map of Australasia and Antarctica (1739), by Philippe Bauche (1700–1773). The original, held by the British Library, is accompanied by text describing the expedition, commanded by Bouvet, which led to the map's production. Copyright © HIP/Art Resource, NY.

ABOUT THE EDITOR

Francis Spufford is the author of *I May Be Some Time*, a cultural history of the British obsession with polar exploring; *The Child that Books Built* and *Backroom Boys*. He has won the Somerset Maugham Award, the *Sunday Times* (London) Young Writer of the Year prize, and the Writers' Guild Award for Best Non-Fiction Book of the Year. He lives in Cambridge, England.

THE ENDS OF THE EARTH

THE ARCTIC

THE ENDS OF THE EARTH

An Anthology of the Finest Writing on the Arctic and the Antarctic

THE ARCTIC

Edited by Elizabeth Kolbert

BLOOMSBURY

Published by Bloomsbury USA, New York
Distributed to the trade by Holtzbrinck Publishers

All papers used by Bloomsbury USA are natural, recyclable products made from wood grown in well-managed forests. The manufacturing processes conform to the environmental regulations of the country of origin.

LIBRARY OF CONGRESS CATALOGING-IN-PUBLICATION DATA HAS BEEN APPLIED FOR

ISBN-10 1-59691-443-2
ISBN-13 978-1-59691-443-8

First U.S. Edition 2007

1 3 5 7 9 10 8 6 4 2

Typeset by Westchester Book Group
Printed in the United States of America by Quebecor World Fairfield

CONTENTS

Cercle Polaire
MER DU NORD
Tropique du Cancer
Terre de Laborador
Baye de Hudson
CANADA
NOUV. FRANCE
AMERIQUE SEPTENTRIONALE
LOUISIANE
FLORIDE
Isles Lucayes
ISLES ANTILLES
CALIFORNIE
MER VERMEILLE
NOUVEAU MEXIQUE
Sonora
MEXIQUE
Golfe du Mexique
Mexico
NOUVELLE ESPAGNE
G. de Honduras
Golfe de Panama
TERRE FERME
AMERIQUE MERIDIONALE
GUIANE
Isles Gallapes
Tropique du Cancer
SUD OU MER PACIFIQUE
Equateur ou Ligne Equinoctiale
230
240
250
290
300
310

MER DE LEN
MER D'AMU
TERRE DE YEÇO
MER ORIENTALE
Niphon
ISLES DU JAPON
R. DE COREE
TARTARIE CHINOISE
CHINE
Tonquin
Cochinchine
Golfe de Siam
Isle de Formose
PHILIPPINES
I. de Samar
Mindanao
I. de Borneo
I. de Sumatra
Isles de Gam ou du Nord
ISLES MARIANES
Coromandel
Isle de Ceylan
GOLFE DE BENGALA
Isles des Andamans
Isles Maldives
les Laquedives
MER DES INDES
ARABIE
Arabie Heureuse
Arabie Deserte
Mer Rouge
PERSE
Tibet
ETATS DU MOGOL
INDE
TARTARIE INDEPENDANTE
GRANDE TARTARIE
Calmoucs
Mogols
Siberie
TARTARIE MOSCOVITE
Mer Caspienne
MEDITERRANEE
BARBARIE
Le Sara ou Desert de Barbarie
Desert de Tegasa
NIGRITIE
AFRIQUE
EGYPTE
NUBIE
ABISSINIE
EUROPE
TURQUIE
70
60
50
120
130

INTRODUCTION

ELIZABETH KOLBERT

IN THE SPRING of 1888, the Norwegian doctor-cum-explorer Fridtjof Nansen set off for Iceland. He boarded a whaling ship in Isafjord and sailed to the east coast of Greenland. Once there, he strapped on a pair of skis and headed out across the ice sheet, a trip of some five hundred miles. Save for some problems with the pemmican—the dried meat, which he had ordered from an outfitter in Copenhagen, was too lean, leading "to a craving for fat which can scarcely be realized by anyone who has not experienced it"—the journey went off without incident. Nevertheless, to get from Christiania—now Oslo—to Godthab—now Nuuk—and back again took Nansen over a year.

In the spring of 2001, I set off for a research station known as the North Greenland Ice-core Project. I boarded a cargo plane in Schenectady, New York, and disembarked in Kangerlussuaq, on the island's west coast, six hours later. Another two-hour flight—this one on board an LC-130 equipped with skis and, for extra propulsion, little rockets—took me to North GRIP, at the very center of the ice sheet, at 75° 06' N. In this way, I completed the first half of Nansen's journey (albeit from the opposite direction) in less than a day. Perhaps if I had spent several weeks at the camp, I would have developed a craving for something; honestly, though, I can't imagine what. The afternoon I arrived, coffee and cake were served in the geodesic dome that doubled as the camp's dining hall. In the evening, there was a cocktail party held in a chamber hollowed out of the ice. Dinner was lamb chops slathered in a tomato cream sauce, accompanied by red wine. As I recall, I skipped dessert that night; I was just too stuffed.

At least twenty scientists were living at North GRIP, and at no point did I wander far enough from the camp to lose sight of the tents. Yet if Arctic travel isn't quite what it used to be, now that the danger, hardship, and solitude have

been sheared away, it is still an other-worldly experience. The white, the cold, the three A.M. sun—the scene was unlike anything I had ever encountered before. Of course, I came home and wrote about it.

This is a book of *writings* about the Arctic, which is not quite the same thing as a book about the Arctic. Almost all the selections are by outsiders to the region—explorers, adventurers, anthropologists, novelists. The predominance of non-natives reflects the fact that Arctic people have, traditionally, transmitted their narratives orally, and also the fact that those who have been drawn to the area have, to an astonishing degree, felt compelled to record their impressions. Even today, the number of people who have traveled to the far north is tiny compared with the number who have traveled to Birmingham, say, or Philadelphia. Yet the literature of the Arctic is immense. Trying to choose the selections for this book, I sometimes felt as if everyone who had ever visited the Arctic had left behind an account of his or her (usually his) experience. In one of his many Klondike tales, "An Odyssey of the North," Jack London compares the Arctic whiteness to "a mighty sheet of foolscap" and a dog team racing across the snow to a line drawn in black pencil. For a writer, the image is reversible, so that the blank page—and all its terrors—can also become a metaphor for the ice.

The Arctic is difficult to define. As Barry Lopez notes, "There is no generally accepted definition for a southern limit" to the region. To use the Arctic Circle as the boundary means including parts of Scandinavia so warmed by the Gulf Stream that they support frog life, while at the same time excluding regions around James Bay, in Canada, that are frequented by polar bears. (The Arctic Circle is designated as 66° 33' N; however, owing to a slight wobble in the earth's axial tilt, the real circle of polar night shifts by as much as fifty feet a year.) The works collected here touch on travels as far south as Iceland, which, except for the tiny tip of a tiny island, lies entirely below the Arctic Circle, and as far north as the pole itself—if, that is, you accept Robert Peary's claims to have made it there.

Though speculation about a mysterious, frozen island known as Thule dates all the way back to the Greek geographer Pytheas, this collection begins in the early nineteenth century. By that point, the search for the Northwest Passage was already well under way and had already claimed dozens of lives. It is sometimes suggested that this search was motivated by commerce and sometimes by nationalism. But neither force seems quite adequate; as the historian Glyn Williams has observed, the quest "became almost mystical in

nature, beyond reasonable explanation." I have chosen to focus on the most celebrated—and most disastrous—attempt to find a passage, that led by Sir John Franklin in 1845. Franklin's disappearance prompted a string of rescue missions, many of which also ended badly, and one of which, organized by a Cincinnati businessman named Charles Francis Hall, may have culminated in murder. In the 1960s, Hall's biographer, Chauncey Loomis, traveled to the shore of Thank God Harbor, in northern Greenland, where Hall had been buried, and had his body exhumed. It was in remarkably good shape, demonstrating the cold, hard truth that in the Arctic nothing is ever really lost.

The next generation of explorers was interested in only one direction: due north. I have included accounts of three efforts to reach the pole—those led by Nansen, Peary, and Salomon August Andrée. Nansen's attempt involved spending a year in an icebound boat—the *Fram,* or "Forward"—and two more traversing the ice by dog sled. The farthest Nansen reached was as 86° 14' N. This was 270 miles short of his goal, but at that point—1895—the highest latitude anyone had ever attained. Nansen's Swedish rival, Andrée, came up with the daring if impractical notion of besting him by balloon. With two companions, Andrée set off for the pole from Spitsbergen on July 11, 1897. Three days later, after having drifted some 200 miles to the northeast, the balloon—the *Eagle*—crashed and had to be abandoned. The three men spent the next ten weeks trekking across the ice, trying—unsuccessfully—to reach first one and then a second food cache that had been left for them. Their bodies were finally found more than thirty years later on the island of Kvitøya, roughly at the same latitude from which they had departed.

In their writings, explorers like Franklin and Andrée tend to treat the Arctic as a set of problems: unfordable rivers, blinding snow, drifting ice. (Peary's accounts of his adventures barely mentions the cozier sort of adventures that would lead, among other things, to at least two half-Inuit offspring.) But even before the race to the pole was over, a different sort of interest in the Arctic had begun to produce a different sort of literature. Though the Arctic comprises some of the most inhospitable terrain on earth, people have been living there for thousands of years. The fact that native communities flourished in an environment often fatal to non-native travelers challenged conventional notions of progress. Starting in the late nineteenth century, the Arctic became a favored destination for what might be called explorers of human nature.

Knud Rasmussen is one of the few indigenous authors included in this volume. Born in Ilulissat to a Danish father and Greenlandic mother, Rasmussen was both a prolific writer and a prolific traveler; he participated in eight Arctic expeditions—the so-called Thule expeditions—which studied the region's archeology, ethnography, geology and botany. *Across Arctic America* is Rasmussen's chronicle of the most famous of these expeditions—the fifth—condensed into a few hundred pages. (The scientific report on the 20,000-mile journey ran to ten volumes.) "Some archeologists have made bold to assert that the Eskimos are surviving remnants of the Stone Age we know, and are, therefore, our contemporary ancestors," Rasmussen writes. "We don't have to go so far to claim kinship with them, however, for we *recognize* them as brothers." Rockwell Kent, the American artist, went to live in northern Greenland for a year in the early 1930s; Gontrans De Poncins, a French count, spent fifteen months among the Canadian Inuit in the second half of the decade. Both men were drawn to life in the Arctic because it was remote and dangerous and other. (A similar attraction would later prompt Tété-Michel Kpomassie to make his way from Togo to Greenland's west coast and Gretel Ehrlich to follow Rasmussen's trail.) Their narratives of discovery are just as significantly stories of *self*-discovery.

Several of the works in this collection are explicitly fictional. "Kasiagsak, the Great Liar," is one of the many native legends collected in the late nineteenth century by Heinrich Rink, a Danish official serving in Greenland. In *The Adventures of Captain Hatteras,* Jules Verne invents a polar explorer easily as obsessed as Peary and Andrée. In *The Voyage of the Narwhal,* Andrea Barrett adds a make-believe ship to the long list of actual ones that went off searching for Franklin and his men. *Independent People,* by the Nobel laureate Halldór Laxness, tells the story of Bjartur of Summerhouses, a sheep-farming, poetry-loving Icelander who endures hardships of an almost comically monstrous variety.

*

The immediate inspiration for this collection is the 2007–2008 International Polar Year, which, to accommodate researchers in Antarctica, actually lasts until March 2009. This is the fourth International Polar Year—previous ones were held in 1882–3, 1932–3, and 1957–8—and, like its predecessors, the current IPY is supposed to direct international attention (and resources)

toward polar research. The 2007–2008 IPY differs from earlier ones, however, in that its focus is on the disappearance of its subject matter.

The impact of global warming, increasingly evident all over the world, is most apparent in the Arctic, thanks to an effect sometimes known as the Arctic amplification. While average global temperatures have risen by about .6 degrees C (1 degree F), in the Arctic they have gone up by roughly twice that amount. The change is particularly striking during the coldest months of the year; in Siberia, for example, wintertime temperatures have risen by as much as 4 degrees C (7 degrees F). Since visiting Greenland in 2001, I've made four more trips to the Arctic to report on how the region is changing. Most of those trips were made in the company of scientists, but along the way I also met many native people who spoke eloquently about what is happening. An Inuit hunter named John Keogak, who lives on Banks Island, in the Inuvik Region of Canada's Northwest Territories, told me that he and his fellow-hunters had started to notice that the climate was changing in the mid-1980s. Then a few years ago, people on the island began to see robins, a bird for which the Inuit in his region have no word.

"We just thought, Oh, gee, it's warming up a little bit," he recalled. "It was good at the start—warmer winters, you know—but now everything is going so fast. The things that we saw coming in the early 1990s they've just multiplied.

"Of the people involved in global warming, I think we're on top of the list of who would be most affected," Keogak went on. "Our way of life, our traditions, maybe our families. Our children may not have a future. I mean, all young people, put it that way. It's not just happening in the Arctic. It's going to happen all over the world. The whole world is going too fast."

The warming that has occurred has by now been sufficient to shrink the Arctic ice cap by almost five hundred thousand square miles, to bring millions of acres of permafrost close to the thawing point, and to cause devastating pest outbreaks in the spruce forests of Canada and Alaska. But this is only the beginning. If current trends continue, temperatures in the Arctic will rise by as much as 5 degrees C (9 degrees F) by the end of this century. Sometime before that point, much of the landscape described in these pages will have vanished. For example, current forecasts suggest that a Northwest Passage could be ice-free, at least in summer, by 2025. The North Pole itself could be open water in summer by 2050. Perhaps most ominously of all, the Greenland ice sheet, which at its center is ten thousand feet thick, could begin to disinte-

grate. While it might take centuries for the ice sheet to disappear entirely, once the process of disintegration gets under way, it will start to feed on itself, most likely becoming irreversible. (The Greenland ice sheet holds enough water to raise global sea levels by more than twenty feet.) In this way, the claim of the Arctic on our imagination has been inverted. A landscape that once symbolized the sublime indifference of nature will, for future generations, come to symbolize its tragic vulnerability.

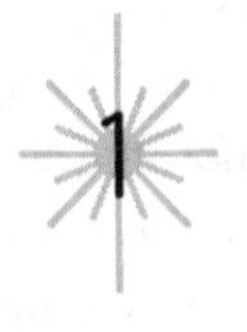

THE EXTREME MISERY OF THE WHOLE PARTY

from *Narrative of a Journey to the Shores of the Polar Sea* (1823)

John Franklin

*Sir John Franklin, an officer in the British Royal Navy, led three Arctic expeditions in the first half of the nineteenth century. The last of these, which resulted not only in his own death, but in the deaths of all 120-plus members of his crew, is the most famous; it prompted several rescue missions, some of which ended no less disastrously. This excerpt is from the increasingly disjointed diaries Franklin kept during his first Arctic expedition (1819–1822), which took him from Hudson Bay to the Coppermine River. Nine of Franklin's men died on the journey—one was shot under mysterious circumstances—and the rest endured desperate privation. Included below is the famous October 4th episode in which Franklin and his men dine on their shoes. (*Tripe de roche, *which Franklin refers to several times, is a kind of lichen.)*

In the afternoon we had a heavy fall of snow, which continued all night. A small quantity of *tripe de roche* was gathered; and Crédit, who had been hunting, brought in the antlers and back bone of a deer which had been killed in the summer. The wolves and birds of prey had picked them clean, but there still remained a quantity of the spinal marrow which they had not been able to extract. This, although putrid, was esteemed a valuable prize, and the spine being divided into portions, was distributed equally. After eating the marrow, which was so acrid as to excoriate the lips, we rendered the bones friable by burning, and ate them also.

On the following morning the ground was covered with snow to the depth of a foot and a half, and the weather was very stormy. These circumstances

rendered the men again extremely despondent; a settled gloom hung over their countenances, and they refused to pick *tripe de roche*, choosing rather to go entirely without eating, than to make any exertion. The party which went for gum returned early in the morning without having found any; but St. Germain said he could still make the canoe with the willows, covered with canvas, and removed with Adam to a clump of willows for that purpose. Mr. Back accompanied them to stimulate his exertion, as we feared the lowness of his spirits would cause him to be slow in his operations. Augustus went to fish at the rapid, but a large trout having carried away his bait, we had nothing to replace it.

The snow-storm continued all the night, and during the forenoon of the 3rd. Having persuaded the people to gather some *tripe de roche*, I partook of a meal with them; and afterwards set out with the intention of going to St. Germain to hasten his operations, but though he was only three-quarters of a mile distant, I spent three hours in a vain attempt to reach him, my strength being unequal to the labour of wading through the deep snow; and I returned quite exhausted, and much shaken by the numerous falls I had got. My associates were all in the same debilitated state, and poor Hood was reduced to a perfect shadow, from the severe bowel complaints which the *tripe de roche* never failed to give him. Back was so feeble as to require the support of a stick in walking; and Dr. Richardson had lameness superadded to weakness. The voyagers were somewhat stronger than ourselves, but more indisposed to exertion, on account of their despondency. The sensation of hunger was no longer felt by any of us, yet we were scarcely able to converse upon any other subject than the pleasures of eating. We were much indebted to Hepburn at this crisis. The officers were unable from weakness to gather *tripe de roche* themselves, and Samandré, who had acted as our cook on the journey from the coast, sharing in the despair of the rest of the Canadians, refused to make the slightest exertion. Hepburn, on the contrary, animated by a firm reliance on the beneficence of the Supreme Being, tempered with resignation to his will, was indefatigable in his exertions to serve us, and daily collected all the *tripe de roche* that was used in the officers' mess. Mr. Hood could not partake of this miserable fare, and a partridge which had been reserved for him was, I lament to say, this day stolen by one of the men.

October 4.—The canoe being finished, it was brought to the encampment, and the whole party being assembled in anxious expectation on the beach, St. Germain embarked, and amidst our prayers for his success, succeeded in reaching the opposite shore. The canoe was then drawn back again,

and another person transported, and in this manner by drawing it backwards and forwards, we were all conveyed over without any serious accident. By these frequent traverses the canoe was materially injured; and latterly it filled each time with water before reaching the shore, so that all our garments and bedding were wet, and there was not a sufficiency of willows upon the side on which we now were, to make a fire to dry them.

[. . .]

It is impossible to imagine a more gratifying change than was produced in our voyagers after we were all safely landed on the southern banks of the river. Their spirits immediately revived, each of them shook the officers cordially by the hand, and declared they now considered the worst of their difficulties over, as they did not doubt of reaching Fort Enterprise in a few days, even in their feeble condition. We had, indeed, every reason to be grateful, and our joy would have been complete had it not been mingled with sincere regret at the separation of our poor Esquimaux, the faithful Junius.

The want of *tripe de roche* caused us to go supperless to bed. Showers of snow fell frequently during the night. The breeze was light next morning, the weather cold and clear. We were all on foot by daybreak, but from the frozen state of our tents and bed-clothes, it was long before the bundles could be made, and as usual, the men lingered over a small fire they had kindled, so that it was eight o'clock before we started. Our advance, from the depth of the snow, was slow, and about noon, coming to a spot where there was some *tripe de roche,* we stopped to collect it, and breakfasted. Mr. Hood, who was now very feeble, and Dr. Richardson, who attached himself to him, walked together at a gentle pace in the rear of the party. I kept with the foremost men, to cause them to halt occasionally, until the stragglers came up. Resuming our march after breakfast, we followed the track of Mr. Back's party, and encamped early, as all of us were much fatigued, particularly Crédit, who having to-day carried the men's tent, it being his turn so to do, was so exhausted, that when he reached the encampment he was unable to stand. The *tripe de roche* disagreed with this man and with Vaillant, in consequence of which, they were the first whose strength totally failed. We had a small quantity of this weed in the evening, and the rest of our supper was made up of scraps of roasted leather. The distance walked to-day was six miles. As Crédit was very weak in the morning, his load was reduced to little more than his personal luggage, consisting of his blanket, shoes, and gun. Previous to setting out, the whole party ate the remains of their old shoes, and whatever scraps of leather they had, to strengthen their stomachs for the fatigue of the

day's journey. We left the encampment at nine, and pursued our route over a range of black hills. The wind having increased to a strong gale in the course of the morning, became piercingly cold, and the drift rendered it difficult for those in the rear to follow the track over the heights; whilst in the valleys, where it was sufficiently marked, from the depth of the snow, the labour of walking was proportionably great. Those in advance made, as usual, frequent halts, yet being unable from the severity of the weather to remain long still, they were obliged to move on before the rear could come up, and the party, of course, straggled very much.

About noon Samandré coming up, informed us that Crédit and Vaillant could advance no further. Some willows being discovered in a valley near us, I proposed to halt the party there, whilst Dr. Richardson went back to visit them. I hoped too, that when the sufferers received the information of a fire being kindled at so short a distance they would be cheered, and use their utmost efforts to reach it, but this proved a vain hope. The Doctor found Vaillant about a mile and a half in the rear, much exhausted with cold and fatigue. Having encouraged him to advance to the fire, after repeated solicitations he made the attempt, but fell down amongst the deep snow at every step.

THE RETURN OF LIGHT

from *Arctic Explorations* (1856)

Elisha Kent Kane

Elisha Kent Kane served as the medical officer on an 1850 American expedition that went in search of John Franklin and his crew. Three years later, he captained a second expedition. Kane was, by all accounts, an ineffectual leader. After spending two winters on their boat, the Advance, *which had frozen into the ice off the west coast of Greenland, he and his men slogged 1,300 miles to the town of Upernavik; amazingly, all but three made it. In this excerpt from Kane's diaries of 1854, spring has just arrived, and with it a profusion of game. This leads Kane to meditate on the hidden richness of the Arctic and on the possibility that some members of Franklin's crew might still be alive.*

We have more fresh meat than we can eat. For the past three weeks we have been living on ptarmigan, rabbits, two reindeer, and seal.

They are fast curing our scurvy. With all these resources,—coming to our relief so suddenly too,—how can my thoughts turn despairingly to poor Franklin and his crew?

. . . Can they have survived? No man can answer with certainty; but no man without presumption can answer in the negative.

If, four months ago,—surrounded by darkness and bowed down by disease,—I had been asked the question, I would have turned toward the black hills and the frozen sea, and responded in sympathy with them, "No." But with the return of light a savage people come down upon us, destitute of any but the rudest appliances of the chase, who were fattening on the most wholesome diet of the region, only forty miles from our anchorage, while I was denouncing its scarcity.

For Franklin, every thing depends upon locality: but, from what I can see

of Arctic exploration thus far, it would be hard to find a circle of fifty miles' diameter entirely destitute of animal resources. The most solid winter-ice is open here and there in pools and patches worn by currents and tides. Such were the open spaces that Parry found in Wellington Channel; such are the stream-holes (stromhols) of the Greenland coast, the polynia of the Russians; and such we have ourselves found in the most rigorous cold of all.

To these spots, the seal, walrus, and the early birds crowd in numbers. One which kept open, as we find from the Esquimaux, at Littleton Island, only forty miles from us, sustained three families last winter until the opening of the north water. Now, if we have been entirely supported for the past three weeks by the hunting of a single man,—seal-meat alone being plentiful enough to subsist us till we turn homeward,—certainly a party of tolerably skilful hunters might lay up an abundant stock for the winter. As it is, we are making caches of meat under the snow, to prevent its spoiling on our hands, in the very spot which a few days ago I described as a Sahara. And, indeed, it was so for nine whole months, when this flood of animal life burst upon us like fountains of water and pastures and date-trees in a southern desert.

I have undergone one change in opinion. It is of the ability of Europeans or Americans to inure themselves to an ultra-Arctic climate. God forbid, indeed, that civilized man should be exposed for successive years to this blighting darkness! But around the Arctic circle, even as high as 72°, where cold and cold only is to be encountered, men may be acclimatized, for there is light enough for out-door labor.

Of the one hundred and thirty-six picked men of Sir John Franklin in 1846, Northern Orkney men, Greenland whalers, so many young and hardy constitutions, with so much intelligent experience to guide them, I cannot realize that some may not yet be alive; that some small squad or squads, aided or not aided by the Esquimaux of the expedition, may not have found a hunting-ground, and laid up from summer to summer enough of fuel and food and seal-skins to brave three or even four more winters in succession.

I speak of the miracle of this bountiful fair season. I could hardly have been much more surprised if these black rocks, instead of sending out upon our solitude the late inroad of yelling Esquimaux, had sent us naturalized Saxons. Two of our party at first fancied they were such.

The mysterious compensations by which we adapt ourselves to climate are more striking here than in the tropics. In the Polar zone the assault is immediate and sudden, and, unlike the insidious fatality of hot countries, produces its results rapidly. It requires hardly a single winter to tell who are to be the

heat-making and acclimatized men. Petersen, for instance, who has resided for two years at Upernavik, seldom enters a room with a fire. Another of our party, George Riley, with a vigorous constitution, established habits of free exposure, and active cheerful temperament, has so inured himself to the cold, that he sleeps on our sledge-journeys without a blanket or any other covering than his walking-suit, while the outside temperature is 30° below zero. The half-breeds of the coast rival the Esquimaux in their powers of endurance.

There must be many such men with Franklin. The North British sailors of the Greenland seal and whale fisheries I look upon as inferior to none in capacity to resist the Arctic climates.

My mind never realizes the complete catastrophe, the destruction of all Franklin's crews. I picture them to myself broken into detachments, and my mind fixes itself on one little group of some thirty, who have found the open spot of some tidal eddy, and under the teachings of an Esquimaux or perhaps one of their own Greenland whalers, have set bravely to work, and trapped the fox, speared the bear, and killed the seal and walrus and whale. I think of them ever with hope. I sicken not to be able to reach them.

It is a year ago to-day since we left New York. I am not as sanguine as I was then: time and experience have chastened me. There is every thing about me to check enthusiasm and moderate hope. I am here in forced inaction, a broken-down man, oppressed by cares, with many dangers before me, and still under the shadow of a hard wearing winter, which has crushed two of my best associates. Here on the spot, after two unavailing expeditions of search, I hold my opinions unchanged; and I record them as a matter of duty upon a manuscript which may speak the truth when I can do so no longer.

MURDER IN THE ARCTIC?

from *Weird and Tragic Shores* (1971)

Chauncey Loomis

In the 1860s, Charles Francis Hall, a Cincinnati businessman-turned-explorer, made two voyages to the Arctic in search of Franklin expedition survivors. Not surprisingly, he found none; nevertheless, his exploits made Hall a popular hero. In 1871, Hall set out on a government-sponsored voyage to the North Pole and within months had died mysteriously. Nearly a hundred years later, his biographer, Chauncey Loomis, traveled to Greenland and exhumed his corpse in the hope of clarifying the circumstances of Hall's death.

For five years after the *Polaris* weighed anchor and steamed through the ice of Thank God Harbor out into Hall Basin, Charles Francis Hall's grave was undisturbed by any human. Eskimos had once hunted the area, but the rings of stone that marked their camp sites were paleolithic; hunting parties had not ventured so far north for hundreds of years. Wind-driven snow and silt blasted the headboard of the grave, but it remained upright, and Hubbard Chester's deep-cut inscription remained sharp and clear. Lemmings burrowed into the mound of the grave, and foxes pawed at its surface, but the coffin beneath was untouched, and the ground willow above remained rooted among the rocks.

In May 1876 Hall's grave had its first human visitors since the departure of the *Polaris.* Members of the British North Polar Expedition led by Captain George Nares arrived with a brass tablet that they had brought from London, knowing they would pass by the gravesite. The tablet was inscribed:

SACRED TO THE MEMORY OF
CAPTAIN C. F. HALL
OF THE U.S. SHIP POLARIS,

WHO SACRIFICED HIS LIFE IN THE ADVANCEMENT OF
SCIENCE
ON NOV^{r} 8th 1871

—

THIS TABLET HAS BEEN ERECTED
BY THE BRITISH POLAR EXPEDITION OF 1875
WHO FOLLOWING IN HIS FOOTSTEPS HAVE PROFITED BY
HIS EXPERIENCE

While twenty-five members of the expedition stood solemnly by, an American flag was hoisted and the tablet was erected at the foot of the grave. The Nares Expedition, like the Polaris Expedition, was not destined to reach the Pole; not long after the ceremony, two of its men, dead of scurvy, were buried only a few hundred yards away from Hall.

Six years later, the grave was visited again. The Greely Expedition, spending the winter thirty miles across Hall Basin at Lady Franklin Bay, came to check on supplies that the Polaris Expedition had cached and to see what was by then known as Hall's Rest. Sergeant William Cross, while rummaging in the wreckage of Bessels's observatory, which had been crudely dismantled before the *Polaris* left, carved his name on one of the boards that lay scattered about. A year later Cross was dead, the first of nineteen men to die in the terrible ordeal of the Greely Expedition. Between 1898 and 1909 Robert Peary passed Thank God Harbor several times aboard the *Roosevelt,* but, with a singlemindedness Hall himself would have admired, did not take time to go ashore. Knud Rasmussen, on his Thule Expedition, arrived in 1917. He found the original headboard lying face down on the ground, perhaps cuffed by the same bear that had bitten deep into the posts supporting the Nares tablet; Rasmussen could plainly see the marks of the animal's teeth in the wood. After Rasmussen's departure, forty years passed before Hall's grave was visited again. In 1958 an American team led by geologist William Davies and assisted by Danish explorer Count Eigel Knuth landed from the icebreaker *Atka,* the first ship to anchor in Thank God Harbor since the *Polaris.* The purpose of "Operation Groundhog" was to locate ice-free aircraft-landing sites as emergency alternatives to the bases at Alert and Thule. For a few weeks the sound of a Jeep was heard on Polaris Promontory. Then the area was returned to its accustomed silence. That silence was broken very briefly a few years later when British geologist Peter Dawes spent a few days in the area.

In August 1968 I arrived at Hall's Rest with three companions. Doctor Franklin Paddock, William Barrett, Thomas Gignoux, and I were flown from Resolute, far to the south in the Arctic Archipelago, by one of Canada's finest bush pilots, W. W. Phipps. The day we arrived was clear; as Weldy Phipps flew his Single Otter low across the hills that border the plain of Polaris Promontory, we could see ahead the deep blue of Hall Basin, the shoreline of Thank God Harbor, and, as we lost altitude, the wreckage of Bessels's observatory close to the beach. We circled, looking for the grave. We could not see it, but knew that it was near the observatory, so Weldy landed on a smooth stretch of plain a mile south of the wreckage. After we had unloaded our equipment, Weldy took off immediately, leaving us standing alone on the plain, dazed in sudden awareness of how isolated we were. He was due to return in two weeks, after he fulfilled some other contracts.

In the course of preparing this biography I had read the government's book on the Polaris Expedition, the journals of its men, the ship's log, the official dispatches, the transcript of the Department of the Navy's inquiry, and masses of other material. My conclusion was, not that Hall certainly had been murdered, not even that he *probably* had been murdered, but only that murder was at least possible and plausible. The conclusion of the Board of Inquiry that he had died of "natural causes, viz, apoplexy," also was possible and plausible, but it had been reached hastily and only by ignoring much of the evidence that the Board itself had wheedled out of witnesses. Secretary Robeson had been under considerable pressure to end investigation; scandal was in the making. That the government was eager to play down the ugly aspects of the affair is indicated by the official book of the expedition, *Narrative of the North Polar Expedition,* written by Rear Admiral C. H. Davis a year after the inquiry. Davis gave the impression that the expedition had been a Boy Scout Jamboree—a bit rough, of course, but enlivened by good cheer and boyish high jinks. The original source materials that Davis had used and distorted show how false that impression was.

I had applied to Denmark's Ministry for Greenland for a permit to travel to Polaris Promontory; arguing that if the case were recent, a court presented with the evidence would order an autopsy, I requested permission to disinter Hall's body and have Frank Paddock perform an autopsy on it. Given the high latitude of its burial, there was a good chance that the body would be well preserved. Approval of my application came only after many letters and finally a trip to Copenhagen, where I met Count Eigel Knuth. An archeologist and old-time sledge traveler, Knuth was one of the last men to have seen

the grave. He also was an adviser to the Ministry on proposed projects in Greenland, and without his agreement I would have no chance of receiving permission. At first it appeared that he would not agree. Hall's grave, he said, was a hallowed place; its remoteness intensified the sense of mystery and beauty associated with any lone grave. The idea of having it disturbed repelled him. After I assured him that I would leave the grave in the condition in which I found it, however, he finally approved.

On the day we arrived at Polaris Promontory we set up camp near the place where Weldy had landed us, about a mile south of the observatory and the grave. We decided not to begin the exhumation until the next day, but after the camp was completed we walked to the gravesite. We could see the Nares tablet first, some distance away across the stony flats. As we approached it, we could see the shape of the mound, covered with large rocks; then a crowbar strangely jutting from the head of the grave; then, finally, Chester's headboard lying face down in the dirt. Change is slow in the cold dry air of the High Arctic; the tablet shone as if it had just been taken from a furnace, and the willow still grew on the mound. Under the benign blue of the sky that day, the place was peaceful—and profoundly still.

We wandered on to the observatory. Here was the litter of man, which is widely strewn throughout the Arctic, all the more noticeable because of the vast inhuman spaces around it. The building no longer stood; its siding lay broken, scattered around its floor as if it had burst from within. Rusted cans, brass nails, iron stoves, a huge davit, an ice saw, shattered glass, and pieces of sailcloth spread out from the observatory like a cancerous growth. Throughout the area were the bowling-ball shapes of ice grenades, packed with black powder that still could explode. While poking through the rubble Tom Gignoux turned over a board, and there was Sergeant Cross's name, carved more than seventy years before when *he* had been doing the poking. Bill Barrett found a broken blue bottle on which the word POISON still could be seen molded into the glass. After momentary, laughing excitement, we realized that it was of little significance: bottles marked POISON could, after all, be contained in a scientific observatory without ominous implications. On the beach I found a Danish shotgun shell, perhaps discharged by Eigel Knuth in 1958. Such sites in the Arctic usually contain layers of history: a hundred yards away were the graves of the two Nares sailors, and not far from them was a paleolithic Eskimo tent-ring.

During the night, under the unsetting sun, the weather changed. When we set out early in the morning to do the job for which we had come, the sky was

suitably lowering, the land suitably bleak. The day before had been too bright for such a morbid piece of work.

For a year I had wondered how I would feel when the coffin was opened. Hall might well have become a skeleton—but in the Arctic air, lying on the permanent frost that had prevented his grave diggers from digging deep, he might have been perfectly preserved. It was impossible to know what was in the coffin, and much as I dreaded finding only a skeleton from which nothing could be proved, I also dreaded finding the man himself, just as he had been. Having spent three years violating his mind by reading his private journals, now I was going to violate his body. I had been haunted by a vision of a rather offended face peering out of the coffin, a face asking, "Is there no limit to what a biographer will do?"

While Frank Paddock, Bill Barrett, and I stood nervously by, Tom Gignoux, not long back from a tour of duty in Vietnam as a Marine, did most of the digging. All of us wanted to be properly solemn, but our nerves short-circuited our sense of awe, and we found ourselves making absurd jokes. As Tom scraped earth off the long coffin lid, revealing pine that was still pale and fresh, Frank looked down at it and said cheerfully, "They didn't build it for the short Hall, did they?" We laughed immoderately. We stopped laughing a minute later when we caught a whiff of decay from within the coffin. During the next ten minutes, while Tom pried carefully at the lid, we stood silent. A piece of the lid broke off, and inside we could see a flag—part of the field of stars—and ice.

I removed the lid after Tom had done all the work, and we stood by the edge of the grave looking down. The body was completely shrouded in a flag. From the waist down, it was covered by opaque ice, but at the base of the coffin a pair of stockinged feet stuck abruptly through it. The front of the torso was clear of the ice, but we could see that its back was frozen into the coffin.

Frank carefully peeled the flag back from the face. It was not the face of an individual, but neither was it yet a skull. There were still flesh, a beard, hair on the head, but the eye sockets were empty, the nose was almost gone, and the mouth was pulled into a smile that a few years hence will become the grin of a death's head. The skin, tanned by time and stained by the flag, was tightening on the skull. He was in a strangely beautiful phase in the process of dust returning to dust. The brown skin, mottled by blue stain and textured by the flag that had pressed against it for almost a hundred years, made him somehow abstract—an icon, or a Rouault portrait.

The autopsy took about three hours. We decided not to try to remove the coffin from the grave or the body from the coffin, embedded as they were in ice. Frank Paddock had to straddle the coffin and lean over to do his work, an agonizing posture to hold for that length of time.

It was very discouraging. At first we thought that the body, still well fleshed, was perfectly preserved, but Frank's scalpel revealed that the internal organs were almost entirely gone, melded into the surrounding flesh. Frank persisted in a meticulous search, but found little that held any hope for analysis. A fingernail and some hair, to be tested for arsenic, were the best samples we had. At last, exhausted, Frank gave up. We put the lid back onto the coffin, and Tom Gignoux shoveled earth back into the grave. After we had piled the rocks back onto the mound, I shoved the strange crowbar into the earth where it had been; later I was to be thankful that I remembered to do it. When we left, Hall's Rest appeared the same as it had been, with only one change that disturbed me: the ground willow planted by the men of the *Polaris* was no longer rooted amid the rocks.

We had almost two weeks to wait before Weldy was to return. We whiled away the time roaming Polaris Promontory, realizing only later that most of our roaming was to the south—the grave lay to the north, and we tended to avoid it. The beach made the best walking. Inland, the shale plain stretched out forty miles deep, depressing in its lifelessness, but along the beach there were life, movement, and sound. Sanderlings, sandpipers, and plovers picked at the waterline; fulmars flew offshore. Occasionally we would disturb nesting Arctic terns, to be delighted by their wheeling and darting attacks on us, their excited and exciting screams. One day when a high wind was blowing on the plain, I found a group of clucking ptarmigans strutting down the protected beach, ridiculous birds having a ladies' club meeting. Our rations were meat bars and dried potatoes; the Danes had forbidden us to live off the land, and only great self-discipline prevented mayhem on the beach that day. Every morning we found the tracks of a fox along the edge of the water, challenging us to catch a glimpse of him. Day after day the little creature eluded us; then one evening while I was sitting quietly, hoping to see him on his nightly route, out of the corner of my eye I saw something move behind me. Foolishly, I jumped up to look. The fox had been stalking me while I was waiting for him. He moved away, not running or even trotting, but keeping his dignity in a stately pace, pretending not to be frightened by what must have been the only human being he had ever seen.

The sky and the light constantly change in the Arctic, because weather systems move rapidly there. Leaden clouds would settle, wind would blow, snow would fall; in a few hours the sky would clear to a deep blue and the wind would calm; then there would be a show of mare's tails and mackerel skies, forecasting another storm; and the cycle would begin again. Local fogs inexplicably would blow in, blow out; we would be walking down the beach in clear air, able to see the mountains of Ellesmere Island thirty miles across Hall Basin, when suddenly we would be plunged into clammy murk, able to see only the ghostly shapes of nearby icebergs.

Hall Basin was clear of ice the day we arrived, but two days later a south wind drove ice up from Kennedy Channel—both floe ice and bergs that had been spawned from glaciers to the south, especially the great Humboldt. Looking at and listening to ice was one of our best diversions. If we stood still and stared out across the basin, we could see it move with the currents, very slowly, very steadily. Many of the small bergs along the beach had eroded into fantastic shapes that changed as we saw them from different angles; some were so smooth that they appeared machine-tooled, others rough-textured; some were in strange animal and birdlike forms, others almost geometrically round, square, trapezoidal. As delightful as the sights of the ice were its sounds. Along the beach, we could hear many: the one that the wind makes when it blows uninterrupted by trees or grass, the lapping of the water on the sand, the cry of birds. But the sounds I remember best were of water steadily dripping from thawing icebergs, and the occasional crack and rumble of big ice breaking out in the bay.

Our two weeks of roaming were well spent. They were lifegiving, images of the beach helping to purge images of the grave from our minds. After those two weeks, I also better understood the man who lay in the grave, and others like him who have felt impelled to travel to the Arctic, yearning for its cold beauty.

*

Weldy picked us up on schedule, and a few days later we were back in the United States. After consultation with specialists in pathology and toxicology, Frank Paddock sent the fingernail and the hair to Toronto's Centre of Forensic Sciences, where they were given a neutron-activation test, a highly sophisticated method of analyzing tiny amounts of material. For some reason I was

not optimistic about our chances of receiving any significant information from the tests, so it came as a surprise when the Centre reported that they had revealed "an intake of considerable amounts of arsenic by C. F. Hall in the last two weeks of his life."

The fingernail had provided the best evidence. Doctor A. K. Perkons of the Centre had sliced it into small segments, working from tip to base, then submitted the segments to the neutron-activation test. The "read-back" from the neutron bombardment indicated increasing amounts of arsenic in the base segments. At the tip, the fingernail contained 24.6 parts per million of arsenic—at the base it contained 76.7 ppm. Assuming a normal growth rate of 0.7 mm a week, Doctor Perkons concluded that the large jump in Hall's body burden of arsenic occurred in the last two weeks of his life. The fact that his arsenic content was high even before the jump could be explained in several ways. Arsenic was often used medicinally in the nineteenth century, and it also was used in hair-dressings, so many persons then had a relatively high content. The normal content today is only 1.5–6.0 ppm. Also, the soil near the grave contained fairly large amounts of arsenic (22.0 ppm); according to Perkons, some could have "migrated" from the soil to the body. "However," Perkons went on in the report, "such migration would not explain the differentially increased arsenic in the sections of both hair and nails toward the root end."

We checked with other authorities, all of whom accepted the accuracy of the Centre's report and agreed with its conclusion: Charles Francis Hall had received toxic amounts of arsenic during the last two weeks of his life.

What conclusions can be drawn? Excited by the report, after I received it I studied my material on the Polaris Expedition in light of the new information, testing various explanations for the arsenic contained in Hall's body. The trouble, as I soon discovered, was that several explanations were possible.

The following list of symptoms of acute arsenic poisoning is quoted from Gleason, Gosselin, Hodge, and Smith, *Clinical Toxicology of Commercial Products* (Baltimore, 1969):

1. Symptoms usually appear ½ to 1 hour after ingestion.
2. Sweetish metallic taste; garlicky odor of breath and stools.
3. Constriction in the throat and difficulty in swallowing. Burning and colicky pains in esophagus, stomach and bowel.
4. Vomiting and profuse painful diarrhea.

5. Dehydration with intense thirst and muscular cramps.
6. Cyanosis, feeble pulse, and cold extremities.
7. Vertigo, frontal headache. In some cases ("cerebral type") vertigo, stupor, delirium, and even mania develop.
8. Syncope, coma, occasionally convulsions, general paralysis, and death.
9. Various skin eruptions, more often as a late manifestation.

As one looks down the list, one sees many of Hall's symptoms: the initial gastrointestinal troubles, the difficulty in swallowing, the dehydration, the stupor, delirium, and mania—even the late manifestations of skin eruption noticed by Chester the day before Hall died. Given the results of the neutron-activation test, this should be no surprise. There is no doubt that Hall received a large amount of arsenic during the period that he showed these symptoms. The question is, How did he receive it?

Arsenic would have been available aboard the ship; in the form of arsenious acid, it was commonly used as a medicine in the nineteenth century. "Arsenious acid," comments the *Dispensatory of the United States* of 1875 in one of its longest entries, "has been exhibited in a great variety of diseases." It was used in the treatment of headaches, ulcers, cancer, gout, chorea, syphilis, even snakebite. In the form of "Fowler's Solution," it was a very popular remedy for fever and for various skin diseases. It was a standard part of any sizable medical kit, and obviously the Polaris Expedition had a large medical kit.

Hall may have dosed himself with it. With such a man, suicide is almost inconceivable, but it would not have to have been suicide; he might have died a victim of the capacity for suspicion that had so often erupted in his life. Platt Evens of the percussion-seal-press lawsuit, William Pomeroy, William Parker Snow, Isaac Hayes, Sidney Budington, Patrick Coleman, and nameless others had aroused his fear of jeopardy and his wrathful self-righteousness. From the beginning, he did not like or trust Emil Bessels. In his sickness, which may indeed have been a stroke, might not he have treated himself rather than put his faith in the "little German dancing master"? Even during the period when he allowed Bessels to treat him, he may have been taking medicine on his own. Bessels testified that Hall had a personal medical kit, containing among other things "patent medicines." Some nineteenth-century patent medicines contained arsenic; although its quantity was not great in any of the medicines that have been tested, Hall might have dosed himself heavily. Or perhaps he gained access to Bessels's kit and took arsenious acid from

it. Tookoolito would have helped him do such a thing, and, quiet Eskimo woman that she was, would not have said a word about it later.

But murder also is possible. The coffee that Hall drank when he boarded the ship after his sledge journey could have been poisoned. Although arsenic is usually tasteless, it can leave a "sweetish metallic taste." Hall complained to Tookoolito about the coffee. "He said the coffee made him sick," she testified. "Too sweet for him." About one half hour after he drank it, he felt pains in his stomach and vomited, symptoms that suggest poisoning.

But the pain and the vomiting could have been caused by a stroke, as could many of the other symptoms listed in *Clinical Toxicology.* If Bessels was telling the truth when he said that Hall also suffered partial paralysis, then a stroke is as satisfactory an explanation of many of them as arsenic poisoning. There is also a possibility that Hall suffered both a stroke and arsenic poisoning. The initial attack could have been a stroke, then the arsenic could have been administered later, during the two weeks of his illness. It should be repeated: there is no doubt that the arsenic was administered. The question remains, How was it administered—and by whom?

The persons who had the most access to Hall during his illness were Sidney Budington, Tookoolito, Ebierbing, Hubbard Chester, William Morton, and Emil Bessels. Others could see him, especially those who shared the cabin with him, but these are the persons who were often with him, treating him or feeding him.

Budington, undergoing a psychological ordeal, drinking heavily, apparently afraid of being so far north, is a suspect, but he actually had less access to Hall than the others. Apparently he seldom approached the sick man or did anything for him. Tookoolito, Ebierbing, Chester, and Morton frequently attended, nursed and fed him, but there is no indication of any possible motive for their doing him injury. Like Budington, they cannot be entirely dismissed as suspects, but they are highly unlikely ones.

If Hall was murdered, Emil Bessels is the prime suspect. A trained scientist, he had the knowledge, and, as ship's surgeon, the material needed to administer arsenic. He had access to Hall much of the time—and when Hall refused Bessels access, his condition improved. Joseph Mauch made a note in his journal on November 1, several days after Hall first refused treatment by Bessels: "Capt. Hall is much better this morning—for the last 2 days he has taken no medicine & today his health is greatly improved, although yet very weak."

When Bessels treated Hall, he gave him some medicines orally, especially cathartics; arsenic could have been mixed with such medicines. He also gave him injections of what he said was quinine; arsenic also could have been injected, as it sometimes was in the treatment of cancers. When Bryan saw Bessels prepare the injections, the process involved the heating of "little white crystals," precisely the way that quinine was usually prepared. But arsenic could be in the form of a white powder, easily mistaken for crystals or mixed with them, and it, too, can be prepared by such heating.

When one considers Bessels as a possible murderer, one notes little things in the transcript of the inquiry that are subject to various interpretations. There is the uncertainty about whether he was in the observatory while Hall was drinking the coffee, as he said he was, or aboard the ship, as Morton and Mauch believed that he might have been. There is his refusal to administer an emetic when Hall was first taken ill; if indeed Hall had suffered a stroke, an emetic would have been dangerous—but an emetic also might have emptied his stomach of poison. There is the persistence of his quinine treatment when Hall's fever had been allayed. And one night, according to Budington's testimony, Bessels came to him complaining that Hall was refusing to take any medicine. Budington volunteered to take the medicine first in front of Hall, like a parent with a child. Bessels refused to let him do so. Small things, straws in the wind.

Bessels had the opportunity, the skill, and probably the material, but why would he do it? He had no apparent rational motive; he would gain nothing concrete by Hall's death. Unlike Budington, for example, he was not afraid of their situation and did not want to retreat south, and therefore Hall's passion to go north was not a threat to him. In fact, Joseph Mauch and Henry Hobby testified that when the *Polaris* was run aground at Etah, Bessels secretively tried to bribe some of the men to return north with him—an ambitious act, and perhaps ambition could be motive enough. With Hall's death, the command actually fell to Budington, but Bessels had more power and independence because Budington was a far weaker man.

But ambition for what? To make major scientific and geographical discoveries and be given full, sole credit for them? This does not seem motive enough for murder. Here we enter the underground streams of mind, the darkness that the Board would not probe. When one of its members asked George Tyson if he thought that "there was any difficulty between Captain Hall and any of the scientific party that would be an inducement for them to do anything toward injuring him," Tyson replied firmly, "No sir." Then

he paused and said, "Unless a man were a monster, he could not do any such thing as that." The Board, not wanting to consider the possibility of monstrosity, moved on to other matters, but perhaps the truth lay precisely in monstrosity.

Joseph Henry had warned Hall that Bessels was "a sensitive man." He must have been very sensitive to justify Henry's making such a comment in a letter that is remarkable for its dry, official tone, and Bessels's behavior on the Polaris Expedition, his quarrels with both Hall and Budington, indeed suggest that he was at least difficult to deal with. Little is known about his later career, but enough is known to indicate that he remained difficult, perhaps abnormally so. For more than ten years he maintained a connection with the Smithsonian. Part of that time was spent compiling the scientific results of the expedition, and he sometimes received needling letters from Baird suggesting that he hasten his work. There is evidence that the Smithsonian was eager to get rid of him.

One reason was his involvement in a controversy in 1880. An International Polar Year was planned for 1882–3, and in 1880 the scientific community in the United States was much concerned with what the country should do about it. Among those who spoke out was Captain Henry Howgate, who had some rather far-fetched ideas about colonizing the Arctic. On February 16, 1880, an interview on the subject with Emil Bessels appeared in the *New York Herald,* an interview that reveals much about the man. The reporter devoted some time to the difficulty of finding Bessels's office in the Smithsonian: "To discover this apartment without a guide would be almost as great an act as reaching the North Pole itself." Then he commented: "When the portals are entered, passing under the heavy folds of green drapery which nearly hide the entrance, the visitor would suppose he had been suddenly translated into the retreat of Faustus." As Bessels was interviewed, he indeed acted like Faustus in his worst manifestations—self-assured to the point of arrogance, scornful of others, convinced that all knowledge was his. He laid down for the reporter what the United States should and should not do during the Polar Year. That much of what he said was correct does not mitigate his irritating condescension:

> "What do you think of the plan originated by Captain Howgate?"
>
> "Howgate's plan? Why, Captain Howgate did not originate any plan whatever. He merely appropriated the ideas of Dr. Hayes and probably those of Lieutenant Weyprecht. As far as these are concerned Captain

Howgate is all right; but with regard to the rest—well, I would prefer to talk about something more rational."

Bessels read aloud a passage of Howgate's writing about the possibility that a superior Eskimo culture already existed somewhere near the Pole; then he made passing reference to the Polaris Expedition: "This even beats Dr. Newman, who wrote for the Polaris Expedition a prayer to be read at the North Pole, consecrating the Pole to liberty, education and religion. I am only astonished that Captain Howgate did not quote him as an Arctic authority." Bessels was hardly being tactful, as Newman was still Chaplain of the Senate.

When the reporter asked him another question, he said: "Let me light a fresh cigar before I answer this question." One can see him, small, natty, his eyes bright with self-assurance, lighting the cigar and leaning back to say, "It amused me to find that a man writing such bombast [Howgate] should have the insolence to point out what caused other expeditions to fail in reaching the Pole, and in what manner they were mismanaged."

When Bessels was asked about an Arctic expedition that Howgate had organized not long before, he snapped, "The sole aim of the thing was to gain cheap reputation and to lay a snare for Congress to appropriate the means for a *real* Arctic expedition." The interviewer asked Bessels why he thought Howgate's expedition had been a failure. The doctor referred to the scientists who had accompanied the expedition without pay "for the mere love of science." Possibly he was thinking of his relationship with Hall when he said, "They had to submit to the orders of an incompetent, harsh skipper, who most seriously interfered with their duties." It should be noted that Howgate was not the skipper—by a wild coincidence, the skipper was George Tyson.

Bessels was right to distrust Captain Howgate. Some years later it was proved that he had taken advantage of his position in the Signal Corps to swindle large sums of money. But the intemperance of Bessels's attack might be explained by something other than his belief that Howgate was a fraud. Henry Howgate had been a member of the Board of Inquiry that investigated the Polaris Expedition. There is no indication that he said or did anything during the inquiry to earn Bessels's enmity, but there is at least a possibility that the doctor held a deep-seated grudge against him for his membership on the Board.

Whatever the cause of Bessels's intemperance, it brought the wrath of Spencer Baird down on his head. Baird wrote him an icy letter, castigating him for his loose mouth. Bessels remained at the Smithsonian for a few more

years, but apparently under a cloud. Past trouble obviously lies behind the terse note that he received from Baird's secretary in 1883:

> Dear Doctor:
>
> We need immediate possession of the room now occupied by you near the north entrance, as we find it necessary to make improved toilet arrangements for visitors. Please therefore remove your property and greatly oblige.
>
> Yours truly,
> Wm. J. Rhees

The tone indicates that this was not the first attempt to dislodge him, but it apparently was the last; his Smithsonian salary soon was stopped, and, presumably, the "retreat of Faustus" became a toilet. Bessels soon after returned to Germany. He died there in 1888—ironically, of apoplexy.

A difficult man—but a monster? Bessels did not seem a monster in the ordinary course of his life, but perhaps he had monstrosity latent within him. The close atmosphere of a wintered-in ship was a test of anyone's mind. Ambitions, dislikes, abnormalities of any kind could be unbearably magnified and intensified, as the whole history of Arctic exploration reveals. On the Polaris Expedition, Budington and others drank, Tyson brooded, the carpenter went insane, young Joseph Mauch, Noah Hayes, and probably most of the men aboard drifted miserably toward paranoia. Hayes's assertion at the conclusion of his journal rings of hard experience and honest self-awareness: "I believe that no man can retain the use of his faculties during one long night to such a degree as to be morally responsible." Bessels scorned Hall, as he apparently scorned many men. Hall was an uneducated boor, but he, Emil Bessels of Heidelberg and Jena, had to serve under him and take his orders. Their relations had been strained at the outset, and Bessels faced at least another year, probably another two years, on that tiny ship, suffering the humiliation of an arrogant man in a subservient position.

Perhaps Bessels murdered Hall. Perhaps. The only certain truth that can be found in this case is a knowledge of the inevitable and final elusiveness of the past. What happened aboard the USS *Polaris* between October 24 and November 8, 1871, can never be entirely known. What went on in the minds of Hall, Bessels, and the others aboard that ship, and what they did furtively on their own, is done, gone, past. The questions that the Board of Inquiry did not ask can be asked today, but many of them cannot be answered.

One way or the other, Charles Francis Hall died, as his friend Penn Clarke said he would, a victim of his own zeal. If a stroke was the primary cause, then he drove himself into it, trying to reach the North Pole at the age of fifty. If Bessels murdered him, then his zeal, which made him strong, also made him an unbearable threat to the doctor's ambitions or a hateful object of the doctor's fears. If he poisoned himself, then it was because the zeal that had made him fiercely independent had also made him fiercely suspicious. The dark side of his independence was his distrust of anyone who seemed in any way to threaten him, his integrity, his desires. Independence is often loneliness, and Hall was a lonely man. Treating himself rather than trusting someone else for treatment would have been a characteristic, almost a symbolic, act.

Will power, energy, and independence are the qualities that made him and perhaps broke him. Nineteenth-century America was filled with the rhetoric of will power, energy, and independence but, as scholars, beginning with Frederick Jackson Turner, have shown, it was an age that increasingly controlled energy and individual will, channeling them into the communal and the cooperative. For better or worse, Hall was the real thing. Pious and patriotic as he was, he had something of the mountain man in him. When he drifted west from New Hampshire, he was looking for wilderness, though he may not have known it. When Cincinnati did not give him what he wanted, then he went north instead of west—not drifting then, but driving with relentless energy and concentrated purpose. In the North, too, he found that full independence was not possible, not among the Eskimos, not aboard the *Polaris.* Hall's voyages to the Arctic were not merely geographical explorations. They were a quest for the kind of independence that was gone from American life—or, closer to the truth, the kind of independence that never existed except in the minds of dreamers like Hall.

*

Just before we left Polaris Promontory after the exhumation, I returned alone to the grave. I had to fulfill Arctic ritual and bury a cannister there with an account of what we had done. After doing the job, I took a last look at the grave, hoping to feel some of the things that I believed I should feel and had not felt during the day of the autopsy. The biographer and the detective still dominated; in spite of myself, all I did was puzzle about the crowbar at the

head of the grave—the crowbar that we had carefully put back in place. After we returned home, while rereading Noah Hayes's journal, I noticed something I had not noticed before; the Indiana farmboy solved at least one mystery and deserves the last word about his hero. The night they buried Hall, he wrote, was too cold, too miserable for them to mount a headboard, so they jammed the crowbar into the mound. "A fit type of his will," wrote Hayes, "an iron monument marks his tomb."

SEE THE ESQUIMAUX

from *The Voyage of the Narwhal* (1998)

Andrea Barrett

The Voyage of the Narwhal *is Andrea Barrett's 1998 fictional version of a Franklin rescue expedition. The captain of the* Narwhal, *Zechariah "Zeke" Voorhees, gets separated from his men and is presumed dead; months later he returns home with the Eskimos, Annie and Tom, who have saved him. In this passage, two old friends of Voorhees, Alexandra Copeland and Erasmus Wells, attend his public exhibition of the rescuers.*

HERE IN THE THEATER'S gallery, near the prostitutes scattered like iridescent fish through the shoals of dark-clothed men, Alexandra felt drab in her brown silk dress. Two seats down from her, a woman in a chartreuse gown with lemon-trimmed flounces was striking a deal with a pleasant-looking man. They would meet on the landing, Alexandra heard them agree. Directly after the lecture. The man's voice dropped and the woman shook her head, shivering the egret feathers woven into her hair. "Twenty dollars," she said. The man nodded and disappeared, leaving Alexandra to marvel at the transaction.

"There must be a thousand people," Erasmus said, scanning the crowd. "Maybe more."

"It's frightening," she said. "How good Zeke is at promoting himself."

All around the city, on lampposts and tavern doors, in merchants' windows and omnibuses, posters advertised the exhibition. A clumsy woodcut showed Zeke holding a harpoon and Annie a string of fish, Tom peeping out from behind her flared boots. In the background were mountains cut by a fjord, and above those a banner headline: MY LIFE AMONG THE

ESQUIMAUX. A caption touted the remarkable discoveries made by Zechariah Voorhees:

Two Fine Specimens of the Native Tribes!
More Exotic than the Sioux and Fox Indians Exhibited by George Catlin in London and Paris!
See the Esquimaux Demonstrate Their Customs!

Zeke had run a smaller version in the newspaper and mailed invitations to hundreds of his family's friends and business associates—organizing this first exhibition, Alexandra thought, like a military campaign. Ahead of him lay Baltimore, Washington, Richmond, New York, Providence, Albany, Boston.

Erasmus said, "Can you see Lavinia?" and Alexandra, scouting the boxes on the second tier, finally spotted her dead center, flanked by Linnaeus and Humboldt and Zeke's parents and sisters. She was touching her hair then her cheek then her brooch then her nose, turning her head from side to side as if the mood of the entire audience were expressing itself through her. Everyone, Alexandra thought, made nervous by this month's chain of disasters. Across the ocean, off the coast of Ireland, the telegraph cable being laid with such fanfare had broken. Two trains had crashed south of Philadelphia, killing several passengers; last week a steamship on its way to New York from Cuba had sunk. Each of these seemed to heighten the financial panic set off by a bank failure in Ohio. Banks were closing everywhere; the stock exchange was in an uproar. The papers were full of news about bankrupt merchants and brokers. Alexandra's own family, who had no money to lose, hadn't been touched so far, and the engraving firm seemed stable. But Erasmus, whose income came primarily from his father's investments, had suffered some losses. And Zeke's father's firm was in trouble, which suddenly made Zeke's future—and Lavinia's as well—uncertain. Suddenly it mattered what Zeke charged for the exhibition tickets, and how many tickets were sold. The theater was full of people desperate for distraction.

In the glow of the gaslights Zeke strode out in full Esquimaux regalia, adjusted the position of two large crates, and took his place at the podium. The roar of applause was startling, as was the ease with which he spoke. If he had notes, Alexandra couldn't see them. Swiftly, eloquently, he sketched for the audience an outline of the voyage of the *Narwhal*, making of the confused first months a spare, dramatic narrative.

Their first sights of Melville Bay and Lancaster Sound, their encounters with the Netsilik and their retrieval of the Franklin relics; the discovery of the *Resolute* and their stormy passage up Ellesmere until they were frozen in; their long winter and the visit of Ootuniah and his companions; the first trip to Anoatok. No mention, Alexandra noticed, of Dr. Boerhaave's death, nor of the other men who'd died: nor of Erasmus. It was "I" all the time, "I" and "me" and "mine;" occasionally "we" or "my men." No names, only him. Beside her, Erasmus fidgeted.

Twenty minutes, she guessed. Twenty minutes for the part of the voyage involving the crew; then another fifteen for Zeke's solo trip north on foot and his return to the empty ship. "Now," Zeke was saying, "now began the most interesting part of my experience in the arctic. I was all alone, and winter was coming. I had to prepare myself."

From the crates he began to pull things. His hunting rifle, sealskins, a tin of ship's biscuit, a jar of dried peas. His black notebook, the sight of which made Erasmus groan. Into his talk he wove some stray lines from that, and then read aloud the section about the arrival of Annie and Nessark and Marumah. "The *angekok* is the tribe's general counselor and advisor," he explained. "As well as its wizard. His chief job is to determine the reason for any misfortune visiting the tribe—and the *angekok* of Annie's tribe determined that the cause of their children's sickness was me. So was my life changed by a superstition. From the day these people arrived I entered into a new life."

He described the journey to Anoatok and his first days there. Then he said, "But you must meet some of the people among whom I stayed." He stepped back from the podium and whistled.

There was rattling backstage, and the crack of a whip. Two dogs appeared—not his huge black hunting dogs but beagles, ludicrous in their harnesses, gamely trotting side by side. Apparently Zeke would not subject his own pets to this. Behind them they pulled a small sledge on wheels, with Tom crouched on the crossbars and Annie grasping the uprights and waving a little whip. Both Annie and Tom wore fur jackets with the hoods pulled up and shadowing their faces. When the sledge reached the front of the podium, Zeke gave a sharp command that stopped the beagles. They sat, drooling eagerly as Zeke held out bits of biscuit, and then lay down in their traces with their chins on their paws. Their eyes followed Zeke as he moved around the stage, but Annie and Tom stared straight out at the audience, shielding their eyes against the glare.

"These are two of the people who rescued me," Zeke said. "The names they use among us are Annie and Tom."

While they stood still he recited some facts. Annie and Tom belonged to the group of people John Ross had discovered in 1818 and called Arctic Highlanders—there were just a few hundred of them, he said, scattered from Cape York to Etah. Fewer each year; their lives were hard and their children sickened; he feared they were dying out. They moved nomadically throughout the seasons, among clusters of huts a day's journey apart and near good hunting sites. All food was shared among them, as if they were one large family. Because no driftwood reached their isolated shores, they had no bows and arrows, nor kayaks, and in this they differed from the Esquimaux of Boothia and southern Greenland. They'd developed their own ways, substituting bone for wood—bone harpoon shafts and sledge parts and tent poles. "A true sledge," Zeke said, "would have bone crosspieces lashed to the runners with thongs, and ivory strips fastened to the runners." He went on to explain how they subsisted largely on animals from the sea.

"The term 'Esquimaux' is French and means 'raw meat eaters,' " Zeke said. "But there's nothing disgusting in this, the body in that violent climate craves blood and the juices of uncooked food." From the nearest crate he took a paper bundle, which he unwrapped to reveal a Delaware shad. A few strokes of a knife yielded three small squares of flesh. Two he held out to Annie and Tom, keeping the third for himself. The beagles whined. Zeke popped the flesh in his mouth and chewed, while Annie and Tom did the same on either side of him. The audience gasped, and Alexandra could see this pleased Zeke enormously.

"With the help of my two friends," he said, "I would like to demonstrate for you some of the elements of daily life among these remarkable people."

Now Alexandra saw the bulk of what the crates contained. Certainly he hadn't carried all these objects home with him; he must have made some here, with Annie's help and whatever supplies he could find. There was a long-handled net, which Tom seized and carried to the top of one crate. He made darting and swooping motions as Zeke described capturing dovekies. "These arrive by the million," Zeke said. "When the hunter's net is full, he kills each bird by pressing its chest with his fingers, until the heart stops."

A soapstone lamp—where had this come from?—with a wick made from moss; Zeke filled it with whale oil and had Annie light it with a sliver of wood he first lit with a match, telling the audience they must imagine lumps of blubber slowly melting. In the huts, he said, with these lamps giving off heat and light, with food cooking and wet clothes drying and children frolicking, it had been warm no matter what the outside temperature. He brought out

more hides and had Annie demonstrate how the women of her tribe scraped off the inner layers to make the hides pliable. "This crescent-shaped knife is an *ulo*," he said, and Annie sat on her knees with her feet tucked beneath her thighs and the skin spread before her, rubbing it with the blade. Beside Alexandra, Erasmus pressed both hands to his ribs.

"Are you all right?" she said. She couldn't take her eyes from the stage.

"That's exactly the way I soften a dried skin before I mount it," Erasmus said. "I have a drawshave I use like her *ulo*."

Zeke said, "The women chew every inch after it's dried, to make it soft," and Annie put a bit of the hide in her mouth and ground her teeth. "I can't show you the threads, which are made from sinews," he said. "But the needles are kept in these charming cases." Annie held up an ivory cylinder, through which passed a bit of hide bristling with needles.

Zeke took Tom's hand and seized a pair of harpoons; then he and Tom lay down and pretended to be inching up on a seal's blowhole, waiting for the seal to surface. As they mimicked the strike Zeke spoke loudly, a flow of vivid words that had the crowd leaning forward. They were seeing what Zeke wanted them to see, Alexandra thought. Not what was really there: not a rickety makeshift sledge, two floppy-eared beagles, a tired woman and a nervous boy moved like mannequins by the force of Zeke's voice. Not them, or a man needing to make a living, but the arctic in all its mystery: unknown landscapes and animals and another race of people.

Her face was wet; was she weeping? As Zeke's antics continued Alexandra found herself thinking of her parents and the last day she'd seen them. Pulling away from the ferry dock, waving good-bye, sure they'd be reunited in a week. Then the noise, the terrible shocking noise. Great plumes of steam and smoke and cinders spinning down to the water—and her parents, everyone, gone. Simply gone.

She turned to Erasmus, who had his face in his hands. Gently she touched him and said, "You have to look."

He raised his head for a second but then returned his gaze to his shoes. "I won't," he said passionately. "I hate this. All my life the thing I've hated most is being *looked at*. I can't bear it when people stare at me. I know just how she feels, all of us peering down at her. It's disgusting. It's worse than disgusting. People stared at me like this when I returned from the Exploring Expedition, and again when I came back without Zeke. Now we're doing the same thing to her."

Had she known this about him? She looked away from him, back at the

stage; she felt a shameful pleasure, herself, in regarding Annie and Tom. She longed to draw them.

Annie had pushed her hood back from her sweating face, while Tom had stretched out on the sledge and was pulling at one of the beagle's ears. From his crate Zeke took a wooden figure clothed in a miniature jacket and pants. "The children play with dolls," Zeke said. "Just as ours do." Tom released the beagle's ear, seizing the doll and pressing it to his chest. Then Zeke was winding string around Annie's fingers, saying, "Among this tribe, a favorite game with the women and children is called *ajarorpok*, which is much like our child's game of cat's cradle, only more complicated."

He said something to Annie and stepped away. Annie's hands darted like birds and paused, holding up a shapely web. "This represents a caribou," Zeke said.

Alexandra tried to see a creature in the loops and whorls, not knowing that, for Annie, it was as if the stage had suddenly filled with beautiful animals. Not knowing that for Annie this evening moved as if the *angekok* who'd brought Zeke to them had bewitched her, putting her into a trance in which she both was and was not on this stage. The *angekok* had shared with her the secret fire that let him see in the dark, to the heart of things. For her Zeke's bird net wasn't a broomstick and knotted cotton but a narwhal's tusk and plaited sinews; on her fingers she felt the fat she'd scraped from the seal. She was home, and she was also here, doing what she'd been told in a dream to do.

She was to watch these people, ranged in tiers above her, and commit them to memory, so that she could bring a vision of them to her people back home. Their pointed faces and bird-colored garments; the way they gathered in great crowds but didn't touch each other or share their food. Their tools, their cooking implements, their huts that couldn't be moved when the weather changed. In a dream she'd heard her mother's voice, singing the song that had risen from her tribe's first sight of the white men.

Her mother had been a small girl on the summer day when floating islands with white wings had appeared by the narrow edge of ice off Cape York. From the islands hung little boats, which were lowered to the water; these spat out sickly men in blue garments, who couldn't make themselves understood but who offered bits of something that looked like ice, which held the image of human faces; round dry tasteless things to eat; parts of their garments, which weren't made of skins.

"At first," her mother had said, "we thought the spirits of the air had come to us." On the floating island her mother had seen a fat, pink, hairless animal,

a man with eyes concealed behind ovals of unmelting ice, bulky objects on which to sit, something like a frozen arm, with which to hit something like a needle. The two men who'd stepped first on the ice had worn hats shaped like cooking pots. Through them, her people had learned they weren't alone in the world.

Much later, when Annie was grown, she'd had her mother's experience to guide her when the other strangers arrived. Kane and his men had taught Annie to understand their ungainly speech, and Annie had learned that the world was larger than she'd understood, though much of it was unfortunate, even cursed. Elsewhere, these visitors said, were lands with no seals, no walrus, no bears; no sheets of colored light singing across the sky. She couldn't understand how these people survived. They'd been like children, dependent on her tribe for clothes, food, sledges, dogs; surrounded by things which were of no use to them and bereft of women. Like children they gave their names to the landscape, pretending to discover places her people had known for generations.

From them she'd gained words for the visions of her mother's childhood: a country called England and another called America; men called officers; ships, sails, mirrors, biscuits, cloth, pig, eyeglasses, chair. Wood, which came from a giant version of the tiny shrubs they knew. Hammer and nails. Later she'd added the words Zeke had taught her while he lived with them; then the names for the vast array of unfamiliar things she'd encountered here. In the dream her mother had given her this task: to look closely at all around her, and to remember everything. To do this while guarding her son.

Her hands darted and formed another shape, which Zeke claimed represented ponds amid hills but in which she saw her home. She felt the warm liver of the freshly killed seal, she tasted sweet blood in her mouth. In the gaslights she saw the moon and the sun, brother and sister who'd quarreled and now chased each other across the sky. At first her mother had thought the strangers must come from these sources of light. Her hands flew in the air.

"Can you see what she's doing?" Alexandra whispered to Erasmus. "I can't see what she's making."

"I have to go," Erasmus said. "We have to go. Can we go?"

MOUNT HATTERAS

from *The Adventures of Captain Hatteras* (1866)

Jules Verne

Jules Verne wrote The Adventures of Captain Hatteras *at the same time that he was working on* Journey to the Center of the Earth. *Hatteras is an obsessed Arctic explorer who resolves that an Englishman must be the first to reach the North Pole. He sets forth on his boat, the* Forward, *and despite the usual hazards—ice, mutiny, hunger—he reaches his goal, where he finds land or, more specifically, a volcano. The passage below chronicles his decision to touch the pole, even if that means leaping into a flaming crater.*

After [a] substantial conversation, all settled down in the grotto as best they could, and dropped off.

All except for Hatteras. Why did this extraordinary man not sleep?

Had the aim of his life not been accomplished? Had he not accomplished the bold project he held dearest? Why did calm not succeed agitation in that burning soul? Was it not natural that, his projects accomplished, Hatteras would fall into a sort of despondency, that his relaxed nerves would need rest? After his success, it would have even been normal to feel that sadness which always follows satisfied desires.

But no. He still appeared over-excited, more so than ever. However, it was not the idea of returning which made him agitated. Did he want to go even further? Did his travel ambition not have any limit, and did he find the world too small because he had been everywhere on it?

Whatever the reason, he could not sleep. And yet, that first night spent at the Pole of the globe was clear and calm. The island was completely uninhabited. Not a single bird in the burning air, not an animal on the cindery

ground, not a fish in the boiling waters. Only the dull and distant rumbling of the mountain, with its head producing dishevelled plumes of incandescent smoke.

When Bell, Johnson, Altamont, and the doctor awoke, Hatteras was no longer with them. Worried, they left the grotto, and found the captain standing on a rock. His eyes remained immutably fixed on the summit of the volcano. He had his instruments in his hand; he had clearly just measured the mountain's position.

The doctor went up to him and spoke several times before he could interrupt his contemplation. Finally, the captain seemed to understand.

"Off we go!" said the doctor, examining him attentively. "Let's explore everywhere on our island; we're ready for our last excursion."

"The last," said Hatteras with the intonation of people dreaming out loud, "yes, the last. But also," he added very keenly, "the most wonderful!"

While he spoke, he rubbed his hands over his forehead as if to calm the boiling inside.

Altamont, Johnson, and Bell joined them; Hatteras now seemed to emerge from his hallucinatory state.

"My friends," he said emotionally, "thank you for your courage; thank you for your perseverance; thank you for your superhuman efforts, which have enabled us to set foot on this land!"

"Captain," said Johnson, "we only followed orders, and the honour belongs to you."

"No, no," replied Hatteras in a violent outpouring; "to all of you as much as to me! To Altamont and all of us and the doctor as well! Oh, may my heart blow its top in your hands! It can no longer contain its joy and gratitude!"

Hatteras seized the hands of his good companions around him. He came, he went, he was no longer in control of himself.

"We only did our duty as Britons," said Bell.

"And friends," added the doctor.

"Yes, but not everyone was able to do his duty. Some failed! However, those who betrayed should be forgiven, like those who let themselves be dragged into betrayal! Poor men! I pardon them, do you hear, doctor?"

"I do," replied Clawbonny, who was seriously worried at Hatteras's exultation.

"So I don't want them to lose the riches they came so far to earn. No, nothing in my arrangements has changed, and they will be wealthy . . . if ever they get back to Britain!"

It would have been difficult to remain unmoved by the tone Hatteras said these words in.

"But, captain," said Johnson, trying to joke, "you'd think you were making your last will and testament."

"Perhaps I am," replied Hatteras gravely.

"But you still have a good long life of fame before you," continued the old sailor.

"Who knows?" said the captain, followed by a long silence.

The doctor did not dare interpret the meaning of these last words.

But Hatteras needed to ensure he was understood, for he added, in a hurried voice he could hardly control:

"My friends, listen to me. We have done a great deal so far, and yet there remains much to do."

The captain's companions looked at each other with astonishment.

"Remember, we're at the land of the Pole, but not the Pole itself!"

"What do you mean?" said Altamont.

"Could you please explain?" exclaimed the doctor, afraid to guess.

"Yes," said Hatteras forcefully; "I said that a Briton would set foot on the Pole of the globe; I said it, and it shall come to pass."

"What . . . ?" said the doctor.

"We are still forty-five seconds away from that mysterious point," said Hatteras with growing emotion, "and to it I shall go!"

"But it's on top of that volcano!" said the doctor.

"I shall go."

"You can't get there!"

"I shall go."

"It's a wide open flaming crater!"

"I shall go."

The energy and conviction with which Hatteras said these last words cannot be conveyed. His friends were stupefied; they looked in terror at the mountain, which was waving its plume of flames in the air.

The doctor then spoke; he insisted; he pressed Hatteras to renounce his projects; he said everything his heart could conjure up, from humble prayers to friendly threats; but he obtained nothing from the unquiet soul of the captain, caught up in an insanity that could be called "polar madness."

There remained only violent means to prevent this madman, who was heading for a fatal fall. But foreseeing that such means would produce serious disorders, the doctor decided to employ them only as a last resort.

In any case, he hoped that physical impossibilities and impenetrable obstacles would prevent Hatteras from executing his project.

"Since such is the case," he said, "we will go with you."

"Good, but only halfway up the mountain! No further! You need to take back to Britain the copy of the affidavit attesting to our discovery if . . ."

"Nevertheless . . ."

"It's decided!" replied Hatteras in an unshakeable tone. Since the prayers of his friends had not changed matters, the captain remained in command.

The doctor did not try to insist further, and a few moments later the little troop had equipped themselves to face difficulties and, led by Duke, had set off.

The sky was shining brightly. The thermometer read fifty-two (11°C). The air was dramatically imbued with the clarity particular to that high latitude. It was eight in the morning.

Hatteras took the lead with his good dog; Bell, Altamont, the doctor, and Johnson followed close behind.

"I'm afraid," said Johnson.

"There's nothing to fear," replied the doctor. "We're right beside you."

What a remarkable island, and how to depict its special physiognomy, which had the unexpected newness of youth! This volcano did not seem to be very old, and geologists would have concluded that it had been formed at a recent date.

The rocks clung to each other, maintaining themselves only through a miracle of equilibrium. The mountain was, so to speak, merely an agglomeration of stones thrown there. No earth, no moss, not the thinnest lichen, no trace of vegetation. The carbonic acid vomited by the crater had not yet had time to join up with hydrogen from the water or ammonia from the clouds, to form organized matter under the effect of the light.

This island, lost in the open sea, was due only to the accumulation of the successive volcanic ejections; several other mountains of the globe were formed in a similar way; what they threw out from their breast was sufficient to build them, like Etna, which has already vomited a volume of lava larger than its own mass, and Mount Nuovo, near Naples, generated by scoria in the short space of forty-eight hours.

The pile of rocks composing Queen's Island had clearly come out of the entrails of the earth; it was plutonian to the utmost degree. Where it now stood, the endless sea had once stretched, formed in the first days by the condensation of water vapour over the cooling earth; but, as the volcanoes of the

old and new worlds went out, or more precisely became dormant, they had to be replaced by new fire-breathing craters.

In effect, the earth can be compared to a vast spherical boiler. Because of the central fire, immense quantities of vapours are generated, stored at a pressure of thousands of atmospheres, and they would blow up the earth if there were no safety valves.

These valves are the volcanoes; when one closes, another opens; and at the Pole, where the terrestrial crust is thinner because of the flattening, it is not surprising that a volcano should have formed, and become visible as its massif rose above the water.

The doctor, while following Hatteras, noticed the strange features; his foot found volcanic tuff, as well as pumiceous slag deposits, ash, and eruptive rocks similar to the syenites and granites of Iceland.

But if he concluded that the islet was of recent formation, this was because the sedimentary terrain had not yet had time to form.

Water was also absent. If Queen's Island had had the advantage of being several centuries old, springs would have been spurting from its breast, as in the neighbourhood of volcanoes. Now, not only was it devoid of liquid molecules, but the steam which rose from the lava streams seemed to be absolutely anhydrous.

This island was thus of recent formation, and just as it had one day appeared, it could easily disappear on another and sink back to the depths of the ocean.

As the men climbed, the going became more and more difficult; the flanks of the mountain were approaching the vertical, and careful precautions had to be taken to avoid landslides. Often tall clouds of ash twisted around the travellers, threatening to suffocate them, and torrents of lava blocked their path. On the few horizontal parts, streams of lava, cooled and solid on the surface, allowed lava to flow bubbling under their hard crusts. Each man therefore had to carefully test the ground in front of him so as not to be suddenly plunged into the rivers of fire.

From time to time the crater vomited out blocks of rock, red hot from the inflamed gases; some of these masses exploded in the air like bombs, and the debris flew enormous distances in all directions.

It can be imagined what countless dangers this ascent of the mountain involved, and how insane one had to be to attempt it.

However, Hatteras climbed with surprising agility and, disdaining the help of his iron-tipped stick, moved up the steepest slopes without hesitation.

Soon he arrived at a circular rock, a sort of plateau about ten feet wide; an incandescent river divided at a ridge of rock higher up, then surrounded it, leaving only a narrow pathway, along which Hatteras boldly moved.

There he stopped, and his companions were able to join him. He seemed to be visually estimating the distance still to be covered; horizontally, he was less than 600 feet from the crater and the mathematical point of the Pole; but vertically, 1,500 feet still remained to be climbed.

The climb had already lasted three hours; Hatteras did not appear tired; his companions were exhausted.

The summit of the volcano could apparently not be reached. The doctor resolved to prevent Hatteras from climbing any more, at any price. He tried first with gentle means, but the captain's exultation had reached the point of delirium; during the march he had shown every sign of increasing insanity, but anyone who knew him, who had followed him through the various stages of life, could hardly be surprised. As Hatteras rose higher above the ocean, his excitement grew; he no longer lived in the realm of men; he was becoming greater than the mountain itself.

"Hatteras," the doctor said to him, "enough! We can't take a single step more."

"Stay then," replied the captain in a strange voice. "I am going higher!"

"No, what you're doing is useless! You're at the Pole of the world!"

"No, no! Higher!"

"My friend, it's me talking to you, Dr Clawbonny. Do you recognize me?"

"Higher, higher!" repeated the madman.

"No! We will not let . . ."

The doctor had not finished before Hatteras made a superhuman effort, jumped across the lava flow, and was out of the reach of his companions.

They uttered a cry; they thought Hatteras had fallen into the torrents of fire; but the captain had landed on the other side, together with his dog Duke, who refused to leave him.

He was hidden behind a curtain of smoke, and his voice could be heard diminishing with the distance.

"Northwards, northwards!" he was crying. "To the top of Mount Hatteras! Remember Mount Hatteras!"

They could not consider joining the captain; they had only a slim chance of repeating his feat, imbued as he was with the strength and skill particular to madmen; it was impossible to cross that stream of fire, impossible also to

go around it. Altamont tried in vain to get over; he nearly perished trying to jump the lava; his companions had to restrain him bodily.

"Hatteras, Hatteras!" the doctor shouted.

But the captain did not reply, and only the barely perceptible barking of Duke echoed over the mountain.

Nevertheless, Hatteras could occasionally be glimpsed through the columns of smoke and ash raining down. Sometimes his arms, sometimes his head emerged from the swirling shapes. Then he would vanish only to reappear higher up, clinging on to rocks. He got smaller at that fantastic speed of objects rising in the air. Half an hour later, he already seemed to be only half his real height.

The air was full of the dull sounds of the volcano; the mountain was vibrating and groaning like an overheated boiler; its flanks could be felt shivering. Hatteras was still climbing. Duke still followed.

Occasionally landslides occurred behind them and enormous boulders, accelerating as they rebounded on the crests, rushed down to drown at the bottom of the polar basin.

Hatteras did not even turn round. He had used his stick as a shaft to hoist the Union Jack. His terrified companions hung on to his every move. He gradually became microscopic, and Duke had shrunk to the size of a large rat.

There came a time when the wind pushed a vast curtain of flame over them. The doctor shouted in anguish; but Hatteras reappeared, still standing, waving his flag.

The vision of this frightening ascent lasted more than an hour. An hour of battle with the loose rocks and the holes full of ash, into which this hero of the impossible often disappeared up to his waist. Sometimes he performed acrobatics, buttressing himself with his knees and back against the projections of the mountain; sometimes he hung by his hands from some ridge, blown in the wind like a dried tuft of dry vegetation.

Finally he reached the top of the volcano, the very mouth of the crater. The hope then came to the doctor that the wretch, having reached his goal, would come back down again, and would only have the dangers of the return to face.

He shouted one last time:

"Hatteras! Hatteras!"

The doctor's appeal was heart-rending and moved the American to the depths of his soul.

"I'll save him!" exclaimed Altamont.

With a single leap across the torrents of fire, almost falling into them, he disappeared amongst the rocks.

Clawbonny had not had time to stop him.

Meanwhile, having reached the peak, Hatteras was advancing over the abyss on an overhanging rock below which there was nothing. Boulders rained down all around. Duke was still with him. The poor animal was apparently already caught up in the vertiginous attraction of the abyss. Hatteras was waving his flag, which was lit up with incandescent reflections, as the red muslin stood out in long folds in the breath from the crater.

Hatteras was shaking the standard with one hand. With the other, he was pointing at the Pole of the celestial globe, directly above him. However, he still seemed to be hesitating. He was still seeking the mathematical point where all the lines of meridian meet, and where, in his sublime obstinacy, he wanted to set foot. All of a sudden, the rock gave way under his feet. He disappeared. A terrible cry from his companions sounded as far as the mountain peak. A second, a century passed! But Altamont was there, and Duke too. The man and the dog had seized the unfortunate creature just as he fell into the chasm.

Hatteras was saved, saved against his will, and half an hour later, the captain of the *Forward* was lying senseless, reposing in the arms of his despairing companions. When he came to again, the doctor examined his eyes in silent distress. But this unconscious regard, like a blind man who looks without seeing, did not respond.

"Good God," said Johnson, "he's blind!"

"No, he isn't. My poor friends, we have saved only Hatteras's body. His soul has remained at the summit of the volcano! His reason is dead!"

"Insane!" cried Johnson and Altamont in dismay.

"Insane," the doctor replied. And large tears ran from his eyes.

THE WINTER NIGHT

from *Farthest North* (1897)

Fridtjof Nansen

In 1888, the Norwegian explorer Fridtjof Nansen completed the first ski crossing of the Greenland ice; five years later, he set out for the North Pole. His ship, the Fram, *or* Forward, *had been designed to withstand the crushing pressure of the ice, and the plan was to drift, as opposed to sail, north. The excerpt below describes the arrival of winter, and a Christmas dinner on board the* Fram. *Nansen eventually reached 86° 14' N, the highest latitude achieved to that point.*

One day differed very little from another on board, and the description of one is, in every particular of any importance, a description of all.

We all turned out at eight, and breakfasted on hard bread (both rye and wheat), cheese (Dutch-clove cheese, Cheddar, Gruyère, and Mysost, or goat's-whey cheese, prepared from dry powder), corned beef or corned mutton, luncheon ham or Chicago tinned tongue or bacon, cod-caviare, anchovy roe; also oatmeal biscuits or English ship-biscuits—with orange marmalade or Frame Food jelly. Three times a week we had fresh-baked bread as well, and often cake of some kind. As for our beverages, we began by having coffee and chocolate day about, but afterwards had coffee only two days a week, tea two, and chocolate three.

After breakfast some men went to attend to the dogs—give them their food, which consisted of half a stock-fish or a couple of dog-biscuits each, let them loose, or do whatever else there was to do for them. The others went all to their different tasks. Each took his turn of a week in the galley—helping the cook to wash up, lay the table, and wait. The cook himself had to arrange his bill of fare for dinner immediately after breakfast, and to set about his

preparations at once. Some of us would take a turn on the floe to get some fresh air, and to examine the state of the ice, its pressure, etc. At 1 o'clock all were assembled for dinner, which generally consisted of three courses—soup, meat, and dessert; or, soup, fish, and meat, or, fish, meat, and dessert; or sometimes only fish and meat. With the meat we always had potatoes, and either green vegetables or macaroni. I think we were all agreed that the fare was good; it would hardly have been better at home; for some of us it would perhaps have been worse. And we looked like fatted pigs; one or two even began to cultivate a double chin and a corporation. As a rule, stories and jokes circulated at table along with the bock-beer.

After dinner the smokers of our company would march off, well fed and contented, into the galley, which was smoking-room as well as kitchen, tobacco being tabooed in the cabins except on festive occasions. Out there they had a good smoke and chat; many a story was told, and not seldom some warm dispute arose. Afterwards came, for most of us, a short siesta. Then each went to his work again until we were summoned to supper at 6 o'clock, when the regulation day's work was done. Supper was almost the same as breakfast, except that tea was always the beverage. Afterwards there was again smoking in the galley, while the saloon was transformed into a silent reading-room. Good use was made of the valuable library presented to the expedition by generous publishers and other friends. If the kind donors could have seen us away up there, sitting round the table at night with heads buried in books or collections of illustrations, and could have understood how invaluable these companions were to us, they would have felt rewarded by the knowledge that they had conferred a real boon—that they had materially assisted in making the *Fram* the little oasis that it was in this vast ice desert. About half-past seven or eight cards or other games were brought out, and we played well on into the night, seated in groups round the saloon table. One or other of us might go to the organ, and, with the assistance of the crank-handle, perform some of our beautiful pieces, or Johansen would bring out the accordion and play many a fine tune. His crowning efforts were "Oh, Susanna!" and "Napoleon's March Across the Alps in an Open Boat." About midnight we turned in, and then the night watch was set. Each man went on for an hour. Their most trying work on watch seems to have been writing their diaries and looking out, when the dogs barked, for any signs of bears at hand. Besides this, every two hours or four hours the watch had to go aloft or on to the ice to take the meteorological observations.

I believe I may safely say that on the whole the time passed pleasantly and imperceptibly, and that we throve in virtue of the regular habits imposed upon us.

My notes will give the best idea of our life, in all its monotony. They are not great events that are here recorded, but in their very bareness they give a true picture. Such, and no other, was our life. I shall give some quotations direct from my diary:

Tuesday, September 26th. Beautiful weather. The sun stands much lower now; it was 9° above the horizon at midday. Winter is rapidly approaching; there are 14½° of frost this evening, but we do not feel it cold. To-day's observations unfortunately show no particular drift northward; according to them we are still in 78° 50' north latitude. I wandered about over the floe towards evening. Nothing more wonderfully beautiful can exist than the Arctic night. It is dreamland, painted in the imagination's most delicate tints; it is color etherealized. One shade melts into the other, so that you cannot tell where one ends and the other begins, and yet they are all there. No forms—it is all faint, dreamy color music, a far-away, long-drawn-out melody on muted strings. Is not all life's beauty high, and delicate, and pure like this night? Give it brighter colors, and it is no longer so beautiful. The sky is like an enormous cupola, blue at the zenith, shading down into green, and then into lilac and violet at the edges. Over the ice-fields there are cold violet-blue shadows, with lighter pink tints where a ridge here and there catches the last reflection of the vanished day. Up in the blue of the cupola shine the stars, speaking peace, as they always do, those unchanging friends. In the south stands a large red-yellow moon, encircled by a yellow ring and light golden clouds floating on the blue background. Presently the aurora borealis shakes over the vault of heaven its veil of glittering silver—changing now to yellow, now to green, now to red. It spreads, it contracts again, in restless change; next it breaks into waving, many-folded bands of shining silver, over which shoot billows of glittering rays, and then the glory vanishes. Presently it shimmers in tongues of flame over the very zenith, and then again it shoots a bright ray right up from the horizon, until the whole melts away in the moonlight, and it is as though one heard the sigh of a departing spirit And all the time this utter stillness, impressive as the symphony of infinitude. I have never been able to grasp the fact that this earth will some day be spent and desolate and empty. To what end, in that case, all this beauty, with not a creature to rejoice in it? Now I begin to divine it. *This* is the coming earth—here are beauty and

death. But to what purpose? Ah, what is the purpose of all these spheres? Read the answer, if you can, in the starry blue firmament.

[. . .]

Friday, October 13th. Now we are in the very midst of what the prophets would have had us dread so much. The ice is pressing and packing round us with a noise like thunder. It is piling itself up into long walls, and heaps high enough to reach a good way up the *Fram*'s rigging; in fact, it is trying its very utmost to grind the *Fram* into powder. But here we sit quite tranquil, not even going up to look at all the hurly-burly, but just chatting and laughing as usual. Last night there was tremendous pressure round our old dog-floe. The ice had towered up higher than the highest point of the floe and hustled down upon it. It had quite spoiled a well, where we till now had found good drinking-water, filling it with brine. Furthermore, it had cast itself over our stern ice-anchor and part of the steel cable which held it, burying them so effectually that we had afterwards to cut the cable. Then it covered our planks and sledges, which stood on the ice. Before long the dogs were in danger, and the watch had to turn out all hands to save them. At last the floe split in two. This morning the ice was one scene of melancholy confusion, gleaming in the most glorious sunshine. Piled up all round us were high, steep ice walls. Strangely enough, we had lain on the very verge of the worst confusion, and had escaped with the loss of an ice-anchor, a piece of steel cable, a few planks and other bits of wood, and half of a Samoyede sledge, all of which might have been saved if we had looked after them in time. But the men have grown so indifferent to the pressure now that they do not even go up to look, let it thunder ever so hard. They feel that the ship can stand it, and so long as that is the case there is nothing to hurt except the ice itself.

In the morning the pressure slackened again, and we were soon lying in a large piece of open water, as we did yesterday. To-day, again, this stretched far away towards the northern horizon, where the same dark atmosphere indicated some extent of open water. I now gave the order to put the engine together again; they told me it could be done in a day and a half or at most two days. We must go north and see what there is up there. I think it possible that it may be the boundary between the ice-drift the *Jeannette* was in and the pack we are now drifting south with—or can it be land?

We had kept company quite long enough with the old, now broken-up floe, so worked ourselves a little way astern after dinner, as the ice was beginning to draw together. Towards evening the pressure began again in earnest, and was especially bad round the remains of our old floe, so that I believe we

may congratulate ourselves on having left it. It is evident that the pressure here stands in connection with, is perhaps caused by, the tidal wave. It occurs with the greatest regularity. The ice slackens twice and packs twice in 24 hours. The pressure has happened about 4, 5, and 6 o'clock in the morning, and almost at exactly the same hour in the afternoon, and in between we have always lain for some part of the time in open water. The very great pressure just now is probably due to the spring-tide; we had new moon on the 9th, which was the first day of the pressure. Then it was just after midday when we noticed it, but it has been later every day, and now it is at 8 P.M.

The theory of the ice-pressure being caused to a considerable extent by the tidal wave has been advanced repeatedly by Arctic explorers. During the *Fram*'s drifting we had better opportunity than most of them to study this phenomenon, and our experience seems to leave no doubt that over a wide region the tide produces movement and pressure of the ice. It occurs especially at the time of the spring-tides, and more at new moon than at full moon. During the intervening periods there was, as a rule, little or no trace of pressure. But these tidal pressures did not occur during the whole time of our drifting. We noticed them especially the first autumn, while we were in the neighborhood of the open sea north of Siberia, and the last year, when the *Fram* was drawing near the open Atlantic Ocean; they were less noticeable while we were in the polar basin. Pressure occurs here more irregularly, and is mainly caused by the wind driving the ice. When one pictures to one's self these enormous ice-masses, drifting in a certain direction, suddenly meeting hinderances—for example, ice-masses drifting from the opposite direction, owing to a change of wind in some more or less distant quarter—it is easy to understand the tremendous pressure that must result.

Such an ice conflict is undeniably a stupendous spectacle. One feels one's self to be in the presence of titanic forces, and it is easy to understand how timid souls may be overawed and feel as if nothing could stand before it. For when the packing begins in earnest it seems as though there could be no spot on the earth's surface left unshaken. The ice cracks on every side of you, and begins to pile itself up; and all of a sudden you too find yourself in the midst of the struggle. There are howlings and thunderings round you; you feel the ice trembling, and hear it rumbling under your feet; there is no peace anywhere. In the semi-darkness you can see it piling and tossing itself up into high ridges nearer and nearer you—floes 10, 12, 15 feet thick, broken, and flung on the top of each other as if they were feather-weights. They are quite near you now, and you jump away to save your life. But the ice splits in front of

you, a black gulf opens, and water streams up. You turn in another direction, but there through the dark you can just see a new ridge of moving ice-blocks coming towards you. You try another direction, but there it is the same. All round there is thundering and roaring, as of some enormous waterfall, with explosions like cannon salvoes. Still nearer you it comes. The floe you are standing on gets smaller and smaller; water pours over it; there can be no escape except by scrambling over the rolling ice-blocks to get to the other side of the pack. But now the disturbance begins to calm down. The noise passes on, and is lost by degrees in the distance.

This is what goes on away there in the north month after month and year after year. The ice is split and piled up into mounds, which extend in every direction. If one could get a bird's-eye view of the ice-fields, they would seem to be cut up into squares or meshes by a network of these packed ridges, or pressure-dikes, as we called them, because they reminded us so much of snow-covered stone dikes at home, such as, in many parts of the country, are used to enclose fields.

Saturday, October 14th. To-day we have got on the rudder; the engine is pretty well in order, and we are clear to start north when the ice opens to-morrow morning. It is still slackening and packing quite regularly twice a day, so that we can calculate on it beforehand. To-day we had the same open channel to the north, and beyond it open sea as far as our view extended. What can this mean? This evening the pressure has been pretty violent. The floes were packed up against the *Fram* on the port side, and were once or twice on the point of toppling over the rail. The ice, however, broke below; they tumbled back again, and had to go under us after all. It is not thick ice, and cannot do much damage; but the force is something enormous. On the masses come incessantly without a pause; they look irresistible; but slowly and surely they are crushed against the *Fram*'s sides. Now (8.30 P.M.) the pressure has at last stopped. Clear evening, sparkling stars, and flaming northern lights.

Sunday, October 15th. To our surprise, the ice did not slacken away much during last night after the violent pressure; and, what was worse, there was no indication of slackening in the morning, now that we were quite ready to go.

Monday, October 16th. Ice quiet and close. Observations on the 12th placed us in 78° 5' north latitude. Steadily southward. This is almost depressing.

Thursday, October 19th. The ice slackened a little last night. In the morning I attempted a drive with six of the dogs. When I had managed to

harness them to the Samoyede sledge, had seated myself on it, and called "Pr-r-r-r, pr-r-r-r!" they went off in quite good style over the ice. But it was not long before we came to some high pack-ice and had to turn. This was hardly done before they were off back to the ship at lightning speed, and they were not to be got away from it again. Round and round it they went, from refuse-heap to refuse-heap. If I started at the gangway on the starboard side, and tried by thrashing them to drive them out over the ice, round the stern they flew to the gangway on the port side. I tugged, swore, and tried everything I could think of, but all to no purpose. I got out and tried to hold the sledge back, but was pulled off my feet, and dragged merrily over the ice in my smooth sealskin breeches, on back, stomach, side—just as it happened. When I managed to stop them at some pieces of pack-ice or a dust-heap, round they went again to the starboard gangway, with me dangling behind, swearing madly that I would break every bone in their bodies when I got at them. This game went on till they probably tired of it, and thought they might as well go my way for a change. So now they went off beautifully across the flat floe until I stopped for a moment's breathing space. But at the first movement I made in the sledge they were off again, tearing wildly back the way we had come. I held on convulsively, pulled, raged, and used the whip; but the more I lashed the faster they went on their own way. At last I got them stopped by sticking my legs down into the snow between the sledge-shafts, and driving a strong seal-hook into it as well. But while I was off my guard for a moment they gave a tug. I lay with my hinder-part where my legs had been, and we went on at lightning speed—that substantial part of my body leaving a deep track in the snow. This sort of thing went on time after time. I lost the board I should have sat on, then the whip, then my gloves, then my cap—these losses not improving my temper. Once or twice I ran round in front of the dogs, and tried to force them to turn by lashing at them with the whip. They jumped to both sides and only tore on the faster; the reins got twisted round my ankles, and I was thrown flat on the sledge, and they went on more wildly than ever. This was my first experience in dog driving on my own account, and I will not pretend that I was proud of it. I inwardly congratulated myself that my feats had been unobserved.

Saturday, October 21st. I have stayed in to-day because of an affection of the muscles, or rheumatism, which I have had for some days on the right side of my body, and for which the doctor is "massaging" me, thereby greatly adding to my sufferings. Have I really grown so old and palsied, or is the whole thing imagination? It is all I can do to limp about; but I just wonder if

I could not get up and run with the best of them if there happened to be any great occasion for it: I almost believe I could. A nice Arctic hero of 32, lying here in my berth! Have had a good time reading home letters, dreaming myself at home, dreaming of the home-coming—in how many years? Successful or unsuccessful, what does that matter?

I had a sounding taken; it showed over 73 fathoms (135 m.), so we are in deeper water again. The sounding-line indicated that we are drifting southwest. I do not understand this steady drift southward. There has not been much wind either lately; there is certainly a little from the north to-day, but not strong. What can be the reason of it? With all my information, all my reasoning, all my putting of two and two together, I cannot account for any south-going current here—there ought to be a north-going one. If the current runs south here, how is that great open sea we steamed north across to be explained? and the bay we ended in farthest north? These could only be produced by the north-going current which I presupposed.

Sunday, October 22d. Henriksen took soundings this morning, and found 70 fathoms (129 m.) of water. "If we are drifting at all," said he, "it is to the east; but there seems to be almost no movement." No wind to-day. I am keeping in my den.

Monday, October 23d. Still in the den. To-day, 5 fathoms shallower than yesterday. The line points southwest, which means that we are drifting north-eastward. Hansen has reckoned out the observation for the 19th, and finds that we must have got 10 minutes farther north, and must be in 78° 15' N. lat. So at last, now that the wind has gone down, the north-going current is making itself felt. Some channels have opened near us, one along the side of the ship, and one ahead, near the old channel. Only slight signs of pressure in the afternoon.

Tuesday, October 24th. Between 4 and 5 A.M. there was strong pressure, and the *Fram* was lifted up a little. It looks as if the pressure were going to begin again; we have spring-tide with full moon. The ice opened so much this morning that the *Fram* was afloat in her cutting; later on it closed again, and about 11 there was some strong pressure; then came a quiet time; but in the afternoon the pressure began once more, and was violent from 4 to 4.30. The *Fram* was shaken and lifted up; didn't mind a bit.

Wednesday, October 25th. We had a horrible pressure last night. I awoke and felt the *Fram* being lifted, shaken, and tossed about, and heard the loud cracking of the ice breaking against her sides. After listening for a little while I fell asleep again, with a snug feeling that it was good to be on board the *Fram*.

It is quickly getting darker. The sun stands lower and lower every time we see it; soon it will disappear altogether, if it has not done so already. The long, dark winter is upon us, and glad shall we be to see the spring; but nothing matters much if we could only begin to move north. There is now southwesterly wind, and the windmill, which has been ready for several days, has been tried at last and works splendidly. We have beautiful electric light to-day, though the wind has not been especially strong (5–8 m. per second). Electric lamps are a grand institution. What a strong influence light has on one's spirits! There was a noticeable brightening-up at the dinner-table to-day; the light acted on our spirits like a draught of good wine. And how festive the saloon looks! We felt it quite a great occasion—drank Oscar Dickson's health, and voted him the best of good fellows.

To-morrow is the *Fram*'s birthday. How many memories it recalls of the launch-day a year ago!

Thursday, October 26th. 54 fathoms (90 m.) of water when the soundings were taken this morning. We are moving quickly north—due north—says Peter. It does look as if things were going better. Great celebration of the day, beginning with target-shooting. Then we had a splendid dinner of four courses, which put our digestive apparatus to a severe test. The *Fram*'s health was drunk amidst great and stormy applause. The proposer's words were echoed by all hearts when he said that she was such an excellent ship for our purpose that we could not imagine a better (great applause), and we therefore wished her, and ourselves with her, long life (hear, hear!).

Sitting here now alone, my thoughts involuntarily turn to the year that has gone since we stood up there on the platform, and she threw the champagne against the bow, saying: "*Fram* is your name!" and the strong, heavy hull began to glide so gently. I held her hand tight; the tears came into eyes and throat, and one could not get out a word. The sturdy hull dived into the glittering water; a sunny haze lay over the whole picture. Never shall I forget the moment we stood there together, looking out over the scene. And to think of all that has happened these four last months! Separated by sea and land and ice; coming years, too, lying between us—it is all just the continuation of what happened that day. But how long is it to last? I have such difficulty in feeling that I am not to see home again soon. When I begin to reflect, I know that it may be long, but I will not believe it.

To-day, moreover, we took solemn farewell of the sun. Half of its disk showed at noon for the last time above the edge of the ice in the south, a flattened body, with a dull red glow, but no heat. Now we are entering the night

of winter. What is it bringing us? Where shall we be when the sun returns? No one can tell.

Friday, October 27th. The soundings this morning showed 52 fathoms (95 m.) of water. According to observations taken yesterday afternoon, we are about 3' farther north and a little farther west than on the 19th. It is disgusting the way we are muddling about here. We must have got into a hole where the ice grinds round and round, and can't get farther. And the time is passing all to no purpose; and goodness only knows how long this sort of thing may go on. If only a good south wind would come and drive us north out of this hobble!

Sunday, October 29th. Peter shot a white fox this morning close in to the ship. For some time lately we have been seeing fox-tracks in the mornings, and one Sunday Mogstad saw the fox itself. It has, no doubt, been coming regularly to feed on the offal of the bears. Shortly after the first one was shot another was seen; it came and smelt its dead comrade, but soon set off again and disappeared. It is remarkable that there should be so many foxes on this drift-ice so far from land.

Monday, October 30th. To-day the temperature has gone down to 18° below zero (–27° C.). I took up the dredge I had put out yesterday. It brought up two pails of mud from the bottom, and I have been busy all day washing this out in the saloon in a large bath, to get the many animals contained in it. They were chiefly starfish, waving starfish, medusæ *(Astrophyton),* sea-slugs, coral insects *(Alcyonaria),* worms, sponges, shell-fish, and crustaceans; and were, of course, all carefully preserved in spirits.

Tuesday, October 31st. Forty-nine fathoms (90 m.) of water to-day, and the current driving us hard to the southwest. We have good wind for the mill now, and the electric lamps burn all day. The arc lamp under the skylight makes us quite forget the want of sun. Oh! light is a glorious thing, and life is fair in spite of all privations! This is Sverdrup's birthday, and we had revolver practice in the morning. Of course a magnificent dinner of five courses—chicken soup, boiled mackerel, reindeer ribs with baked cauliflower and potatoes, macaroni pudding, and stewed pears with milk—Ringnes ale to wash it down.

Sunday, November 5th. A great race on the ice was advertised for to-day. The course was measured, marked off, and decorated with flags. The cook had prepared the prizes—cakes, numbered, and properly graduated in size. The expectation was great; but it turned out that, from excessive training during the few last days, the whole crew were so stiff in the legs that they were not able to move. We got our prizes all the same.

So it is Sunday once more. How the days drag past! I work, read, think, and

dream; strum a little on the organ; go for a walk on the ice in the dark. Low on the horizon in the southwest there is the flush of the sun—a dark fierce red, as if of blood aglow with all life's smouldering longings—low and far-off, like the dreamland of youth. I can sit and gaze and gaze, my eyes entranced by the dream-glow yonder in the west, where the moon's thin, pale, silver sickle is dipping its point into the blood; and my soul is borne beyond the glow, to the sun, so far off now—and to the home-coming! Our task accomplished, we are making our way up the fjord as fast as sail and steam can carry us. On both sides of us the homeland lies smiling in the sun; and then . . . the sufferings of a thousand days and hours melt into a moment's inexpressible joy. Ugh! that was a bitter gust—I jump up and walk on. What am I dreaming about! so far yet from the goal—hundreds and hundreds of miles between us, ice and land and ice again. And we are drifting round and round in a ring, bewildered, attaining nothing, only waiting, always waiting, for what?

Wednesday, November 8th. The storm (which we had had the two previous days) is quite gone down; not even enough breeze for the mill. We tried letting the dogs sleep on the ice last night, instead of bringing them on board in the evening, as we have been doing lately. The result was that another dog was torn to pieces during the night. It was "Ulabrand," the old brown, toothless fellow, that went this time. "Job" and "Moses" had gone the same way before. Yesterday evening's observations place us in 77° 43' north latitude and 138° 8' east longitude. This is farther south than we have been yet. No help for it; but it is a sorry state of matters; and that we are farther east than ever before is only a poor consolation.

Here I sit in the still winter night on the drifting ice-floe, and see only stars above me. Far off I see the threads of life twisting themselves into the intricate web which stretches unbroken from life's sweet morning dawn to the eternal death-stillness of the ice. Thought follows thought—you pick the whole to pieces, and it seems so small—but high above all towers one form . . . *Why did you take this voyage?* . . . Could I do otherwise? Can the river arrest its course and run up hill? My plan has come to nothing. That palace of theory which I reared, in pride and self-confidence, high above all silly objections has fallen like a house of cards at the first breath of wind. Build up the most ingenious theories and you may be sure of one thing—that fact will defy them all. Was I so very sure? Yes, at times; but that was self-deception, intoxication. A secret doubt lurked behind all the reasoning. It seemed as though the longer I defended my theory, the nearer I came to doubting it. But *no,* there is no getting over the evidence of that Siberian drift-wood.

But if, after all, we are on the wrong track, what then? Only disappointed human hopes, nothing more. And even if we perish, what will it matter in the endless cycles of eternity?

[. . .]

Monday, December 25th (Christmas-day). Thermometer at 36° Fahr. below zero (–38° C.). I took a walk south in the beautiful light of the full moon. At a newly made crack I went through the fresh ice with one leg and got soaked; but such an accident matters very little in this frost. The water immediately stiffens into ice; it does not make one very cold, and one feels dry again soon.

They will be thinking much of us just now at home and giving many a pitying sigh over all the hardships we are enduring in this cold, cheerless, icy region. But I am afraid their compassion would cool if they could look in upon us, hear the merriment that goes on, and see all our comforts and good cheer. They can hardly be better off at home. I myself have certainly never lived a more sybaritic life, and have never had more reason to fear the consequences it brings in its train. Just listen to to-day's dinner menu:

1. Ox-tail soup;
2. Fish-pudding, with potatoes and melted butter;
3. Roast of reindeer, with pease, French beans, potatoes, and cranberry jam;
4. Cloudberries with cream;
5. Cake and marchpane (a welcome present from the baker to the expedition; we blessed that man).

And along with all this that Ringnes bock-beer which is so famous in our part of the world. Was this the sort of dinner for men who are to be hardened against the horrors of the Arctic night?

Tuesday, December 26th. 36° Fahr. below zero (–38° C.). This (the same as yesterday's) is the greatest cold we have had yet. I went a long way north to-day; found a big lane covered with newly frozen ice, with a quite open piece of water in the middle. The ice rocked up and down under my steps, sending waves out into the open pool. It was strange once more to see the moonlight playing on the coal-black waves, and awakened a remembrance of well-known scenes. I followed this lane far to the north, seemed to see the outlines of high land in the hazy light below the moon, and went on and on; but in the end it turned out to be a bank of clouds behind the moonlit vapor rising from the open water. I saw from a high hummock that this opening stretched north as far as the eye could reach.

The same luxurious living as yesterday; a dinner of four courses. Shooting with darts at a target for cigarettes has been the great excitement of the day. Darts and target are Johansen's Christmas present from Miss Fougner.

Wednesday, December 27th. Wind began to blow this afternoon, 19½ to 26 feet per second; the windmill is going again, and the arc lamp once more brightens our lives. Johansen gave notice of "a shooting-match by electric light, with free concert," for the evening. It was a pity for himself that he did, for he and several others were shot into bankruptcy and beggary, and had to retire one after the other, leaving their cigarettes behind them.

Thursday, December 28th. A little forward of the *Fram* there is a broad, newly formed open lane, in which she could lie crossways. It was covered with last night's ice, in which slight pressure began to-day. It is strange how indifferent we are to this pressure, which was the cause of such great trouble to many earlier Arctic navigators. We have not so much as made the smallest preparation for possible accident, no provisions on deck, no tent, no clothing in readiness. This may seem like recklessness, but in reality there is not the slightest prospect of the pressure harming us; we know now what the *Fram* can bear. Proud of our splendid, strong ship, we stand on her deck watching the ice come hurtling against her sides, being crushed and broken there and having to go down below her, while new ice-masses tumble upon her out of the dark, to meet the same fate. Here and there, amid deafening noise, some great mass rises up and launches itself threateningly upon the bulwarks, only to sink down suddenly, dragged the same way as the others. But at times when one hears the roaring of tremendous pressure in the night, as a rule so deathly still, one cannot but call to mind the disasters that this uncontrollable power has wrought.

I am reading the story of Kane's expedition just now. Unfortunate man, his preparations were miserably inadequate; it seems to me to have been a reckless, unjustifiable proceeding to set out with such equipments. Almost all the dogs died of bad food; all the men had scurvy from the same cause, with snow-blindness, frost-bites, and all kinds of miseries. He learned a wholesome awe of the Arctic night, and one can hardly wonder at it.

I am almost ashamed of the life we lead, with none of those darkly painted sufferings of the long winter night which are indispensable to a properly exciting Arctic expedition. We shall have nothing to write about when we get home.

ANDRÉE'S SECOND DIARY

from *Andrée's Story* (1930)

Salomon August Andrée

Salomon August Andrée was a Swedish engineer who worked in the Stockholm patent office. His plan to reach the North Pole by hydrogen balloon held great appeal for the Swedes, who feared they were losing the polar competition to Fridtjof Nansen and the Norwegians. With two companions, Andrée took off from Spitsbergen on July 11, 1897. There is a good deal of evidence that Andrée knew, but kept secret, the fact that his balloon, the Eagle, *was leaking gas and therefore could not make it to the Pole. On July 14, 1897, 200 miles from its departure point, the* Eagle *crashed. Andrée spent the next ten weeks trekking over the ice, finally reaching a tiny island in the Svalbard Archipelago. The excerpt below is all that was decipherable from the final pages of Andrée's diaries, which were discovered thirty years later along with his body and those of his companions.*

Page 1 ……… 897
…… with cutti[ng] ……
…… … … beginning of a
…… … the hut hung
…… … … the day passed
……… … … … … …
…… imp … … … vation
…… … … … … … to
…… … … … … all too
…… … we there as a matter of
…… a not unimportant
…… of the island correctly. But
…… … … … could with

…… … the lowland. The question
…… … … … … here with everything
…… … … t … … …we first
…… … … … … to reach
…… … … … … … …
……… of
…… … … … should
…… … … and up
…… … on the island.

Page 2 In the evening 5 b ……………………
Riders or geese ………………………
5th in the morning……………………
the previously mentioned …………..
we had ………………………………
lucky that we …………………………
there and ……………………………
ing it …… … …………….. ………..
I …… … t …
along the glacier ………….. ………..
from the glacier ………….. ………..
our hard … … not ……….. ………..
even if late at night ……….. ………..
the day's energetic labour … ………..
middle of the night ……….. ………..
……… for the (flaming) ……… outside …
northern lights neither……… ………..
warmed … … ………….. ………..
…… …… k …………….. ………..
…… …… my ………….. ………..
We christened o[n] [acc]ount ………..
of this the district ……….. ………..
the place to "M ……….. place."

Page 3 ………….. …… during the day the 6th
………….. heavy wind w
………….. …… … could not ……… much
………….. undertook however a short

.... we at last
.... Swedes
.... to be the
.... icy
.... at once interes-
.... we ... high
.... from the sea found
.... All the ground
.... stone-brash
.... of the gravel was
.... Granite lay partly
.... great walls [ridges?]
.... which however were
.... If ... could
.... whole
.... (large)
.... in darkness.
.... in the snow-
hut .... transport of the goods
the .... neighbourhood. This was
a heavy .... was done

Page 4 ...... g was busy ... . at ...
feared that ... .
such with ....
which we f ....
of it. ....
rings ....
cier. ....
had set foot ....
if it possibly ....
to look at tr........ ....
the glacier ....
ought I think ....
than one ....
the visible ....
the sea but .... k. ... .

not since ……….. …… … … ……………………
to. Our ……….. …… … … ……………………
s … … ……….. …… … … ……………………
worth … ……….. …… … … ……………………
……mon with … …… … … ……………………
and intestine …… …… …… envious
now give … ……… impression … innocent
white doves but of ……… …… l carrion
birds ……….. …… … … ……………………

Page 5 ……….. …… …… bad weather and we fear ……
……….. …… we keep in the tent the whole day
……….. …… …… so that we could
……….. …… …… …… on the hut.
……….. …… …… …… to escape
……….. …… …… …… …… like
……….. …… …… …… out on the sea
……….. crash … …… …… grating
……….. …… …… …… … driftwood
……….. …… …… …… to move about a little
……….. …… …… …… ermits

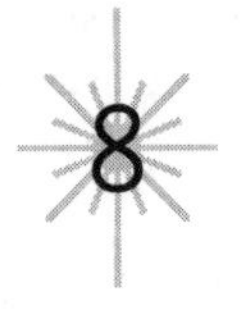

WE REACH THE POLE

from *The North Pole* (1910)

Robert Peary

The American explorer Robert Peary claimed to have reached the North Pole on April 6, 1909, after having wintered on Ellesmere Island. This claim has been much disputed. Critics have noted that no one who accompanied Peary on the final leg of his journey was trained in navigation, and that the pace the explorer supposedly set after his last support party headed back was several times faster than that set during the rest of the journey. (Disturbed by Peary's lack of corroboration, Congress even held hearings into the matter.) In 1989, National Geographic, *which had been a major backer of Peary's 1909 expedition, concluded that if he did miss the Pole, it was not by more than five miles. This finding, too, has been disputed.*

THE LAST MARCH NORTHWARD ended at ten o'clock on the forenoon of April 6. I had now made the five marches planned from the point at which Bartlett turned back, and my reckoning showed that we were in the immediate neighborhood of the goal of all our striving. After the usual arrangements for going into camp, at approximate local noon, of the Columbia meridian, I made the first observation at our polar camp. It indicated our position as 89° 57'.

We were now at the end of the last long march of the upward journey. Yet with the Pole actually in sight I was too weary to take the last few steps. The accumulated weariness of all those days and nights of forced marches and insufficient sleep, constant peril and anxiety, seemed to roll across me all at once. I was actually too exhausted to realize at the moment that my life's purpose had been achieved. As soon as our igloos had been completed and we had eaten our dinner and double-rationed the dogs, I turned in for a few hours of absolutely necessary sleep, Henson and the Eskimos having unloaded the

sledges and got them in readiness for such repairs as were necessary. But, weary though I was, I could not sleep long. It was, therefore, only a few hours later when I woke. The first thing I did after awaking was to write these words in my diary: "The Pole at last. The prize of three centuries. My dream and goal for twenty years. Mine at last! I cannot bring myself to realize it. It seems all so simple and commonplace."

Everything was in readiness for an observation* at 6 P.M., Columbia meridian time, in case the sky should be clear, but at that hour it was, unfortunately, still overcast. But as there were indications that it would clear before long, two of the Eskimos and myself made ready a light sledge carrying only the instruments, a tin of pemmican, and one or two skins; and drawn by a double team of dogs, we pushed on an estimated distance of ten miles. While we traveled, the sky cleared, and at the end of the journey, I was able to get a satisfactory series of observations at Columbia meridian midnight. These observations indicated that our position was then beyond the Pole.

Nearly everything in the circumstances which then surrounded us seemed too strange to be thoroughly realized; but one of the strangest of those circumstances seemed to me to be the fact that, in a march of only a few hours, I had passed from the western to the eastern hemisphere and had verified my

---

* The instruments used in taking observations for latitude may be either a sextant and an artificial horizon, or a small theodolite. Both these instruments were taken on the sledge journey; but the theodolite was not used, owing to the low altitude of the sun. Had the expedition been delayed on the return until May or June, the theodolite would then have been of value in determining position and variation of the compass.

The method of taking meridian observations with a sextant and an artificial horizon on a polar sledge journey is as follows: if there is any wind, a semicircular wind-guard of snow blocks, two tiers high, is put up, opening to the south. If there is no wind, this is not necessary.

The instrument box is firmly bedded in the snow, which is packed down to a firm bearing and snow is packed around the box. Then something, usually a skin, is thrown over the snow, partly to prevent any possible warmth from the sun melting the snow and shifting the bearing of the box; partly to protect the eyes of the observer from the intense reflected glare of light from the snow.

The mercury trough of the artificial horizon is placed on top of the level box, and the mercury, which has been thoroughly warmed in the igloo, is poured into the trough until it is full. In the case of the special wooden trough devised and used on the last expedition, it was possible to bring the surface of the mercury level with the edges of the trough, thus enabling us to read angles very close to the horizon.

The mercury trough is covered with what is called the roof—a metal framework carrying two pieces of very accurately ground glass, set inclined, like the opposite sides of the

position at the summit of the world. It was hard to realize that, in the first miles of this brief march, we had been traveling due north, while, on the last few miles of the same march, we had been traveling south, although we had all the time been traveling precisely in the same direction. It would be difficult to imagine a better illustration of the fact that most things are relative. Again, please consider the uncommon circumstance that, in order to return to our camp, it now became necessary to turn and go north again for a few miles and then to go directly south, all the time traveling in the same direction.

As we passed back along that trail which none had ever seen before or would ever see again, certain reflections intruded themselves which, I think, may fairly be called unique. East, west, and north had disappeared for us. Only one direction remained and that was south. Every breeze which could possibly blow upon us, no matter from what point of the horizon, must be a south wind. Where we were, one day and one night constituted a year, a hundred such days and nights constituted a century. Had we stood in that spot during the six months of the arctic winter night, we should have seen every star of the northern hemisphere circling the sky at the same distance from the horizon, with Polaris (the North Star) practically in the zenith.

All during our march back to camp the sun was swinging around in its ever-moving circle. At six o'clock on the morning of April 7, having again

---

roof of a house. The object of this roof is to prevent any slightest breath of wind disturbing the surface of the mercury and so distorting the sun's image in it, and also to keep out any fine snow or frost crystals that may be in the atmosphere. In placing the trough and the roof on the top of the instrument box, the trough is placed so that its longer diameter will be directed toward the sun.

A skin is then thrown down on the snow close to the box and north of it, and the observer lies down flat on his stomach on this, with his head to the south, and head and sextant close to the artificial horizon. He rests both elbows on the snow, holding the sextant firmly in both hands, and moving his head and the instrument until the image or part of the image of the sun is seen reflected on the surface of the mercury.

The principle on which the latitude of the observer is obtained from the altitude of the sun at noon is very simple. It is this: that the latitude of the observer is equal to the distance of the center of the sun from the zenith, plus the declination of the sun for that day and hour.

The declination of the sun for any place at any hour may be obtained from tables prepared for that purpose, which give the declination for noon of every day on the Greenwich meridian, and the hourly change in the declination.

Such tables for the months of February, March, April, May, June, and July, together with the ordinary tables for refraction to minus 10° Fahrenheit, I had with me on pages torn from the "Nautical Almanac and Navigator."

arrived at Camp Jesup, I took another series of observations. These indicated our position as being four or five miles from the Pole, towards Bering Strait. Therefore, with a double team of dogs and a light sledge, I traveled directly towards the sun an estimated distance of eight miles. Again I returned to the camp in time for a final and completely satisfactory series of observations on April 7 at noon, Columbia meridian time. These observations gave results essentially the same as those made at the same spot twenty-four hours before.

I had now taken in all thirteen single, or six and one-half double, altitudes of the sun, at two different stations, in three different directions, at four different times. All were under satisfactory conditions, except for the first single altitude on the sixth. The temperature during these observations had been from minus 11° Fahrenheit to minus 30° Fahrenheit, with clear sky and calm weather (except as already noted for the single observation on the sixth).

[. . .]

In traversing the ice in these various directions as I had done, I had allowed approximately ten miles for possible errors in my observations, and at some moment during these marches and countermarches, I had passed over or very near the point* where north and south and east and west blend into one.

Of course there were some more or less informal ceremonies connected

---

* Ignorance and misconception of all polar matters seem so widespread and comprehensive that it appears advisable to introduce here a few a b c paragraphs. Anyone interested can supplement these by reading the introductory parts of any good elementary school geography or astronomy.

The North Pole (that is, the geographical pole as distinguished from the magnetic pole, and this appears to be the first and most general stumbling block of the ignorant) is simply the point where that imaginary line known as the earth's axis—that is, the line on which the earth revolves in its daily motion—intersects the earth's surface.

Some of the recent sober discussions as to the size of the North Pole, whether it was as big as a quarter, or a hat, or a township, have been intensely ludicrous.

Precisely speaking, the North Pole is simply a mathematical point, and therefore, in accordance with the mathematical definition of a point, it has neither length, breadth, nor thickness.

If the question is asked, how closely can the Pole be determined (this is the point which has muddled some of the ignorant wiseacres), the answer will be: That depends upon the character of the instruments used, the ability of the observer using them, and the number of observations taken.

If there were land at the Pole, and powerful instruments of great precision, such as are used in the world's great observatories, were mounted there on suitable foundations and used by practised observers for repeated observations extending over years, then it would be possible to determine the position of the Pole with great precision.

with our arrival at our dfficult destination, but they were not of a very elaborate character. We planted five flags at the top of the world. The first one was a silk American flag which Mrs. Peary gave me fifteen years ago. That flag has done more traveling in high latitudes than any other ever made. I carried it wrapped about my body on every one of my expeditions northward after it came into my possession, and I left a fragment of it at each of my successive "farthest norths": Cape Morris K. Jesup, the northernmost point of land in the known world; Cape Thomas Hubbard, the northernmost known point of Jesup Land, west of Grant Land; Cape Columbia, the northernmost point of North American lands; and my farthest north in 1906, latitude 87° 6' in the ice of the polar sea. By the time it actually reached the Pole, therefore, it was somewhat worn and discolored.

A broad diagonal section of this ensign would now mark the farthest goal of earth—the place where I and my dusky companions stood.

It was also considered appropriate to raise the colors of the Delta Kappa Epsilon fraternity, in which I was initiated a member while an undergraduate student at Bowdoin College, the "World's Ensign of Liberty and Peace," with its red, white, and blue in a field of white, the Navy League flag, and the Red Cross flag.

After I had planted the American flag in the ice, I told Henson to time the

---

With ordinary field instruments, transit, theodolite, or sextant, an extended series of observations by an expert observer should permit the determination of the Pole within entirely satisfactory limits, but not with the same precision as by the first method.

A single observation at sea with sextant and the natural horizon, as usually taken by the master of a ship, is assumed under ordinary satisfactory conditions to give the observer's position within about a mile.

In regard to the difficulties of taking observations in the arctic regions, I have found a tendency on the part of experts who, however, have not had practical experience in the arctic regions themselves, to overestimate and exaggerate the difficulties and drawbacks of making these observations due to the cold.

My personal experience has been that, to an experienced observer, dressed in furs and taking observations in calm weather, in temperatures not exceeding say 40° below zero Fahrenheit, the difficulties of the work resulting from cold alone are not serious. The amount and character of errors due to the effect of cold upon the instrument might perhaps be a subject for discussion, and for distinct differences of opinion.

My personal experience has been that my most serious trouble was with the eyes.

To eyes which have been subjected to brilliant and unremitting daylight for days and weeks, and to the strain of continually setting a course with the compass, and traveling towards a fixed point in such light, the taking of a series of observations is usually a nightmare; and the strain of focusing, of getting precise contact of the sun's images, and

Eskimos for three rousing cheers, which they gave with the greatest enthusiasm. Thereupon, I shook hands with each member of the party—surely a sufficiently unceremonious affair to meet with the approval of the most democratic. The Eskimos were childishly delighted with our success. While, of course, they did not realize its importance fully, or its world-wide significance, they did understand that it meant the final achievement of a task upon which they had seen me engaged for many years.

Then, in a space between the ice blocks of a pressure ridge, I deposited a glass bottle containing a diagonal strip of my flag and records of which the following is a copy:

90 N. Lat., North Pole,
April 6, 1909.

Arrived here to-day, 27 marches from C. Columbia.

I have with me 5 men, Matthew Henson, colored, Ootah, Egingwah, Seegloo, and Ookeah, Eskimos; 5 sledges and 38 dogs. My ship, the S. S. *Roosevelt,* is in winter quarters at C. Sheridan, 90 miles east of Columbia.

The expedition under my command which has succeeded in reaching the Pole is under the auspices of the Peary Arctic Club of New York City, and has been fitted out and sent north by the members and friends of the

---

of reading the vernier, all in the blinding light of which only those who have taken observations in bright sunlight on an unbroken snow expanse in the arctic regions can form any conception, usually leaves the eyes bloodshot and smarting for hours afterwards.

The continued series of observations in the vicinity of the Pole, noted above, left me with eyes that were, for two or three days, useless for anything requiring careful vision, and had it been necessary for me to set a course during the first two or three days of our return I should have found it extremely trying.

Snow goggles, as worn by us continually during the march, while helping, do not entirely relieve the eyes from strain, and during a series of observations the eyes become extremely tired and at times uncertain.

Various authorities will give different estimates of the probable error in observations taken at the Pole. I am personally inclined to think that an allowance of five miles is an equitable one.

No one, except those entirely ignorant of such matters, has imagined for a moment that I was able to determine with my instruments the precise position of the Pole, but after having determined its position approximately, then setting an arbitrary allowance of about ten miles for possible errors of the instruments and myself as observer, and then crossing and recrossing that ten mile area in various directions, no one except the most ignorant will have any doubt but what, at some time, I had passed close to the precise point, and had, perhaps, actually passed over it.

club for the purpose of securing this geographical prize, if possible, for the honor and prestige of the United States of America.

The officers of the club are Thomas H. Hubbard, of New York, President; Zenas Crane, of Mass., Vice-president; Herbert L. Bridgman, of New York, Secretary and Treasurer.

I start back for Cape Columbia to-morrow.

Robert E. Peary,
United States Navy.

90 N. Lat., North Pole,
April 6, 1909.

I have to-day hoisted the national ensign of the United States of America at this place, which my observations indicate to be the North Polar axis of the earth, and have formally taken possession of the entire region, and adjacent, for and in the name of the President of the United States of America.

I leave this record and United States flag in possession.

Robert E. Peary,
United States Navy.

If it were possible for a man to arrive at 90° north latitude without being utterly exhausted, body and brain, he would doubtless enjoy a series of unique sensations and reflections. But the attainment of the Pole was the culmination of days and weeks of forced marches, physical discomfort, insufficient sleep, and racking anxiety. It is a wise provision of nature that the human consciousness can grasp only such degree of intense feeling as the brain can endure, and the grim guardians of earth's remotest spot will accept no man as guest until he has been tried and tested by the severest ordeal.

Perhaps it ought not to have been so, but when I knew for a certainty that we had reached the goal, there was not a thing in the world I wanted but sleep. But after I had a few hours of it, there succeeded a condition of mental exaltation which made further rest impossible. For more than a score of years that point on the earth's surface had been the object of my every effort. To its attainment my whole being, physical, mental, and moral, had been dedicated. Many times my own life and the lives of those with me had been risked. My own material and forces and those of my friends had been devoted to this object. This journey was my eighth into the arctic wilderness. In that wilderness I had spent

nearly twelve years out of the twenty-three between my thirtieth and my fifty-third year, and the intervening time spent in civilized communities during that period had been mainly occupied with preparations for returning to the wilderness. The determination to reach the Pole had become so much a part of my being that, strange as it may seem, I long ago ceased to think of myself save as an instrument for the attainment of that end. To the layman this may seem strange, but an inventor can understand it, or an artist, or anyone who has devoted himself for years upon years to the service of an idea.

But though my mind was busy at intervals during those thirty hours spent at the Pole with the exhilarating thought that my dream had come true, there was one recollection of other times that, now and then, intruded itself with startling distinctness. It was the recollection of a day three years before, April 21, 1906, when after making a fight with ice, open water, and storms, the expedition which I commanded had been forced to turn back from 87° 6' north latitude because our supply of food would carry us no further. And the contrast between the terrible depression of that day and the exaltation of the present moment was not the least pleasant feature of our brief stay at the Pole. During the dark moments of that return journey in 1906, I had told myself that I was only one in a long list of arctic explorers, dating back through the centuries, all the way from Henry Hudson to the Duke of the Abruzzi, and including Franklin, Kane, and Melville—a long list of valiant men who had striven and failed. I told myself that I had only succeeded, at the price of the best years of my life, in adding a few links to the chain that led from the parallels of civilization towards the polar center, but that, after all, at the end the only word I had to write was failure.

But now, while quartering the ice in various directions from our camp, I tried to realize that, after twenty-three years of struggles and discouragement, I had at last succeeded in placing the flag of my country at the goal of the world's desire. It is not easy to write about such a thing, but I knew that we were going back to civilization with the last of the great adventure stories—a story the world had been waiting to hear for nearly four hundred years, a story which was to be told at last under the folds of the Stars and Stripes, the flag that during a lonely and isolated life had come to be for me the symbol of home and everything I loved—and might never see again.

The thirty hours at the Pole, what with my marchings and countermarchings, together with the observations and records, were pretty well crowded. I found time, however, to write to Mrs. Peary on a United States postal card which I had found on the ship during the winter. It had been my custom at

various important stages of the journey northward to write such a note in order that, if anything serious happened to me, these brief communications might ultimately reach her at the hands of survivors. This was the card, which later reached Mrs. Peary at Sydney:—

"90 North Latitude, April 7th.

"*My dear Jo,*

"I have won out at last. Have been here a day. I start for home and you in an hour. Love to the 'kidsies.'

"Bert."

In the afternoon of the 7th, after flying our flags and taking our photographs, we went into our igloos and tried to sleep a little, before starting south again.

I could not sleep and my two Eskimos, Seegloo and Egingwah, who occupied the igloo with me, seemed equally restless. They turned from side to side, and when they were quiet I could tell from their uneven breathing that they were not asleep. Though they had not been specially excited the day before when I told them that we had reached the goal, yet they also seemed to be under the same exhilarating influence which made sleep impossible for me.

Finally I rose, and telling my men and the three men in the other igloo, who were equally wakeful, that we would try to make our last camp, some thirty miles to the south, before we slept, I gave orders to hitch up the dogs and be off. It seemed unwise to waste such perfect traveling weather in tossing about on the sleeping platforms of our igloos.

Neither Henson nor the Eskimos required any urging to take to the trail again. They were naturally anxious to get back to the land as soon as possible—now that our work was done. And about four o'clock on the afternoon of the 7th of April we turned our backs upon the camp at the North Pole.

Though intensely conscious of what I was leaving, I did not wait for any lingering farewell of my life's goal. The event of human beings standing at the hitherto inaccessible summit of the earth was accomplished, and my work now lay to the south, where four hundred and thirteen nautical miles of ice-floes and possibly open leads still lay between us and the north coast of Grant Land. One backward glance I gave—then turned my face toward the south and toward the future.

KASIAGSAK, THE GREAT LIAR

from *Tales and Traditions of the Eskimo* (1875)

Hinrich Rink

Hinrich Rink served as Royal Inspector of South Greenland for the Danish government. During his years on the island, he made many scientific and ethnographic studies of the region. His Tales and Traditions of the Eskimo, *first published in 1875, contains several dozen stories, mostly from Greenland, which Rink transcribed in Danish and then translated into English.*

KASIAGSAK, WHO WAS LIVING with a great many skilful seal-hunters, always returned in the evening without a catch of his own. When he was out, his wife, named Kitlagsuak, was always restless and fidgety, running out and in looking out for him, in the hope that he might be bringing home something; but he generally returned empty-handed. One day, being out in his kayak, he observed a black spot on a piece of ice, and it soon turned to be a little seal. His first intention was to harpoon it, but he changed his mind, and broke out, saying, "Poor little thing! it is almost a pity. Perhaps it has already been wounded by somebody else; perhaps it will slide down in the water when I approach it, and then I need only take hold of it with my hands." So saying he gave a shout, at which the seal was not slow to get down. Presently it appeared close before the point of his kayak; but he called out still louder than before, and the seal went on diving up and down quite close to him. At length he made up his mind to chase and harpoon it; but somehow it always rose at a greater distance, and was soon entirely lost to him. Kasiagsak now put back, merely observing, "Ye silly thing! ye are not easy to get at; but just wait till next time."

Another day he went seaward in bright, fine weather. Looking towards land he got sight of the other kayakers, and observed that one of them had

just harpooned a seal, and that the others were all hurrying on to his assistance. As to himself, he never stirred, but remained quite unconcerned in his former place. He also noticed that the one who had caught the seal tugged it to the shore, and made it fast to a rock on the beach, intending to return in pursuit of others. He instantly put further out to sea; but when he had got quite out of sight he returned to the beach by a roundabout way, and made straight for the other man's seal, and carried it off. The towing-line was all around ornamented with walrus-teeth, and he was greatly delighted at the prospect of getting home with this prize. Meanwhile his wife had been wandering about in expectation of him, and looking out for the returning kayakers. She at length cried out, "There is a kayak!"—at which more people came running out; and shading her eyes with her hand, she continued, "It looks like Kasiagsak, and he moves his arms like one tugging something along with him. Well, I suppose it will now be my turn to give you a share, and ye shall all get a nice piece of blubber." As soon as he landed she hastened to ask him, "Where didst thou get that beautiful tugging-line?" He answered, "This morning at setting out I thought it might come in handy, as I was bent on having a catch, and so I brought it out with me; I have kept it in store this long time." "Hast thou, indeed?" she rejoined, and then began the flensing and carving business. She put the head, the back, and the skin aside; all the rest, as well as the blubber, she intended to make a grand feast upon. The other kayakers successively returned, and she took care to inform each of them separately that a seal was already brought home; and when some of the women came back from a ramble on the beach, she repeated the whole thing over to them. But while they were sitting down to supper in the evening, a boy entered, saying, "I have been sent to ask for the towing-line; as to the seal, that is no matter." Turning to Kasiagsak, his wife now put in, "Didst thou tell me an untruth?" He only answered, "To be sure I did"; whereto his wife remarked, "What a shame it is that Kasiagsak behaves thus!" but he only made a wry face, saying, "Bah!" which made her quite frightened; and when they lay down to rest he went on pinching her and whistling until they both fell asleep.

Another day, rowing about in his kayak, he happened to observe a black spot away on a flake of ice. On nearing it he made it out to be only a stone. He glanced round towards the other kayakers, and then suddenly feigned to be rowing hard up to a seal, at the same time lifting the harpoon ready to lance it; but presently went to hide himself behind a projecting point of the ice, from which he managed to climb it and roll the stone into the sea with a

splash, making it all froth and foam. Meanwhile he got into his kayak again, making a great roar in order to call the others to his assistance. When they came up to him they observed that he had no bladder, and he said, "A walrus has just gone down with my bladder; do help me to catch sight of him; meantime I will turn back and tell that I have lanced a walrus." He hurried landwards, and his wife, who happened to be on the look-out, again shouted, "A kayaker!" He called out that he had made a lucky hit. "I almost do believe it is Kasiagsak; do ye hear him in there?" Meantime he had approached the shore, and said, "In chasing a walrus I lost my bladder; I only came home to tell you this." His wife now came running into the house, but being in such a hurry she broke the handle of her knife. However, she did not mind this, but merely said, "Now I can get a handle of walrus-tooth for my knife, and a new hook for my kettle." In the evening Kasiagsak had chosen a seat on the hindermost part of the ledge, so that only his heels were to be seen. The other kayakers stayed out rather long; but the last of them on entering brought a harpoon-line and a bladder along with him, and turning to Kasiagsak observed, "I think it is thine; it must have been tied round some stone and have slipped off; here it is." His wife exclaimed, "Hast thou been telling us new lies?" at which he only answered her, "Why, yes; I wanted to play you a trick, you see."

Another day, when he was kayaking along the coast, he remarked some loose pieces of ice away on a sandy beach at some distance; he rowed up to them and went ashore. Two women, gathering berries, watched his doings all along. They saw him fill his kayak with bits of broken ice; and this done, he waded down into the water till it reached his very neck, and then turned back and got upon the beach, where he set to hammering his kayak all over with stones; and having finally stuffed his coat with ice, he turned towards home. At some distance he commenced shrieking aloud and crying, "Ah me! a big iceberg went calving (bursting and capsizing) right across my kayak, and came down on the top of me"; and his wife repeated his ejaculations, adding, "I must go and see about some dry clothes for him." At last they got him up on shore, and large bits of ice came tumbling out of his clothes, while he went on lamenting and groaning as if with pain, saying, "I had a very narrow escape." His wife repeated the tale of his misfortunes to every kayaker on his return home; but at last it so happened that the two women who had seen him likewise returned, and they at once exclaimed, "Is not that he whom we saw down below the sand-cliffs, stuffing his clothes with ice." On this, the wife cried out, "Dear me! has Kasiagsak again been lying to us?" Subsequently Kasiagsak went

to pay a visit to his father-in-law. On entering the house he exclaimed, "Why, what's the matter with you that your lamps are not burning, and ye are boiling dog's flesh?" "Alas!" answered the master, pointing to his little son, "he was hungry, poor fellow! and having nothing else to eat we killed the dog." Kasiagsak boastingly answered him, "Yesterday we had a hard job at home. One of the women and I had our hands full with the great heaps of seals and walruses that have been caught. I have got both my storehouses chokefull with them; my arms are quite sore with the work." The father-in-law now rejoined, "Who would ever have thought that the poor little orphan boy Kasiagsak should turn out such a rich man!" and so saying, he began crying with emotion; and Kasiagsak feigned crying likewise. On parting from them the following day, he proposed that his little brother-in-law should accompany him in order to bring back some victuals, adding, "I will see thee home again"; and his father said, "Well, dostn't thou hear what thy brother-in-law is saying? thou hadst better go." On reaching home, Kasiagsak took hold of a string and brought it into the house, where he busied himself in making a trap, and taking some scraps of frizzled blubber from his wife's lamp, he thrust them out as baits for the ravens. Suddenly he gave a pull at the string, crying out, "Two!—alas! one made its escape;" and then he ran out and brought back a raven, which his wife skinned and boiled. But his brother-in-law had to look to the other people for some food; and at his departure the next day, he likewise received all his presents from them, and not from Kasiagsak.

Another day he set off in his kayak to visit some people at a neighbouring station. Having entered one of the houses, he soon noticed that some of the inmates were mourning the loss of some one deceased. He questioned the others, and on hearing that they had lost a little daughter named Nepisanguak, he hastened in a loud voice to state, "We have just got a little daughter at home, whom we have called Nepisanguak"; on which the mourning parents and relations exclaimed, "Thanks be to thee that ye have called her by that name"; and then they wept, and Kasiagsak also made believe to be weeping; but he peeped through his fingers all the while. Later in the day they treated him richly with plenty of good things to eat. Kasiagsak went on saying, "Our little daughter cannot speak plainly as yet; she only cries 'apangaja!'" but the others said, "She surely means 'sapangaja'" (*sapangat*, beads); "we will give thee some for her"; and at his departure he was loaded with gifts—such as beads, a plate, and some seal-paws. Just as he was going to start, one of the men cried out to him, "I would fain buy a kayak, and I can pay it back with a good pot; make it known to the people in thy place." But

Kasiagsak said, "Give it to me; I have got a new kayak, but it is a little too narrow for my size." At length he started along with his presents, and the pot stuck upon the front part of his kayak. At home he said, "Such a dreadful accident! a boat must surely have been lost; all these things I bring you here, I have found tossed about on the ice" and his wife hastened into the house to give her cracked old pot a smash, and threw away the shoulder-blades that till now had served her instead of plates, and ornamented her coat with beads, and proudly walked to and fro to make the pearls rattle. The next day a great many kayakers were announced. Kasiagsak instantly kept as far back on the ledge as possible. As soon as the kayakers put in to shore, they called out, "Tell Kasiagsak to come down and fetch off some victuals we have brought for their little daughter"; but all the reply was, "Why, they have got no daughter at all." Another of the men now put in, "Go and ask Kasiagsak for the new kayak I bought of him"; but the answer was, "He certainly has no new kayak." At this information they quickly got up to the house, which they entered, taking their several gifts back, and last of all cutting the flaps ornamented with beads away from the wife's jacket. When the strangers were gone she said as before, "Kasiagsak has indeed been telling a lie again." His last invention was this: he one day found a small bit of whale-skin floating on the top of the water, and bringing it home he said, "I have found the carcass of a whale; follow me and I will show you it": and the boat was got out, and they started. After a good while they asked him, "Whereabout is it?" but he merely answered them, "Away yonder"; and then a little bit further, "we shall soon get at it." But when they had gone a long way from home without seeing anything like a floating whale, they got tired of Kasiagsak, and put a stop to all his fibs by killing him then and there.

SONGS OF THE INUIT

from *Across Arctic America* (1927)

Knud Rasmussen

Knud Rasmussen, an explorer and anthropologist, was born in 1879 in what is now Ilulissat, Greenland, to a Danish father and an Inuit mother. His knowledge of the Greenlandic language allowed him to communicate with native people across the region. Among the books Rasmussen wrote was Across Arctic America: Narrative of the Fifth Thule Expedition, *which chronicled his three-year, 20,000-mile trek, via dogsled, from Greenland to Siberia. This excerpt is from a section of the book devoted to a song festival of Inuit living on the shore of a lake in northern Canada. The singing takes place in the tent of Rasmussen's host, Igjugarjuk. (An* angakoq *is a shaman.)*

WOMEN DO NOT as a rule sing their own songs. No woman is expected to sing unless expressly invited by an *angakoq*. As a rule, they sing songs made by the men. Should it happen, however, that a woman feels a spirit impelling her to sing, she may step forth from the chorus and follow her own inspiration. Among the women here, only two were thus favored by the spirits; one was Igjugarjuk's first wife, Kivkarjuk, now dethroned, and the other Akjartoq, the mother of Kinalik.

KIVKARJUK'S SONG

I am but a little woman
Very willing to toil,
Very willing and happy
To work and slave.
And in my eagerness

To be of use,
I pluck the furry buds of willow
Buds like beard of wolf.

I love to go walking far and far away,
And my soles are worn through
As I pluck the buds of willow,
That are furry like the great wolf's beard.

AKJARTOQ'S SONG

I draw a deep breath,
But my breath comes heavily
As I call forth the song.

There are ill rumors abroad,
Of some who starve in the far places,
And can find no meat.

I call forth the song
From above,
Hayaya—haya.

And now I forget
How hard it was to breathe,
Remembering old times,
When I had strength
To cut and flay great beasts.
Three great beasts could I cut up
While the sun slowly went his way
Across the sky.

In addition to ordinary hunting songs and lyrics there are songs of derision, satires with a mercilessly personal address; two men will stand up in turn and accuse each other before the assembled neighbors. These accusations, even when well founded, are received with surprising calmness, whereas "evil or angry words" may have far more serious effects.

I give here Utahania's impeachment of one Kanaijuaq who had quarrelled

with his wife and attempted to desert her, leaving her to her fate out in the wilds; the woman, however, had proved not only able to stand up for herself in a rough-and-tumble, but left her husband of her own accord and went to shift for herself, taking her son with her.

Something was whispered
Of man and wife
Who could not agree.
And what was it all about?
A wife who in rightful anger
Tore her husband's furs across,
Took their canoe
And rowed away with her son.
Ay—ay, all who listen,
What do you think of him,
Poor sort of man?
Is he to be envied,
Who is great in his anger
But faint in strength,
Blubbering helplessly
Properly chastised?
Though it was he who foolishly proud
Started the quarrel with stupid words.

Kanaijuaq retorted with a song accusing Utahania of improper behavior at home; his hard words however, seemed to make no difference to their friendship. Far more serious was the effect of malicious words in the case of Utahania's foster-son who was once upbraided by his foster-father as follows:

"I wish you were dead! You are not worth the food you eat." And the young man took the words so deeply to heart that he declared he would never eat again. To make his sufferings as brief as possible, he lay down the same night stark naked on the bare snow, and was frozen to death.

Halfway through the festival it was announced that Kinalik, the woman angakoq, would invoke her helping spirits and clear the way of all dangers ahead. Sila was to be called in to aid one who could not help himself. All the singing now ceased, and Kinalik stood forth alone with her eyes tightly closed. She uttered no incantation, but stood trembling all over, and her face twitched from time to time as if in pain. This was her way of "looking inward," and

penetrating the veil of the future; the great thing was to concentrate all one's force intently on the one idea, of calling forth good for those about to set out on their journey.

Igjugarjuk, who never let slip an opportunity of exalting his own tribe at the expense of the "salt water Eskimo," informed me at this juncture that their angakoqs never danced about doing tricks, nor did they have recourse to particular forms of speech; the one essential was truth and earnestness—all the rest was mere trickwork designed to impress the vulgar.

When Kinalik had reached the utmost limit of her concentration, I was requested to go outside the tent and stand on a spot where there were no footmarks, remaining there until I was called in. Here, on the untrodden snow, I was to present myself before Sila, standing silent and humble, and desiring sky and air and all the forces of nature to look upon me and show me goodwill.

It was a peculiar form of worship or devotion, which I now encountered for the first time; it was the first time, also, that I had seen Sila represented as a benign power.

After I had stood thus for a time, I was called in again. Kinalik had now resumed her natural expression, and was beaming all over. She assured me that the Great Spirit had heard her prayer, and that all dangers should be removed from our path; also, that we should have success in our hunting whenever we needed meat.

This prophecy was greeted with applause and general satisfaction; it was plain to see that these good folk, in their simple, innocent fashion, gave us their blessing and had done all they could to render it effective. There was no doubting the sincerity of their goodwill.

On the following night: we were racing at full speed over the wintry surface of Lake Hikoligjuaq. The firm ice was spread with a thin layer of soft, moist snow, acting as a soft carpet to the dogs' paws, and the long rest in complete idleness with plenty of fresh caribou meat had given them a degree of vitality that made it a pleasure to be out once more. We had two lads with us as guides, who had borrowed Igjugarjuk's dogs, but it was not long before they were hopelessly out-distanced, and we had to content ourselves with a guess at our direction.

Early in the morning, before the sun was fairly warm, we reached the southern shore of the lake and camped in a pleasant little valley, fastening the dogs in a thicket of young willow that stood bursting in bud to greet the spring.

In the course of the day we went out to reconnoitre. And it was not long before we came upon a solitary caribou hunter observing us from a little hill. He was just taking to flight when the two lads from the last village, who had now come up, recognized him and called him by name, when he walked up smiling to meet them. He informed us that there was a village of five tents a couple of hours' journey farther inland, and that we could reach the place without difficulty, although the ground was bare. We tried to persuade him to come back with us to the camp, but he preferred to go on ahead and tell his comrades of the strange meeting. And before we had gone far, the whole party came down and overtook us, they had been too impatient to wait for our arrival. It was hard work for the dogs to get the sledge over the numerous hills, and even the level ground was difficult going, sodden as it was with water and broken by tussocks and pools. There were plenty of willing hands, however, and we made our way, albeit slowly, with a great deal of merriment. Miteq and I had to face an endless rain of questions. These inland folk look upon the sea as something wonderful and mysterious, far beyond their ken; and when we explained that we had had to cross many seas in coming from our own land to theirs, they regarded our coming in itself as something of a marvel. And we agreed with them in their surprise at our being able to understand one another's speech.

Suddenly speech and laughter died away; the dogs pricked up their ears, and a strange silence fell upon all. There, full in our way, lay the body of a woman prone on the ground. We stood for a moment at a loss. Then the men went forward, while we held back our dogs. The figure still lay motionless. A loud wailing came from the party ahead, and Miteq and I stood vaguely horrified, not knowing what it meant. Then one of the men came back and explained that we had found the corpse of a woman who had been lost in a blizzard the winter before—and he pointed to one of those bending over her; that was her husband.

It had been a hard winter, and just when the cold was most severe, six of those in the village had died of hunger. A man named Atangagjuaq then determined to set out for a neighboring village in search of aid, and his wife, fearing lest, weak as he was, he might be unable to complete the journey, had followed after him. She herself, however, had been lost in the snow before coming up with him. They had searched for her that winter, and in the following spring, but without result; and now here she lay, discovered by the merest accident right athwart our course.

I walked forward to view the body of this woman who had lost her life in a vain attempt to help her husband. There was nothing repulsive in the sight;

she just lay there, with limbs extended, and an expression of unspeakable weariness on her face. It was plain to see that she had walked on and on, struggling against the blizzard till she could go no farther, and sank exhausted, while the snow swiftly covered her, leaving no trace.

The body was left lying as it was; no one touched it. We drove on, and in an hour's time reached the Eskimo camp.

These people are quick to change from one extreme of feeling to another. We had not gone far on our way before the dead woman, to all seeming, was forgotten, and the merriment that had met with so sudden a check broke out afresh. As soon as we had put up our tent, the men got hold of our ski, and went off to try them in a good deep snowdrift that still lay in a gap. They had never seen ski before, and great shouts of laughter greeted the first attempts of those venturesome enough to try them. One of the gayest of the party was Atangagjuaq, who but a few minutes earlier had stood weeping beside the body of his wife.

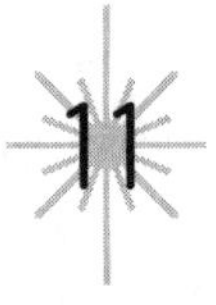

THE GARDEN OF EDEN

from *Salamina* (1935)

Rockwell Kent

In Salamina, *the American artist Rockwell Kent chronicles the year (1932–33) that he spent living and painting on Greenland. This selection describes a five-day stay in an isolated house on the Karrat Fjord. Accompanying Kent is a native Greenlander named Pauline, who has been delivered to him—along with rice, coffee, and oatmeal—by his local provisioner, David.*

Sunlight to see by, ice to travel on, and work to do. The work was painting. It was for that that I had come to Greenland; by that, and maybe for that, that I lived and found it *almost* good most anywhere, alone. "If a man," said Socrates, "sees a thing when he is alone, he goes about straightway seeking until he finds someone to whom he may show his discoveries, and who may confirm him in them." So does the artist then seek solitude that, seeing when he is alone, he shall be constrained to such utterance as may endure to seek and find his friends. They will confirm him.

"Discoveries"—are we discoverers then, we writers, poets, sculptors, picture-painters? It is all anyone at most can be. Leif Ericsson, Magellan, Cook, the architect of the first pyramid, the builder of the first arch, Homer, Shakespeare, Euclid, Newton, Einstein: all are discoverers, revealers, of what was and is, of continents, of natural law, of the human soul. God, let us say, made Adam. It was for Michelangelo to discover, as though for the first time, how beautiful God's Adam was. And it remains for all of us, forever, to discover as though for the first time how beautiful the sunrise is, and the moon, and night, and plain and mountain, land and sea, and man and woman; how *beautiful* life is. And whether we pursue discovery in the environment at home which is familiar to us all, or abroad in the remoter

and less-known regions of the earth, we'll find the field still unexplored and rich in undiscovered beauty.

[. . .]

[A]bout five miles from Nugatsiak is the island of Karrat, which, though one of the smaller islands of that archipelago, is an imposing landmark by reason of its comparative isolation and the noble architecture of its mountain mass. With towers and buttressed walls reared high upon a steep escarpment, it has the dignity of a great citadel standing to guard the gateway to the glamorous region of Umiamako. I'd thought of some day camping there, to paint. So that when, having arrived at Nugatsiak on this trip with David, it came to my ears that there stood an untenanted house on the western end of Karrat I had at once but one idea: to look it over with a view to staying there. I promptly called upon the owner, a Nugatsiak man, crawled in to him, bowed low to him and stood with bended head in deference to his ceiling and my head, sat with him and his numerous family in one of the smallest, lowliest, dirtiest, and most friendly houses I'd had cause to enter, and got at once his glad consent to use his Karrat house. "You'll need the key," he said. "It is standing in the lock."

That I went next day to Karrat with supplies for only overnight was sheer stupidity; for no sooner had I seen the cove where stood the house, and had one glimpse of its stupendous views, than it was settled in my mind to stay my time out there. "You'll leave me here," I said to David, "go straight back to Nugatsiak for the night, bring me more canvases and food tomorrow, and then go hunting where you please; and stay—five days."

The cove, three sides surrounded by the steep hillsides and ledges of the foreland, lay beautifully sheltered from most winds. Its background was the donjon keep of Karrat; it faced the mountainous environs of the mouth of Kangerdlugsuak. One would breathe deep and fast who lived in such a place.

The house I might not readily have found had I been there alone, so dwarfed were man-sized things. David, who knew the place, drove straight in to the head of the cove, whipped the dogs up the steep and high embankment of a knoll, and stopped. An edge of turf across a mound of snow: that was the house.

It took us but a minute or two to clear away the drifted snow from around the low doorway, to turn the key (the owner had been right, the house was locked), pass through the low turf passageway, and enter. We found ourselves inside a sort of ice cave, dimly and glamorously illuminated by the cold daylight which filtered through the snowbank at the window and through a

snowed-up stovepipe hole in the roof. Where, through this hole, and through innumerable leaks in the flat roof, water had trickled there hung great glittering ice stalactites that were at variance with my intentions there. And directly under the stovepipe hole—there was no stove—was an accumulated mound of ice a foot thick at its crest. Walls, ceiling, floor; ice, ice: it wasn't as you'd picture home. No matter; it would do. We brought my things indoors, I set the primus up and lighted it and put on snow to melt; in twenty minutes we were drinking coffee. The sound of dripping water filled the house.

And now, with David waiting to depart, I took pencil and paper and drew up a list of those things of which I'd have need in the days to follow. It was a problem to express my wants in such a way as Pavia would understand: I supplemented pigeon Eskimo by art. Rice, oatmeal, coffee, a Greenland halibut—that's all. Oh, no! a postscript just for fun, to make old Pavia laugh. "And,"—*"ama,"* I wrote, *"niviassak pinakak."* And I drew a picture of the thing: a pretty girl. Then, instructing David to bring me from my own supplies a quantity of canvases, I sent him on his way.

Who hasn't, over and over again in the course of his life, fixed up places to live in? Houses out of chairs and shawls, houses in trees, houses of snow, caves, lairs in impenetrable thickets, tents, log houses, lofts, abandoned houses, sheds, boats, homes: the thrill of making them is never lost. I looked my ice cave kindly in the face; it wept encouragement. Enough. And with an ardor worthy of the task I set to knocking the stalactites from its hoary brow, scraping the ice from its incrusted cheeks, digging the ice up from—we'll have to drop our physiognomy—the floor, and melting it, as much as my small pocket primus would, from everywhere. And at bedtime I had the deep satisfaction of having converted a dry cold ice cave into an ice-cold sump. Such, we may say, is progress. I crawled into my sleeping-bag and slept.

Just as a painter doesn't *have* to have a duplex studio, north light, and inside balcony with silk brocades draped over it, so does he not have to have an easel. Outdoors they're such a nuisance that I never use them. A stick to prop the canvas with and stones to hold it down: that's good most any time of year. But in *deep* snow, a couch; I found one in my Karrat house. It was a rather elegant affair, homemade, of course, just wood, but shapely, with a sort of arm or back at one end of it. With this couch planted to its belly in the snow, my canvas propped up at the arm-embellished end, my palette flat in front of it, I sat next morning on the hill and worked. My theme was mountains, and its foreground, snow, the snow plain of the frozen fiord. So

hours passed. It must have been near noon when, looking up, I saw that there had crept into my foreground plane a minute sledge propelled by insect dogs. They *did* look small in that immense environment. I left my work and went to meet them at the house.

David had brought my things, I made that out as he drew near. But what a lot of stuff! My canvases, they bulked up large; but it was hard to see just what he had, against the sun's glare on the snow. The hill now hid both sledge and dogs. I heard the whip's report, the shrill yap of a dog; *"Eu, eu!"* the voice of David. Heads down, tails up, the dogs appear, the dogs, the sledge, the driver—David, and—by God—the girl. My postscript in the flesh.

"So you brought everything," said I to David as we unpacked.

"Yes," said he, "I think so."

"Then come indoors; we'll all have coffee."

✳

"Five days. Remember, David, and come back."

"I will," said David, and drove off.

Pauline—that was her name—and I watched from the hilltop till he turned the point. We waved good-by.

She was a pleasant, quiet-wayed, mature young woman of twenty. She was dumpy, round-cheeked, good-looking only if one fancied Greenland looks. She was a normal, healthy Greenland Eve. But for the respect that I had formed for Spartan Greenland womanhood I would have felt misgivings about the Eden that I had brought her to. She let me put such thoughts away. And whether, during the days that followed, she was occupied with chipping ice from the ceiling, or with digging out the rotting bones and filth from the frozen bog which was the floor, or shivering in idleness, or wading in the snowdrifts to keep warm, she was absolutely, peacefully, content and happy. Entering that dripping house I'd shudder and, by way of cheer, ask, *"Ajorpa?"*—"Are things too bad?" She'd look up from her work and, smiling, say, *"Ajungulak"*—"It's good." We praise the lark for singing under summer skies; Pauline would stand there singing by the hour in that cave.

We hadn't much to eat; I hadn't figured on two mouths to feed. For lunch I cooked a mess of rice and pemmican or fish, for supper, oatmeal porridge—eaten plain. We had no dishes, so we ate by turns out of the pot; we had no tableware, so I whittled a wooden spoon from a piece of board. *"Mamapok,"* said Pauline, tasting the porridge. That means "Delicious."

It never got warm in the house, the primus was too small. And I couldn't burn it continuously, for I had little petroleum. Within an hour of putting out the stove the floor and lower walls would freeze again. Pauline had little on. "Where are your clothes, Pauline?" I asked. "There," she answered, pointing to a pair of kamiks and an anorak.

There was no way for us to sleep but in my sleeping-bag. If out of gallantry I'd given it to her, I would out of misery during the night have crawled into it beside her. And if I hadn't, no one would believe it. We both got in the bag. We tried to get in with our clothes on, and we couldn't. Then we took them off. Two fingers in one finger of a glove: that tight, we managed it. The nights were hell. We shared the privilege by turns of working out an arm and leaving it outside to cool and almost freeze. That was our one relief. She slept a little, but she didn't sing.

I could have sent Pauline back home the day she came: not for the world would I have sent her back. I'd written for a girl: well, here she was. And whether Pavia was to be credited with a rare sense of humor or with incredible efficiency as a storekeeper is beside the point. *Sent home!* Pauline could never have lived down the ignominy. She knew that well; and stayed.

By day she'd clean the house, and wash the pot and spoon, and stroll around and sing. And then at night, following the example of my intentioned shamelessness, she'd strip her pauper's garments off and slither into the warm reindeer bag beside me. Poor little shivering, ice-cold, uncomplaining Eskimo: how almost numb with cold she was! Then warmth would come and with it, sometimes, sleep.

The days were uneventful; they were mild and fair. Over our heads the dark deep vault of blue, before our eyes huge snow-incrusted mountain walls and one restricted vista far across the snow-covered plain of the sea to the distant peaks and ranges of Nugssuak. The peak of Karrat towered over us; its dark-red rocks were gold against the zenith sky. We seldom walked far, for the snow was deep and heavy, but from the summit of the foreland what a view there was! One saw, near by, the marvelously corrugated mountain sides of Kekertarssuak, and then the whole broad tossing panorama of the northern ranges. Pale gold of sun-illumined snow, and blue; with here and there a patch of bare black mountain side to prove the blinding pitch of all the rest. Snow-blind: perhaps our eyes are stricken to preserve our souls.

We rose at earliest daylight, and turned in at dark; almost all day I worked. We had two visitors. The first, a hunter coming out of Kangerdlugsuak, saw me at work and came to pay a call. He was a most unhappy man. He told me

that his wife had died, and how his house stood empty now. So good a house, glass in the windows, and a stove. Yes, there it stood in Nugatsiak, no one in it. He sat a long time in silence. "There were once many houses on Karrat," he continued. "One house stood there, another there. Yes, there were many houses here one time. Now, only one. The houses, people—gone." I thought he'd weep. He used, he said, to catch a lot of seals; now he caught none. Hard days, hard times. And he had nothing now to eat. Would I lend him twenty-five öre? If he got a seal, he would pay me; if he got none, he wouldn't. He told me that his name was Jakob.

Our second visitor was named Abraham. He brought me a letter from Salamina. And he brought me two ptarmigan; of these we made a feast.

Two visitors and a brace of ptarmigan: these were the events of five days in the life of Pauline. The routine of her island days, and those events, might be the type of what her life would be for forty years. A house, a man to do his work, whatever it might be, apart from her, the household chores, long idle hours at the window or outdoors, a little food, a little warmth: these make the pattern of a Greenland woman's life. And children as a by-product. Yes, Pauline in the course of years as she'd stand idly gazing from the window and singing softly to herself would rock her body soothingly to lull her child to sleep. There'd be that difference.

I wonder what it would be like, that island and Pauline, for life. I wonder whether a white man would have sense enough to let things be, just let the days and the relationship run on and not go prying into her exotic soul, nor even thinking that she had a soul, nor caring. The silly fool would doubtless fall in love with her and spoil it all; display, to her astonishment, the strange behavior of romantic love; insist on finding in her placid face some revelation of unfathomed pagan depths; get staring at her disconcertingly, annoying her. If at such antics Pauline didn't get, at last, the poor oaf's measure, size up what he *was* good for in cash expressed in clothes and beads, in *Danish* clothes, in leisure for herself and servants to support it, if she didn't contrive that her fatuous adorer should leave the island and move with her into the great city of Umanak, she'd be a fool—or a philosopher. Pauline, I think, was neither. And if a romantic white man lured by the visionary primitive did fall in love with her, she'd serve him in her own behalf as other "primitives" have served such men.

Therefore, when on the fifth day David came we packed our household goods and art, and drove away. The people of Nugatsiak crowded around us as we came to land. "How did you like the pretty girl?" asked Pavia. "The pretty girl," I answered him, "was swell."

KABLOONA

from *Kabloona* (1941)

Gontran De Poncins

In 1938, Gontran De Poncins, a French count, set off to live among the Inuit of King William Island, in the Canadian Arctic. This passage from his book Kabloona*—the title refers to a local term for Europeans—recounts De Poncins' arrival at the camp where he would spend much of the next fifteen months. His host is Utak, who has agreed to assist him in return for trade goods equal to the value of one white fox.*

NIGHT WAS FALLING WHEN of a sudden three glimmering points too faint to be called lights pricked the grey scene. The igloos! Through the translucent snow of which these houses are built the feeble gleam of seal-oil lamps was visible, bespeaking the breath of life and the presence of man on this pallid ocean of ice. I crept through a winding tunnel so low that I went on all fours and knocked in the dark against wet and wriggling hairy bodies. These were the dogs. They had taken shelter in the freezing porch against the greater cold outside. Not for an empire would they have stirred out of my path, and over and among them I crawled until I emerged into the igloo.

But was this an igloo? This witch's cave black on one side with the smoke of the lamp and sweating out on the other the damp exudation caused by the warmth of lamp and human bodies! Within, nothing was white save an occasional line that marked the fitting of block to block; and the odor was inconceivable. In the vague light of the lamp shapeless things, men and women, were stirring obscurely. If you wanted a hierarchy of light you might say that before electricity there was the gas-jet, before the gas-jet the lamp, before the lamp the wax taper, before the wax taper the tallow candle, and before the tallow candle

the seal-oil vessel. I was in a brown bear's lair, a troglodyte's cave. What would elsewhere be the stone age was here the ice age.

I was too newly come from Outside to see in the igloo anything but filth: the charnel heap of frozen meat piled on the ground behind the lamp; the gnawed fish-heads strewn everywhere; the sordid rags on the lumpish flesh, as if these Eskimos had worn their party clothes to the Post and were here revealing their true selves, the maculate bodies they covered with skin and fur to hide the truth from the White. And to heighten the horror of the scene, one of these Eskimos would fling himself from time to time into the porch—as the tunnel is called through which I had crawled—to drive out the dogs; and a howling would resound as of murder committed in a subterranean chamber.

Even to-day, as I write, it is still difficult for me to explain how it happened that I was able to accustom myself to this life, so that within a month a description like this would seem to me stupid, would seem a recital of non-essentials and a neglect of everything consequent in Eskimo existence.

Fortunately, I was too overcome with weariness to be able to think. Details met my eye and offended it, but they could not reach as far as my brain. My box had been dragged in, and like an automaton I opened it in order to find something to eat, something "white" that would preserve me from all this. My soup was not there! Had I forgotten it? Probably; and for the reason that I had thought about it too much not to forget it. What was the Eskimo word for "soup"? I thumbed through my dictionary without a thought that the Eskimo might never eat soup, and there might be no word for it. Instead, I cursed the dictionary with the curse usual the world over—that a dictionary never contains the words we need. I could not explain to Utak what was missing; but as he saw me hunting, turning my effects over and over, he too—and this was the only comic note of the evening—he too began to hunt, though he knew not what he was hunting. What was I to eat? That frozen fish? That repellent snow-covered thing I could hear grating in their teeth as they chewed?

The household stared at me, and I needed no word of Eskimo to understand what they were thinking: not only had this white man no titbits to offer to them, he had not even brought his own grub. They said nothing, but their disapproval was unmistakable. Sick at heart, I crept into my bag and fell asleep without a morsel of food.

We slept six in a row, squeezed together in an igloo built to hold three, our heads turned towards the porch. The men lay naked in their caribou

sleeping-bags. I kept my clothes on, and it was as if I were sleeping in a cage with wild beasts. All night long something dripped from the ceiling upon my face, and though each drop sent a twinge of pain through me, I could not evade it because we were squeezed too tightly together. All night long, too, my neighbor, Utak's brother, made use of the tin that served as chamber-pot, and each time he would hold it out at arm's length without stirring, and empty it under my nose. In a corner an old woman spat the whole night through, and between the one and the other, in a spirit of the deepest gloom of heart, through which the two or three images of warmth and comfort that I summoned were unable to make their way, I fell finally asleep.

✳

When I awoke the igloo was empty except for the old woman: the men had gone fishing.

I crawled out of doors and had a look round. It wanted almost an effort to identify the igloos in this landscape. There were four in all, four molehills made of snow; and had it not been for the harpoons and other accoutrements sticking up like vertical black lines drawn on white paper, I should not have seen them. These strokes were the only signs of the existence of a camp in this white infinity.

The camp was deserted. Nothing stirred. Here and there a puppy lay half buried in the snow. The men had gone with their sleds and their dogs. Every day was for them a day of work and travel: every morning they awoke to the same seasonal chores: ice the runners, harness the dogs, unleash the 40-foot serpentine whip with its 12-inch handle, and go off to the fishing or the hunt.

The camp was built on the flank of a ridge, doubtless because the snow here was more plentiful. Below me I could see a wide flat surface which was a lake. Three out of the five men in the camp had gone ice-fishing on this lake; the other two had preferred to go off to another lake, twenty-five miles distant, on pretext that the fish there were bigger. At this time of year the ice was only two feet thick, and fishing was still easy.

✳

Utak came up from the lake before the rest in order to build me an igloo. It was not to be separated altogether from his own, but would be a sort of lean-to opening into his igloo, and through this opening he and his wife would be

able to keep an eye on my tin of biscuits. However, I should at least sleep alone this night.

One hour sufficed Utak for the erection of my spiral shelter, and it was no sooner finished than soiled. The dogs climbed and ran all over it on the outside, as is their habit, and yellowed its dome and sides. Ohudlerk hastened to pay me a visit as soon as I had installed myself. With a great deal of hawking and spitting he explained to me that the igloo was perfect—from which I was to understand how great was my debt to Utak. And Utak himself, by way of creating a fitting atmosphere, came in with the gift of a heap of rotted fish.

An igloo is very pretty when it is new, when it has just been finished and the *iglerk,* the flat couch of snow that rises about fifteen inches from the floor, has been smoothed down. It is so pretty, so white, so pure with its little heaps of powdered snow at the base of the meeting of the blocks, that one is afraid to move in it for fear of soiling it. But the miraculous industry of the Eskimo soon removes this sense of caution and daintiness. In less than a day the igloo is made cosy and homelike: everything is spattered and maculated; the heaps of objects brought inside create great black spots where they lie; the ground is strewn with the debris of fish spat forth in the course of eating; everywhere there are stains of seal blood and droppings of puppies (puppies are allowed indoors).

I am told that there are Eskimos who keep their igloos clean, scraping the floor daily and sprinkling fresh snow over it to cover the stains. This is not the case with the Netsilik of King William Land, who seem to feel the most profound indifference, indeed contempt, for cleanliness. As for my igloo, they invaded it as if in conquered territory; and after all, it was their igloo, I was their guest, they had doubtless the right to treat it as their own. There they sat on my *iglerk,* belching and laughing, picking out a morsel of the fish that lay on the ground—our food and the dogs' as well—as if they had come each time upon something particularly savory, and spitting the bones out straight in front of them.

I say again that I was too green to have any notion of Eskimo values. Every instinct in me prompted resistance, impelled me to throw these men out,—to do things which would have been stupid since they would have astonished my Eskimos fully as much as they might have angered them. I knew nothing, for example, of the variant of communism they practised, and which I later learned was the explanation of their taking possession of me, their shameless sharing among themselves of my goods, which on this occasion made me think of them as inconceivably impudent, filled with effrontery, and of myself

as helpless and in a hopeless situation. They were the masters, I the captive, I said to myself. You wanted to live with the Eskimos, did you? I said. Well, here you are, you silly ass.

Thus, my beginnings went very badly. Worse than the pillage was the fact that two days later my hands froze.

"Una-i-kto!"—It is cold, Utak had said on waking that morning. But we had gone off together on his sled to fish on the great lake whose name I had by now learned. It was called Kakivok-tar-vik, "the place where we fish with the three-pronged harpoon."

Half a mile out from shore Utak began by clearing the snow off the surface of the lake with his native shovel in a circle about twelve feet in diameter. Then he knelt down, a hand shading his eyes, his nose to the ice, and tried to judge whether or not the depth of the lake here was what it should be. I did as he did, and could see the bottom of the lake perfectly, the grasses waving and the fish moving past in their tranquil world. As soon as he spied the fish, Utak became feverish. He ran to the sled, which with the dogs had been left a hundred feet off, came back with an ice chisel, and now the ice was flying in an upward rain of chips. He was cutting out a hole, and it was incredible with what speed and precision he worked. I have seen Eskimos go through five feet of ice with one of these chisels in ten minutes. He would stop at every four or five inches, send down a sort of ladle made of bone, and slowly and cautiously bring up the chips.

When the hole had been pierced through, the water flowed in and brought to the surface the odd chips that still remained, which were carefully ladled off. Then, on the far side of the hole, Utak built a wind-screen of three snow blocks, one set straight ahead of him and each of the others serving as wings. This done, he spread a caribou skin, and knelt on it. With his left hand he unrolled a long cord at the end of which hung a small fish made of bone, with two fins. He let the decoy down into the water, and when he jigged, or pulled on the cord, which he did with the regularity of a clock, the fins beat. The little bone fish was like a water-bug swimming. In his right hand, held very near the hole, was the *kakivok,* the great three-pronged harpoon. When the fish, lured by the decoy, came swimming beneath Utak, he would lower his harpoon gently into the hole, and at the proper moment he would strike, and the fish would be speared.

Nothing was more comical than the silhouette of Utak, his bottom in the air, his nose literally scraping the ice, his eyes fixed on the moving water, his whole being as motionless as a deer at the moment when it takes fright and is

about to run. At first I had knelt beside him. Then, my hands freezing and my muscles stiff, I stood up to stretch. He became furious, for a man walking round the hole frightens away the fish. But one could hum as much as one pleased without disturbing them, and as Utak peered into the hole he kept up a monotonous humming. I came back to where he crouched, for I was fascinated by what he was doing. This seemed to please him, and undoubtedly it did. The Eskimo is very proud of everything that he does, and to see a white man imitating him is for him the highest flattery.

With what patience that left hand, as regular as a metronome, rose and fell while the hours went by! And what passion the Eskimo put into this form of the chase! What intensity was in his gaze! The tiniest fish that passed drew from him muttered words, and it was clear that the game absorbed him, that time and space had fled leaving him only this hole in the ice over which he would peer for days if necessary. As far as the eye could see in every direction the scene was void of life; and in the midst of this immensity a single man, who might have been alone in the world, was absorbed with a scientist's concentration upon . . . upon what? Upon the art of filling his belly.

Had I not been tortured by the cold, I should have been content to watch for hours this admirable adjustment of primitive man to his element. But, although it could not have been more than fifteen degrees below zero, I was freezing. Doubtless my skin had not yet become adapted to this climate. My fingers burned in my gloves, and I was too vain to speak of it. But while I knelt there, thinking of nothing else, suddenly—a fish! Utak's right hand was closing over the handle of the *kakivok,* and before I could see what had happened, the thing was done, the fish was gasping on the ice, had flung itself twice in the air and then lay still, frozen almost on the spot. And Utak was back in the same posture, absorbed again in his chase.

We had been out several hours, and the pain in my fingers became so unbearable that I could have screamed. The heel of my hands also had begun to harden. When, finally, we stood up, I took off my gloves to have a look and saw that my fingers were waxen. I had frozen my eight fingertips.

*

Three days later my fingers were still useless: hard as wood, very painful, whenever I touched anything with them they burned, and I could not so much as roll myself a cigarette. Rubbing them with snow did no good. Dipping them in coal-oil merely produced in them a sensation of cold. There was

no remedy, and the best I could do was to hope they were not permanently frozen. Meanwhile, I was chained to the igloo like a hospital patient to his bed.

From my *iglerk,* my couch, I watched the life of the women through the opening in the wall between our igloos. Unarnak, Utak's wife, was industriously at work with her *kumak-sheun,* her louse-catcher, a long caribou bone with a tuft of polar-bear hair glued to the end. The hairs must have had an extraordinary attraction for the lice, for this species of hunting was always successful. It was a treat—though I agree, of a special kind—to see Unarnak pull three lice in succession off the hairs and crack them in her teeth.

On the skins that covered their *iglerk* the little boy was naked at play. He strutted, grimaced, chattered, and held behind him a looking-glass while he peered round to see in it the reflection of his bottom.

When the child forgot himself on the caribou skin his mother put out a casual hand and scooped the brine off the couch. The hand of the Eskimo is always busy, and it serves him in a thousand ways: for example, to pick the nose and then carefully place the catch in the mouth—a detail for which I beg to be excused, but nothing is more typically Eskimo than this. It is with her hand that Unarnak trims the wick of the seal-oil lamp; and when she has finished she sucks the oil from her fingers, or else wipes her fingers in her hair—though the latter means a less thorough job. I have never yearned to find myself lord of a harem of native mistresses; but the sight of Unarnak would deprive any white man of the temptation to make her dishonorable proposals.

At the other end of the *iglerk* Utak's mother, Niakognaluk, sat in her habitual seat. Squatting beside the completely shapeless old woman was a yellow bitch with flopping ears, rendered equally shapeless by the fact that she was heavy with a coming litter. The old woman sat all day long scraping skins—a task that never ends in the life of the Eskimo, for weather, snow, and water are constantly soaking and hardening the clothes he wears and the skins he lies on, and it is only by this process of continual scraping that the hides can be softened again and made wearable and usable.

Niakognaluk is the only completely bald woman I have ever seen. She sat in her corner wearing an old woollen bonnet, dressed in hides so worn that all fur and hair was long gone out of them and they were as black and shiny as a blacksmith's leather apron. Bowed over the lamp, working with misshapen hands, her feet folded beneath her, she scraped and scraped; and as she worked tirelessly on she would murmur words which for all I knew might have been addressed to the lamp, to the dog, to herself. When a skin was finished she

flung it against the igloo wall with an air of weariness and indifference and got up to get another, holding up her caribou trousers with both hands—the dress of these men and women is much alike—as she staggered across to the pile of skins, bent stiffly down, fumbled in the heap, and reeled back to her corner to squat again over her work.

She had two or three different scrapers to work with, but the real softening was done with her teeth. I have said before, I believe, that the Eskimo's teeth serve him as a third hand, and though I had demonstrations of this again and again, yet each time it was as marvellous in my eyes as a turn at the circus. The miracle was that when Niakognaluk had finished a skin it was really white and as supple as a glove.

Among the Eskimos as with the humble of every land, the Old Woman seemed to express the sum of experience, of hardship, of wisdom. She was symbolic; she was permanence; she was She Who Stays Behind. The others leave or die: she is always there. Each death, each winter, adds its burden to her load of life, bends and bows her a little more, but it does not achieve the breaking of her, and she goes on living. She mutters and seems to grumble, merely because she is old; but because she is old, also, her heart is kind. She makes no demands, and when you make her a little gift she sends forth a worn smile that is warm with friendliness.

Utak's mother was like this. She mumbled constantly over her work. She pretended endlessly that the child, who lived part of the day in the deep hood that hung down her back, would never leave her in peace. Tyrannical as are all Eskimo children, he rode her as if she were a spavined old mare, shook her as if she were a plum-tree; and while she complained her patience was limitless.

Generally the child was out with his mother, and the old woman sat alone with her dog. There was a sort of resemblance between the two. Two slanting slits were all the eyes one saw in the old woman's face, and the same was true of the dog. The bitch's coat and the old woman's covering were of the same color and the same state of decrepitude. Both were worn out by life, neither had any strength left; and when the old woman took up a whip-handle and beat the dog to drive her away, it was feebly and without conviction that she did so. The old dog would moan, but it would not stir. "You see," the dog seemed to say; "you try to beat me, but at bottom you don't even want to. And I don't want to go away. We belong here together." They would sit motionless, looking at each other, the dog with its flopping ears and bowed legs, the old woman with her rounded back and her misshapen hands. I could imagine their dialogue.

"You ought to be ashamed of yourself!" the old woman seemed to say. "To go on having puppies at your age! Will you never stop? And I'm sure their father is that Arluk, that useless hound who howls in harness before you've even laid a whip to him. Kigiarna, aren't you ashamed?"

The bitch would flatten herself out, cringing.

"And of course you intend to drop your litter in the warmth of the igloo. Naturally. I'll find the puppies one of these days in my sleeping-bag, and I'll be the one to bring them up. Not for the first time, either. You weren't so concerned about these things when you were younger. Time was when your puppies were born in the porch; out in the snow, even."

And Kigiarna would approve every word, pitiably.

✳

Among these Netsilik Eskimos, these *Inuit* or "men, preeminently," as they boasted themselves, the routine upon waking in the igloo never varied. It went like this.

First, hawk and spit for at least half an hour.

Second, grumble and mutter until your wife, having crawled out of the *krepik,* the deerskin bag, has taken up the circular knife and cut off a great piece of the frozen fish that lies on the ground.

Third, eat the fish, panting and grumbling meanwhile because wife and child are stirring in the *krepik* and getting into the way of your free arm.

Fourth, between each bite, suck your fingers noisily and tell a story or recount your dream, a satisfied appetite having put you in a good humor.

Fifth, with great deal of puffing and snorting, light the Primus stove. If it refuses to go, fling it across the igloo and slide down growling into the *krepik,* after which silence is restored in the igloo. If the Primus should catch with little trouble,

sixth, brew tea, gulp down two or three mugs, and say *"Una-i-kto,"*—"It is cold,"—so that the Kabloona (the white man) may be seized with compassion and get up to prepare his grub for you. After each mug of tea, wipe up the leaves with your fingers and eat them.

Finally, having eaten and drunk and woken the entire household, come up out of your *krepik,* ready to be off fishing.

✳

The men came in from their jigging and the silent igloo was suddenly filled with stampings, threshings and snortings as of beasts in a stable. Voices and laughter broke forth; the constant and horrible coughing and spitting began that seems always to attack the Eskimo indoors. The tea, which had been boiling all day long above the seal-oil lamp, was poured into mugs and bowls and its steam rose from between their hands in an odor of seal while the air of the igloo became a vapor in which the bodies were seen as shapeless blurs.

Almost immediately an incident took place that gave me a great fright.

Ohudlerk's son, Kakokto, who had been away fishing on the distant lake, had come back to visit his father. He and his wife were standing before me in my igloo, and as it was time to eat and they seemed to be expecting something, I fed them. Whether it was jealousy or not I do not know, but while I was talking to the young couple, Unarnak came in, picked up my sack of flour, and took it into her igloo. There she proceeded to bake an impressive quantity of *baneks*—a sort of flat bread—which she distributed to everybody present. I watched her out of the corner of my eye and observed that she had been lacking in the first article of courtesy, which was to offer the *baneks* first to me. I waited a moment; then, seeing that the sack was not restored to its place, I told her quietly to put it back where she had found it.

Up to that moment everybody had been in splendid humor—those in my igloo because they had supped handsomely, those in Utak's because they had received an unexpected offering. As soon as I had spoken—in the hearing of every one—silence fell. Unarnak, knowing that I considered her at fault, gathered together all the *baneks* and placed them without a word on my *iglerk,* with an air of complete disinterestedness. But her husband would not take it thus. Like a true Asiatic, this Eskimo conceived his finest vengeance to lie in ridicule. Refusing contemptuously the tea I offered him, he let himself back on his couch and then, the igloo being full of people, smoke, and laughter, lying back, and with an almost casual air, he began to tell them how I had frozen my fingers. It was not hard to guess his story.

"And there was the Kabloona," he said sarcastically, "walking in a circle and stamping his feet, blowing on his fingers, making noise enough to frighten away all the fish in the lake, and saying, *'Aiie! Aiie!* My fingers are frozen!' till he looked like an unhappy fish, like the littlest fish in the world."

All of them roared with laughter, for a game of this sort is always played collectively. Besides, they knew that if they laughed the wonderful story would go on. And it did. I could feel the tone of its rise, could see that Utak

had become excited from the very fact of having a tale to tell, and was making it as daring and as cutting as he could.

"Finally we started back," he went on. "It was all he could do to drag himself back to his couch, and there he's been lying these three days past, whimpering and showing his fingers and moaning: *'Una-i-kto!'* "

And Utak, with wonderful mimicry, counterfeited not only my gestures but the very timbre of my voice.

"And to-night, finally, for a couple of grains of flour . . ."

The rest found all this very funny, and each time that a burst of laughter greeted one of his sallies Utak would turn towards me to see how I was taking it. Some of the others, indeed, rose from their places and came round to have a good look at me as I sat there, ill and half stupefied with fever,—for it was curious that my frozen fingers had raised my temperature.

I went on talking to Kakokto and his wife as if I had no notion what was towards, but at bottom I was extremely upset by the sudden turn which things had taken. And they did not entirely stop there. For when Utak had finished, he and his wife got up and went triumphantly out to tell their story in the next igloo.

Remembering what Paddy Gibson had said to me about Utak, I slept with one eye open. This fellow is subject to fits of temper, I thought. He killed his stepfather. He had to leave his own family and come to live on this side of the island because of it. I was uneasy. Never before had I heard that word for white man—Kabloona—pronounced with such contempt; and I suspected that this contempt came into an Eskimo's mouth only when he felt positively aggressive. I seemed to myself imprisoned in a disquieting atmosphere and had no notion what might come of this. I knew only one thing—that I could not retreat from the position of indifference and dignity which I had adopted.

I dropped off to sleep, and suddenly I awoke. I had no notion of the time and could hear the child crying. Utak was standing smoking a cigarette, his wife was stirring about, and all three were clothed. Unarnak came into my igloo and, thinking me asleep, picked up swiftly a pile of skins which belonged to them and had been stored with me for want of room in their igloo. They are going to strike camp and desert me, I said to myself, still with my eye on them. They seemed to be consulting each other, to hesitate. Finally they went to bed. Weariness sent me back to sleep—and in the morning it was all over. The first thing I saw on opening my eyes was Utak bending over me, grinning and offering me a mug of tea, a peace token.

A GREENLAND CHRISTMAS

from *An African in Greenland* (1981)

Tété-Michel Kpomassie

Tété-Michel Kpomassie was born in Togo in 1941. As a teenager, he opened a book called The Eskimos from Greenland to Alaska *and began, in his words, "dreaming of eternal cold." Improbably enough, ten years later he had made his way to Kangerlussuaq. In this excerpt, he is staying in the small village of Rodebay, on the west coast of the island. Hans and Cecilia are his hosts.*

All Greenland adults, both men and women, love cigars, which they smoke in a very curious fashion, not wasting even the ash . . . After taking a few puffs on a lighted cigar, they put it out by spitting on it, and immediately put the burnt end in their mouths and bite off the glowing ash along with a slice of raw tobacco. They turn it rapidly over and over on the tongue—the cigar end is sometimes still red-hot—and chew it, rolling their eyes with ineffable delight as they swallow the hot ash mixed with shreds of tobacco, then spit out black saliva and announce, "*Mam-mâk!*" (It's good!) The rest of the cigar is stowed away in a pocket, to be repeatedly lit, smoked, extinguished, and consumed all over again. According to them, this warms your face and improves your circulation when you're out in a kayak in the cold.

So Hans seemed very happy with my packet of cigars and, wanting to give me a present in return, that evening brought me a seashell he'd picked up on the shore, and a black ballpoint pen, property of the Rodebay school, whose name was engraved on it.

At half-past six I visited Thue, who offered me coffee and a big, hard biscuit. Hendrik and Marianna had moved out with their children and gone to live with Søren Petersen; the house was quieter, but sad and empty. So there

was Thue, alone now with his two youngest children and Saqaq, who had not yet left him. Apart from them, he had his kayak—which never brought back any seal and which he wouldn't be able to use for several months while the sea was frozen over—and his sledge, which had several wooden slats missing. And nobody, nobody in the village offered to help him!

Arkaluk no longer had a shirt of his own, and had to wear one of his father's. The shirt was far too big for him, so, to make it fit better, the boy, standing in front of me with the shirt on, started to poke new buttonholes alongside the old ones, using a knife. As he cut the cloth, he bent his head down over his chest and kept sniffing in the snot that dripped from his nostrils.

*

Friday, December 31. Starting at one in the afternoon, groups of children made the rounds of the houses, though they didn't sing as they had at Christmas. When they came to us, Cecilia doled out cakes and they left with thanks.

That day produced the strangest sight I'd seen since the start of the festivities. Young men disguised as spirits roamed the village in the endless night. They generally gathered by the roadside in front of the unlit house used as a communal workshop for building kayaks and sleds. These "spirits" looked like great bundles of clothes; they were called *midartut* (*midartôk* in the singular). I was oddly reminded of those grotesque unmasked creatures called *zangbéto*, who in the villages of my native land loom up out of the darkness, creep into huts, and by their unexpected and terrifying presence silence little children who cry in the night. But unlike the *zangbéto* of Togo, the *midartut* of Greenland play no useful role in society. They spring out and give you a fright, run after passers-by, perform a ponderous dance, sometimes roll on the ground at your feet, and almost never say a word.

We were at Jorgensen and Augustina's, clasping hands and singing carols round the Christmas tree with the children, when a *midartôk* arrived. He was covered in such an array of skins that he must have had trouble breathing; and he had a stick in his hand. As soon as he came in, the terrified children scattered and hid—some under the bed, some behind their father or mother. Hans forgot the pious hymns we'd been singing and broke into a lively, lilting song. The *midartôk* answered with his own wild dance, leaping in the air and banging on the floor with his stick. At the end they gave him a cake, and he left saying, "*kuyanâk*"—the one word he had spoken during the whole performance.

I left Jorgensen's to fetch my tape recorder, so as to record those rousing

tunes to which the *midartut* dance. But on my way home I was stopped by a man named Eliassen and his wife. He told me, "You can see the *midartut* another time. Come and have coffee with us."

I followed them back to their house, which turned out to be full of people drinking and enjoying themselves. Their pretty little thirteen-year-old daughter was serving the visitors, helped by her sister, Adina, who was six.

The guests were already half stewed. Eliassen, behaving as if his own wife wasn't there, started to take great liberties with one girl: sitting in front of her, he kept teasing her with shouts of "*Tui!*" as he darted his hand up her skirt. Meanwhile his wife sat down opposite me and gave me a long, slow look; then suddenly her face lit up with a sugary smile, exposing her upper gums and three missing teeth, while blue veins pulsed on her neck and flat bosom, where a crucifix gleamed. With the same bedroom eyes, she told me three times, "*Assavakit*" (I love you), in the hearing of her husband, who burst out laughing at the sight of my embarrassment.

I don't know if *assavakit* has some more respectable meaning; apart from the girls, Hans and several other men had often said it to me as a sign of friendship. Probably it also means simply "I like you."

We all toasted one another with one glass and then another in quick succession, for, besides cups of coffee, we all had before us three glasses containing different alcoholic drinks, and these were filled as soon as we emptied them. In the living room, women enticed me with come-hither stares and suggestive movements.

In the thick of all this uproar, the door was flung open and in came Thue, quite well dressed. How people detested him! They all blamed him for letting me leave his home. They didn't throw him out, but they wouldn't give him anything to drink. He managed a tight-lipped smile and gave each of the guests a long stare. Then, as if asserting his rights, he suddenly grabbed a glass three-quarters full of *imiak* just as we were shouting "*Skol!*"—but he didn't raise the glass to anyone before he started drinking.

Then the *midartut* arrived, and among them I recognized Maria; some of them had pulled old pink stockings over their heads, weirdly flattening their noses. But soon we heard the church bell ringing: it was midnight, and all these drunken people set off to the church for the service. No *midartôk*, however, took part.

This is when you should see Rodebay—after the New Year's Eve midnight mass. Exhilarated with drink and song, chanting as they roamed through the village, delirious villagers started on another round of visits. I followed the

midartut into about ten houses, then finally stopped at Knud's, where there was still light in the windows. He was with a little man in a white anorak. As he got up to greet me, Knud signaled to me not to speak too loudly: a woman was curled up asleep on the couch in a dim corner of the living room. I took the glass of beer he handed me to celebrate the New Year, and five minutes later set off home to bed.

As I was crossing the yard, I met a man who kept shining his torch right in my eyes. In my country this is considered rude. Who could it be? I turned my own light on him: it was Johan Dorf. Both of us lowered our torches. Thinking no more of it, I went on my way.

I was right by the door of the outside larder when Dorf attacked me from behind and knocked me to the ground. I got up with grazed fingers, and Dorf, like a vicious mongrel that runs off after creeping up and pouncing on a man, scuttled off to Knud's house. I followed him there to have it out with him at once; as soon as I opened the door he swung around, aimed a kick at my belly, and slammed the door in my face. I heard him screaming inside like a madman, "Get out of here, you rotten nigger!"

It was the first time I'd ever been called that, though I'd long ago realized that when someone having a dispute with a black man calls him "rotten nigger" or "filthy nigger" or some such name, it's always some embittered neurotic trying to work off frustrations that have nothing at all to do with the "nigger." In this case Johan Dorf, who was envious by nature, was jealous of my success with the women there.

I managed to get the door open again, but he gave me another kick in the stomach and heaved it shut again. Shouting for Knud, I pulled at the door again, and the bastard gave me a third kick. Trying to dodge it, I slipped and fell on the steps. I was just getting up to open the door yet again when Knud came flying outside.

"Run for it," he shouted. "They're fighting inside."

Winded, and suddenly weak at the knees, feebly I went along.

"Let's go and hide in the school. We're on our own. We're friends, Michel!" Knud panted as we stumbled through the deep snow, floundering into drifts.

We burst into the school building, and Knud double-locked the door. We sat down and he started to cry. Only then did he explain to me that the woman sleeping in his house, so dead drunk that she couldn't walk home, was Johan Dorf's wife.

"But what's that got to do with me?"

All the while, I was trying to control my temper. Knud took my torch to go

and see if things had calmed down at his place. He came back very agitated, then went out again. Finally he returned and said we could go back now.

In the yard he had me wait near a shed by the house while he went into the living room. A minute later he came back and called my name.

Two men who had nothing to do with the affair were sitting calmly at the table. The woman in question hadn't stirred: she was still fast asleep. We talked about Dorf's strange behavior, and I informed Knud and his visitors that I intended to report the attack to the police representative; every village has a man with the imposing title of *Usted bestyrer*—literally, "village administrator"—whose main function is to keep the police informed. Detectives visiting a community where a crime has been committed go and see him first of all. The one in Rodebay was called Knud Jørgensen.

They told me Knud was falling-down drunk, so seeing that the whole village was now unprotected by the law, I decided not to be made a fool of: the next time I saw him, I'd teach Johan Dorf a lesson. I was twenty-four. Johan, strongly built and fat as a pig, was nearly thirty-eight. So he wasn't old—he could put up a fight.

Suddenly the door flew open. Dorf came in and made for his wife. I called his name but he just scowled at me, so I grabbed him by the collar. Trying to hold me back, the others hung onto my coat so hard that it ripped, but I took it off and started fighting. The table overturned. Dorf managed to throw himself on top of me, but I pounded his face with my fists, then felt sick at the sight of the bloody mess I made of him. My *kamiks* slipped in the beer; everybody shouted at once. Finally we stopped, and they pulled Dorf to one side. Knud asked him why he had knocked me down and kicked me in the stomach.

"I don't remember," he said as he slumped gasping onto the couch.

This idiotic answer made me go for him again. I hit him in the face, and we were at it again, till he sat down, winded once more. Suddenly, in came the little man in the white anorak who had been at Knud's an hour or so earlier; I let this little puppet stand there and wave his arms. Saqaq arrived, learned what was happening, and wanted to give me a hand. Gradually, however, everything calmed down. They showed Dorf the door. He turned back. Knud shouted "Out!" and gave him a kick in the rear. Some people came to take him and his drunken wife away.

Many of the villagers were afraid of this bully Dorf who, because he ran the Danish shop and handled the trading company's operations in the village (buying seal, fox, and other skins), liked to be feared and to play the top dog. But that evening, his fat face puffed and bleeding, he slunk through the village

with lowered head. They congratulated me on teaching him a lesson, and Knud suggested a snack.

A group of children came and sang at the window to wish Knud a happy New Year. He invited them in and handed out cakes. I calmed down at the sight of these children with their peaceful faces.

But that evening still had surprises in store. When the children had left, the five of us sat down round the living-room table and began to pitch into some cold meat. Besides myself and Knud, there were his houseboy Évat, Saqaq, and the young man Paviâ. A dispute soon broke out among the three young Greenlanders over the quality of Knud's *imiak*. Saqaq found it too weak, therefore second-rate. Knud did not reply, but his houseboy Évat couldn't allow his boss's *imiak* to be slandered like this, so he and his friend Paviâ saw Saqaq out, but first, to my surprise, they calmly asked him to apologize and to say thanks by shaking hands with everyone, beginning with Knud. And so he did, before he left!

✳

Saturday, January 1. We split up at eight in the morning. What blackness outside! When it was time to leave I couldn't find my torch; someone had been seen picking it up after the fight, but nobody remembered who. I went off flanked by Évat and Paviâ, who saw me all the way to Hans's door.

"It's for your protection," they told me.

They were afraid of possible reprisals by Dorf's friends.

Our *Usted bestyrer* never asked me anything about the fight. Recovering from his night on the bottle, he remarked: "Dorf looks as if he fell out of a *timissatôq* (helicopter)."

As for Dorf himself, far from making any more trouble, he kept out of my way. And from that day on, every time he had a row with someone, they'd say: "So you think you can push me around because I'm small and weak? I won't fight—you're stronger than me. But try it with Mikilissuak and see what happens!"

✳

In the evening there was a big New Year's dance at the *Forsamlinghus* (the village hall). I went with Hans, who was not sure where Cecilia had got to. We each took a kroner, the admission price.

The village hall, a small clapboard building, was lit inside by an oil lamp. Hendrik stood at the far end of the packed room and played the accordion: he was the only musician, but everyone joined in the singing.

At first, with pride they sang *Nunarput utorkarssuángoravit* (Our Old Country), a fine poem written by Hendrik Lund around 1912, which soon became Greenland's national anthem to a tune by Jonathan Petersen. The poet sings the praises of the *Kalâtdit*, the Greenland man, who survives today and will live on forever, because he can make the most of his country's resources. Then everybody roared the refrain:

> But we must set ourselves new aims,
> And go on growing, that one day we may be
> The respected equals of all other nations.

Hendrik kept squeezing his accordion, but it's not easy dancing to a national anthem! So someone started singing "Narssaq," a song that salutes the most fertile region of the south, where cows, ponies, and sheep graze peacefully in the meadows among clapboard houses. Narssaq abounds with forage and food and green fields lying at the foot of the mountains! The song goes on to compare the delights of Narssaq with those of the Biblical lands of Lebanon and Sharon.

During this dance, Søren and his partner Amalia, who had had a few drinks too many, fell flat on the floor. Far from asking if they had hurt themselves, everyone gathered around and held their sides with laughter.

The next song was "Sunia": When Sunia appears at the far end of the fjord, towing a whale behind his boat, there is great joy in the village. The Inuit happily divide the animal among themselves, and every villager has as much meat and blubber as he wants! But Niels, Lukas, and Markers down so much *mattak* that it sticks in their throats. So they have to be thumped on the back to make all the *mattak* go down! The singers end with the wish that some day Greenlanders will be sailing real whaling ships of their own.

But the liveliest air, started by Hans, was called "Aanayarak ayagtugaussok," Snow White! For the Inuit really have a sung version of this fairy tale. In this song the wicked stepmother, jealous of Snow White's beauty, throws her out in the snow. But the fox, the crow, the snow goose, and the arctic hare lead her to the turf cottage of the Seven Dwarfs. When Snow White gets there she puts the house to rights, scrubs the floor, trims the wick on the

seal-oil lamp, and prepares a tasty meal of dried fish, *mattak*, and reindeer meat. Then the Seven Dwarfs come back from the coal mine. Delighted to find their cottage spick and span, they adopt Snow White; overjoyed, they begin to dance the rhythmic steps of the *sisamak*, the strange dance we ourselves were now dancing. We formed a big circle, and to the music of the accordion took four steps to the left, four to the right, then swung giddily round with our partners. It was a reel, a cheerful fifteenth-century Scots country dance that the natives had learned from European whalers.

Despite all the liveliness and gaiety, I felt a little disappointed. In their amusements, the inhabitants of this western coast have retained hardly anything of their own cultural heritage, nothing that really belongs to them. The accordion which Hendrik tirelessly played was a foreign instrument. As for the Eskimo drums, made of a circular wooden frame covered with a stretched membrane which is tapped on the edge with a slender stick (strangely enough, never on the membrane itself), nowadays they can be found only at the National Museum in Copenhagen! I rather missed the New Year festivities in my home village, where our dances, not copied from anyone else's, are cadenced to the rhythms of the tom-toms.

To give Hendrik a breather, they brought out records and a small battery-powered record player, and we danced till morning.

On various pretexts, the holiday continued all through the first week of January. On Monday the 3rd, the whole village assembled at the pastor's, where we drank till five in the morning: it was his birthday. The next day was Knud's birthday, and his house never emptied. The following day it was someone else's turn, and so on.

Those who had enough provisions spent only an hour or two a day repairing their outbuildings or kayaks. Only Thue went hunting every day, but only from ten until noon. Today he came back with an eider and a little guillemot. In the dark of their living room, I could dimly see Maria boiling the birds, and Saqaq squatting down. In the other room the oil lamp was burning feebly, and Thue was sitting on the floor drinking. I had a cup of tea with him while he got drunk on "Patria," a terrible Danish wine.

January 5. This evening—unusually—we were all home, and everyone was busy with some kind of work. Cecilia sat on the floor with a thimble on her finger, mending the soles of a pair of *kamiks* with a bodkin, some seal sinew

thread, and scraps of new skin. She made two patches on each sole, at the heel and toe. Poyo worked on Hans's seal net, while Hans himself concentrated on making a dog whip, which he plaited in a curious way with a single leather thong, so that it resembled a pigtail. The lash was about seven meters long, tapering at the end. Hans started the braiding about fifteen centimeters from the thicker end, that would later be attached to a wooden handle. In the tough leather thong he made a slit through which he passed the other end of the lash, then pulled on it tightly to make a knot. Then he made the next slit, the next knot, and so on. The knots were very tight and close together. It was hard work, but the result was a series of beautiful interlacing patterns, like little superimposed triangles with the apex of one entering the base of the next, while at the same time a number of elegant wavy lines appeared on either side of the leather thong. To pull the knots as tight as possible, Hans used not only his hands and knees but also his teeth, worn down to the gums: he tugged with them and chewed at the leather, with dizzying jerks of his head. He made sixteen knots in all, and then added to the end of the whip a finer lash of very smooth skin that came from a baby narwhal found in its mother's womb. It is this flexible and hard-wearing section which is used to discipline the huskies. After three days' work, the whip was ready. The thick end was attached to the wooden handle by several layers of cord passing over and under the leather. The whip was nine to ten meters long, and very clean and white now. But wait until it had been used for one winter!

Then Hans made an ice chopper. The handle, two meters long, would later be shortened by twenty centimeters. The blade, flat and rather long, was inserted halfway into the handle, which was split at one end; then it was lashed tight with coiled twine, and reinforced lower down by a nailed iron plate, with six nails on each of the handle's four sides. The beveled blade was sharpened with a file.

*

January 6. There were still a few *midartut* to be seen in the village. It all came to an end that evening—it was the last day, the last night, the end of the festivities! Tomorrow the Christmas decorations would be taken down in all the houses . . .

How quiet we all were at home that night! The oil lamp was turned up high, and several candles had been lit at once, as if their combined brightness could drive out our crushing boredom. Hans, who all day long had been

alternately sullen and morose, then touchy and surly, slumped in a chair and heaved a loud sigh. He had an Eskimo book in his hand, but he kept nodding off. Cecilia shuttled silently between the living room and the kitchen, where Poyo sat quiet as a mouse. Was it post-Christmas fatigue—here big celebrations are nearly always followed by deep depressions—or was it just the weather? It was piercingly cold now. The Greenland cold, strangely enough, didn't make you shiver or cause your teeth to chatter, for it wasn't just all around you, it was *inside* you. It permeated everything: houses, clothing, people, things. You were reluctant to touch a plate, a pan, a cigarette lighter in your pocket, a watch left at your bedside overnight, and so on.

Yet the arctic cold is less intense and harsh during the months when the ground is still covered by a thick layer of snow than in March and April when that snow is transformed into ice. Today the temperature, which was minus seventeen degrees in the morning, dropped to minus nineteen. The weather had been mild up to now, considering it was the middle of the arctic winter. But by January I already felt as if I were living in a refrigerator.

Well now . . . A few days before, old Sophia and her husband Knud had introduced me to one of their daughters, Else, who had recently arrived for the holidays from Jakobshavn, where she worked as a nurse. I spent that night in their house, in bed with Else. The coal stove went out as soon as everyone was in bed, and the inside temperature plummeted. At two in the morning, I still couldn't get to sleep. How cold it was, my God, how cold! A woman's warmth doesn't protect you completely from that sort of cold! Yet we had two blankets and an eiderdown over us, and I was wearing a pull-over, too! Else wore nothing!

The next day, January 7, when school started again, the temperature dropped to minus twenty-two degrees. I looked on as poor little children emerged from houses whose faintly glowing windows were curtained with frost, and watched them trudge to school mutely through the frigid whiteness. A great silence reigned in the village. That return to school after the holidays was one of the saddest sights I have ever seen: not one voice, not one sound, not a single dog barking.

The local men, who also seemed to have taken a long Christmas holiday, returned to their hunting that same day. Clad in animal skins, with their ice picks on their shoulders, they headed for the bay, which gave off a grayish haze.

The men had started work again, but not for long, you can be sure. On Thursday, January 13, towards one o'clock in the afternoon, the villagers climbed a mountain behind the village to witness, so they told me, the first

appearance of the sun: we did in fact see a faint yellow glow on the horizon, but the sun itself would remain invisible for another few months.

This "return" of the sun, which seemed to me completely imaginary, gave rise that night to the most curious of the Greenland celebrations, a dance in the town hall that only adult males and their wives may attend—young people are not admitted.

Before going to this affair, Jørgensen and his wife Augustina looked in at our house. Hans, who was getting dressed, appeared in his underclothes and called to his friend's wife. Augustina got up at once and joined him in the bedroom, while her husband stayed with Cecilia in the living room, and never lost his smile. The couple reappeared soon afterward. This time Hans was dressed and wearing boots—of different colors. Apparently this was done on purpose, because he looked at his feet, clicked his tongue, and gave a sly smile. I couldn't account for this freak of fancy, but Jørgensen and some other men going to the dance had apparently had the same odd notion and were also wearing boots that didn't match.

Hans and Jørgensen took a quick drink and set off with their wives. I followed the two couples as far as the village hall, where a man standing guard at the door refused to admit me. "You're not married," he said. But the pastor spotted me from inside and asked that I be let in.

For the first time, I saw coffee being served in the village hall during a dance. The women were no longer separated from the men, as they usually were, but sitting with them. I felt out of place, with everybody paired off from the start. After midnight, with great uproar and loud singing, and faces pouring with sweat, people began to exchange wives. I had read that those present at these public wife-swappings practiced what was referred to as "dowsing the lamps," but that didn't happen on this occasion. You just saw a man get up and go sit beside another couple. He talked for a few moments, then simply left the hall with the other man's wife. Sometimes this happened while the other partner was dancing. The husband thus relieved of his wife looked around him, spotted another woman who might or might not be sitting with her husband, and went up to her. Soon after that, he too went off with a woman who was not his wife. Though the husbands relieved of their mates in this way gave the impression of not being too upset, it looked to me as if most of the women, if you watched closely, were only half willing. Still, like the co-wives in my native land, they seemed resigned to an age-old tradition.

Cecilia went off on Jørgensen's arm! I ran into Hans, who was heading

towards a small group at the other end of the hall. Seeing his friend lead his wife away, he told me jokingly: "You should have got cracking with her, but now she's gone—so hard luck!"

Around one in the morning I left the hall, which was now three-quarters empty. At home I was getting ready for bed when the door opened: it was Hans, coming home with Augustina.

She didn't leave until nine o'clock the next morning. Cecilia didn't get home until eleven.

Hans simply asked her: "Were you cold there in the night?"

"*Namik!*"

"That's good."

And that was all! Life resumed or just continued as if nothing had happened between the two of them, who had just openly swapped partners.

So Cecilia was Jørgensen's standby wife, while his wife Augustina was Hans's. Jørgensen was Hans's best friend. Apparently this exchange of wives took place only among friends, obeyed precise rules, and united the two men by an unbreakable tie. Though this practice brought them fresh pleasure, it also involved strict duties, and was not to be compared with the brief exchanges of girlfriends by the young unmarried men. Neither the Jørgensen family nor Hans's would ever go short of food if one of the two men returned empty-handed from hunting, or fell ill, so long as the other had killed a seal.

Motives of survival, then, have given rise to the strange custom of wife-swapping in the Far North. In this light, certain details of behavior which had made no sense to me a few weeks earlier now appeared in a different perspective. For instance, I remembered Cecilia protesting sharply when Jørgensen told me that his wife Augustina and I should become close friends. A third man—particularly a foreigner—admitted into this union of two couples could endanger, and even destroy, the alliance that would protect one of the two wives in the event (since women don't hunt) that her husband got killed. In the same way, Thue, practically a widower, couldn't enter into the firm and sacred friendship to be formed between two men only by means of their wives, and which laid the basis of unfailing mutual help between the two families.

But if the exchange of wives was a matter only for the couples concerned, why make a public exhibition of it by organizing a dance before the exchange takes place? Probably so as to have no secrets from the village. Hans packed away his boots of different colors, whose significance may be found in the exchange itself, and said to me: "They fit me like a pair of gloves, but they're sometimes hard to wear!"

For my part, I had grasped the meaning of exchanging wives in the Arctic. All the same, I couldn't help wondering what my father and uncles, my brothers, and above all my grandfather would have to say about it, when I told the tale back home. Perhaps they'd simply think I'd been living among madmen.

"But you're a strange lot, too," Hans shot back, "with your eight wives and more under one roof, when it's hard enough living with just one!"

IN A FAR COUNTRY (1899)

Jack London

In 1897, Jack London joined the Klondike gold rush. While in the Yukon, he developed scurvy and, as a result, eventually lost his four front teeth. "In a Far Country" is one of the first stories London published. He sold it to Overland Monthly *in 1899, reportedly for $7.50.*

When a man journeys into a far country, he must be prepared to forget many of the things he has learned, and to acquire such customs as are inherent with existence in the new land; he must abandon the old ideals and the old gods, and oftentimes he must reverse the very codes by which his conduct has hitherto been shaped. To those who have the protean faculty of adaptability, the novelty of such change may even be a source of pleasure; but to those who happen to be hardened to the ruts in which they were created, the pressure of the altered environment is unbearable, and they chafe in body and in spirit under the new restrictions which they do not understand. This chafing is bound to act and react, producing divers evils and leading to various misfortunes. It were better for the man who cannot fit himself to the new groove to return to his own country; if he delay too long, he will surely die.

The man who turns his back upon the comforts of an elder civilization, to face the savage youth, the primordial simplicity of the North, may estimate success at an inverse ratio to the quantity and quality of his hopelessly fixed habits. He will soon discover, if he be a fit candidate, that the material habits are the less important. The exchange of such things as a dainty menu for rough fare, of the stiff leather shoe for the soft, shapeless moccasin, of the feather bed for a couch in the snow, is after all a very easy matter. But his pinch will come in learning properly to shape his mind's attitude toward all things, and especially toward his fellow man. For the courtesies of ordinary

life, he must substitute unselfishness, forbearance, and tolerance. Thus, and thus only, can he gain that pearl of great price,—true comradeship. He must not say "Thank you"; he must mean it without opening his mouth, and prove it by responding in kind. In short, he must substitute the deed for the word, the spirit for the letter.

When the world rang with the tale of Arctic gold, and the lure of the North gripped the heartstrings of men, Carter Weatherbee threw up his snug clerkship, turned the half of his savings over to his wife, and with the remainder bought an outfit. There was no romance in his nature,—the bondage of commerce had crushed all that; he was simply tired of the ceaseless grind, and wished to risk great hazards in view of corresponding returns. Like many another fool, disdaining the old trails used by the Northland pioneers for a score of years, he hurried to Edmonton in the spring of the year; and there, unluckily for his soul's welfare, he allied himself with a party of men.

There was nothing unusual about this party, except its plans. Even its goal, like that of all other parties, was the Klondike. But the route it had mapped out to attain that goal took away the breath of the hardiest native, born and bred to the vicissitudes of the Northwest. Even Jacques Baptiste, born of a Chippewa woman and a renegade *voyageur* (having raised his first whimpers in a deerskin lodge north of the sixty-fifth parallel, and had the same hushed by blissful sucks of raw tallow), was surprised. Though he sold his services to them and agreed to travel even to the never-opening ice, he shook his head ominously whenever his advice was asked.

Percy Cuthfert's evil star must have been in the ascendant, for he, too, joined this company of argonauts. He was an ordinary man, with a bank account as deep as his culture, which is saying a good deal. He had no reason to embark on such a venture,—no reason in the world, save that he suffered from an abnormal development of sentimentality. He mistook this for the true spirit of romance and adventure. Many another man has done the like, and made as fatal a mistake.

The first break-up of spring found the party following the ice-run of Elk River. It was an imposing fleet, for the outfit was large, and they were accompanied by a disreputable contingent of half-breed *voyageurs* with their women and children. Day in and day out, they labored with the bateaux and canoes, fought mosquitoes and other kindred pests, or sweated and swore at the portages. Severe toil like this lays a man naked to the very roots of his soul, and ere Lake Athabasca was lost in the south, each member of the party had hoisted his true colors.

The two shirks and chronic grumblers were Carter Weatherbee and Percy Cuthfert. The whole party complained less of its aches and pains than did either of them. Not once did they volunteer for the thousand and one petty duties of the camp. A bucket of water to be brought, an extra armful of wood to be chopped, the dishes to be washed and wiped, a search to be made through the outfit for some suddenly indispensable article,—and these two effete scions of civilization discovered sprains or blisters requiring instant attention. They were the first to turn in at night, with a score of tasks yet undone; the last to turn out in the morning, when the start should be in readiness before the breakfast was begun. They were the first to fall to at meal-time, the last to have a hand in the cooking; the first to dive for a slim delicacy, the last to discover they had added to their own another man's share. If they toiled at the oars, they slyly cut the water at each stroke and allowed the boat's momentum to float up the blade. They thought nobody noticed; but their comrades swore under their breaths and grew to hate them, while Jacques Baptiste sneered openly and damned them from morning till night. But Jacques Baptiste was no gentleman.

At the Great Slave, Hudson Bay dogs were purchased, and the fleet sank to the guards with its added burden of dried fish and pemmican. Then canoe and bateau answered to the swift current of the Mackenzie, and they plunged into the Great Barren Ground. Every likely-looking "feeder" was prospected, but the elusive "pay-dirt" danced ever to the north. At the Great Bear, overcome by the common dread of the Unknown Lands, their *voyageurs* began to desert, and Fort of Good Hope saw the last and bravest bending to the towlines as they bucked the current down which they had so treacherously glided. Jacques Baptiste alone remained. Had he not sworn to travel even to the never-opening ice?

The lying charts, compiled in main from hearsay, were now constantly consulted. And they felt the need of hurry, for the sun had already passed its northern solstice and was leading the winter south again. Skirting the shores of the bay, where the Mackenzie disembogues into the Arctic Ocean, they entered the mouth of the Little Peel River. Then began the arduous upstream toil, and the two Incapables fared worse than ever. Tow-line and pole, paddle and tump-line, rapids and portages,—such tortures served to give the one a deep disgust for great hazards, and printed for the other a fiery text on the true romance of adventure. One day they waxed mutinous, and being vilely cursed by Jacques Baptiste, turned, as worms sometimes will. But the half-breed thrashed the twain, and sent them, bruised and bleeding, about their work. It was the first time either had been man-handled.

Abandoning their river craft at the headwaters of the Little Peel, they consumed the rest of the summer in the great portage over the Mackenzie watershed to the West Rat. This little stream fed the Porcupine, which in turn joined the Yukon where that mighty highway of the North countermarches on the Arctic Circle. But they had lost in the race with winter, and one day they tied their rafts to the thick eddy-ice and hurried their goods ashore. That night the river jammed and broke several times; the following morning it had fallen asleep for good.

*

"We can't be more'n four hundred miles from the Yukon," concluded Sloper, multiplying his thumb nails by the scale of the map. The council, in which the two Incapables had whined to excellent disadvantage, was drawing to a close.

"Hudson Bay Post, long time ago. No use um now." Jacques Baptiste's father had made the trip for the Fur Company in the old days, incidentally marking the trail with a couple of frozen toes.

"Sufferin' cracky!" cried another of the party. "No whites?"

"Nary white," Sloper sententiously affirmed; "but it's only five hundred more up the Yukon to Dawson. Call it a rough thousand from here."

Weatherbee and Cuthfert groaned in chorus.

"How long 'll that take, Baptiste?"

The half-breed figured for a moment. "Workum like hell, no man play out, ten—twenty—forty—fifty days. Um babies come" (designating the Incapables), "no can tell. Mebbe when hell freeze over; mebbe not then."

The manufacture of snowshoes and moccasins ceased. Somebody called the name of an absent member, who came out of an ancient cabin at the edge of the camp-fire and joined them. The cabin was one of the many mysteries which lurk in the vast recesses of the North. Built when and by whom, no man could tell. Two graves in the open, piled high with stones, perhaps contained the secret of those early wanderers. But whose hand had piled the stones?

The moment had come. Jacques Baptiste paused in the fitting of a harness and pinned the struggling dog in the snow. The cook made mute protest for delay, threw a handful of bacon into a noisy pot of beans, then came to attention. Sloper rose to his feet. His body was a ludicrous contrast to the healthy physiques of the Incapables. Yellow and weak, fleeing from a South American fever-hole, he had not broken his flight across the zones, and was

still able to toil with men. His weight was probably ninety pounds, with the heavy hunting-knife thrown in, and his grizzled hair told of a prime which had ceased to be. The fresh young muscles of either Weatherbee or Cuthfert were equal to ten times the endeavor of his; yet he could walk them into the earth in a day's journey. And all this day he had whipped his stronger comrades into venturing a thousand miles of the stiffest hardship man can conceive. He was the incarnation of the unrest of his race, and the old Teutonic stubbornness, dashed with the quick grasp and action of the Yankee, held the flesh in the bondage of the spirit.

"All those in favor of going on with the dogs as soon as the ice sets, say ay."

"Ay!" rang out eight voices,—voices destined to string a trail of oaths along many a hundred miles of pain.

"Contrary minded?"

"No!" For the first time the Incapables were united without some compromise of personal interests.

"And what are you going to do about it?" Weatherbee added belligerently.

"Majority rule! Majority rule!" clamored the rest of the party.

"I know the expedition is liable to fall through if you don't come," Sloper replied sweetly; "but I guess, if we try real hard, we can manage to do without you. What do you say, boys?"

The sentiment was cheered to the echo.

"But I say, you know," Cuthfert ventured apprehensively; "what's a chap like me to do?"

"Ain't you coming with us?"

"No-o."

"Then do as you damn well please. We won't have nothing to say."

"Kind o' calkilate yuh might settle it with that canoodlin' pardner of yourn," suggested a heavy-going Westerner from the Dakotas, at the same time pointing out Weatherbee. "He'll be shore to ask yuh what yur a-goin" to do when it comes to cookin' an' gatherin' the wood."

"Then we'll consider it all arranged," concluded Sloper. "We'll pull out tomorrow, if we camp within five miles,—just to get everything in running order and remember if we've forgotten anything."

The sleds groaned by on their steel-shod runners, and the dogs strained low in the harnesses in which they were born to die. Jacques Baptiste paused by

the side of Sloper to get a last glimpse of the cabin. The smoke curled up pathetically from the Yukon stove-pipe. The two Incapables were watching them from the doorway.

Sloper laid his hand on the other's shoulder.

"Jacques Baptiste, did you ever hear of the Kilkenny cats?"

The half-breed shook his head.

"Well, my friend and good comrade, the Kilkenny cats fought till neither hide, nor hair, nor yowl, was left. You understand?—till nothing was left. Very good. Now, these two men don't like work. They won't work. We know that. They'll be all alone in that cabin all winter,—a mighty long, dark winter. Kilkenny cats,—well?"

The Frenchman in Baptiste shrugged his shoulders, but the Indian in him was silent. Nevertheless, it was an eloquent shrug, pregnant with prophecy.

Things prospered in the little cabin at first. The rough badinage of their comrades had made Weatherbee and Cuthfert conscious of the mutual responsibility which had devolved upon them; besides, there was not so much work after all for two healthy men. And the removal of the cruel whip-hand, or in other words the bulldozing half-breed, had brought with it a joyous reaction. At first, each strove to outdo the other, and they performed petty tasks with an unction which would have opened the eyes of their comrades who were now wearing out bodies and souls on the Long Trail.

All care was banished. The forest, which shouldered in upon them from three sides, was an inexhaustible woodyard. A few yards from their door slept the Porcupine, and a hole through its winter robe formed a bubbling spring of water, crystal clear and painfully cold. But they soon grew to find fault with even that. The hole would persist in freezing up, and thus gave them many a miserable hour of ice-chopping. The unknown builders of the cabin had extended the side-logs so as to support a cache at the rear. In this was stored the bulk of the party's provisions. Food there was, without stint, for three times the men who were fated to live upon it. But the most of it was of the kind which built up brawn and sinew, but did not tickle the palate. True, there was sugar in plenty for two ordinary men; but these two were little else than children. They early discovered the virtues of hot water judiciously saturated with sugar, and they prodigally swam their flapjacks and soaked their crusts in the rich, white syrup. Then coffee and tea, and especially the dried

fruits, made disastrous inroads upon it. The first words they had were over the sugar question. And it is a really serious thing when two men, wholly dependent upon each other for company, begin to quarrel.

Weatherbee loved to discourse blatantly on politics, while Cuthfert, who had been prone to clip his coupons and let the commonwealth jog on as best it might, either ignored the subject or delivered himself of startling epigrams. But the clerk was too obtuse to appreciate the clever shaping of thought, and this waste of ammunition irritated Cuthfert. He had been used to blinding people by his brilliancy, and it worked him quite a hardship, this loss of an audience. He felt personally aggrieved and unconsciously held his muttonhead companion responsible for it.

Save existence, they had nothing in common,—came in touch on no single point. Weatherbee was a clerk who had known naught but clerking all his life; Cuthfert was a master of arts, a dabbler in oils, and had written not a little. The one was a lower-class man who considered himself a gentleman, and the other was a gentleman who knew himself to be such. From this it may be remarked that a man can be a gentleman without possessing the first instinct of true comradeship. The clerk was as sensuous as the other was æsthetic, and his love adventures, told at great length and chiefly coined from his imagination, affected the supersensitive master of arts in the same way as so many whiffs of sewer gas. He deemed the clerk a filthy, uncultured brute, whose place was in the muck with the swine, and told him so; and he was reciprocally informed that he was a milk-and-water sissy and a cad. Weatherbee could not have defined "cad" for his life; but it satisfied its purpose, which after all seems the main point in life.

Weatherbee flatted every third note and sang such songs as "The Boston Burglar" and "The Handsome Cabin Boy," for hours at a time, while Cuthfert wept with rage, till he could stand it no longer and fled into the outer cold. But there was no escape. The intense frost could not be endured for long at a time, and the little cabin crowded them—beds, stove, table, and all—into a space of ten by twelve. The very presence of either became a personal affront to the other, and they lapsed into sullen silences which increased in length and strength as the days went by. Occasionally, the flash of an eye or the curl of a lip got the better of them, though they strove to wholly ignore each other during these mute periods. And a great wonder sprang up in the breast of each, as to how God had ever come to create the other.

With little to do, time became an intolerable burden to them. This naturally

made them still lazier. They sank into a physical lethargy which there was no escaping, and which made them rebel at the performance of the smallest chore. One morning when it was his turn to cook the common breakfast, Weatherbee rolled out of his blankets, and to the snoring of his companion, lighted first the slush-lamp and then the fire. The kettles were frozen hard, and there was no water in the cabin with which to wash. But he did not mind that. Waiting for it to thaw, he sliced the bacon and plunged into the hateful task of bread-making. Cuthfert had been slyly watching through his half-closed lids. Consequently there was a scene, in which they fervently blessed each other, and agreed, thenceforth, that each do his own cooking. A week later, Cuthfert neglected his morning ablutions, but none the less complacently ate the meal which he had cooked. Weatherbee grinned. After that the foolish custom of washing passed out of their lives.

As the sugar-pile and other little luxuries dwindled, they began to be afraid they were not getting their proper shares, and in order that they might not be robbed, they fell to gorging themselves. The luxuries suffered in this gluttonous contest, as did also the men. In the absence of fresh vegetables and exercise, their blood became impoverished, and a loathsome, purplish rash crept over their bodies. Yet they refused to heed the warning. Next, their muscles and joints began to swell, the flesh turning black, while their mouths, gums, and lips took on the color of rich cream. Instead of being drawn together by their misery, each gloated over the other's symptoms as the scurvy took its course.

They lost all regard for personal appearance, and for that matter, common decency. The cabin became a pigpen, and never once were the beds made or fresh pine boughs laid underneath. Yet they could not keep to their blankets, as they would have wished; for the frost was inexorable, and the fire box consumed much fuel. The hair of their heads and faces grew long and shaggy, while their garments would have disgusted a ragpicker. But they did not care. They were sick, and there was no one to see; besides, it was very painful to move about.

To all this was added a new trouble,—the Fear of the North. This Fear was the joint child of the Great Cold and the Great Silence, and was born in the darkness of December, when the sun dipped below the southern horizon for good. It affected them according to their natures. Weatherbee fell prey to the grosser superstitions, and did his best to resurrect the spirits which slept in the forgotten graves. It was a fascinating thing, and in his dreams they

came to him from out of the cold, and snuggled into his blankets, and told him of their toils and troubles ere they died. He shrank away from the clammy contact as they drew closer and twined their frozen limbs about him, and when they whispered in his ear of things to come, the cabin rang with his frightened shrieks. Cuthfert did not understand,—for they no longer spoke,—and when thus awakened he invariably grabbed for his revolver. Then he would sit up in bed, shivering nervously, with the weapon trained on the unconscious dreamer. Cuthfert deemed the man going mad, and so came to fear for his life.

His own malady assumed a less concrete form. The mysterious artisan who had laid the cabin, log by log, had pegged a wind-vane to the ridge-pole. Cuthfert noticed it always pointed south, and one day, irritated by its steadfastness of purpose, he turned it toward the east. He watched eagerly, but never a breath came by to disturb it. Then he turned the vane to the north, swearing never again to touch it till the wind did blow. But the air frightened him with its unearthly calm, and he often rose in the middle of the night to see if the vane had veered,—ten degrees would have satisfied him. But no, it poised above him as unchangeable as fate. His imagination ran riot, till it became to him a fetich. Sometimes he followed the path it pointed across the dismal dominions, and allowed his soul to become saturated with the Fear. He dwelt upon the unseen and the unknown till the burden of eternity appeared to be crushing him. Everything in the Northland had that crushing effect,—the absence of life and motion; the darkness; the infinite peace of the brooding land; the ghastly silence, which made the echo of each heart-beat a sacrilege; the solemn forest which seemed to guard an awful, inexpressible something, which neither word nor thought could compass.

The world he had so recently left, with its busy nations and great enterprises, seemed very far away. Recollections occasionally obtruded,—recollections of marts and galleries and crowded thoroughfares, of evening dress and social functions, of good men and dear women he had known,—but they were dim memories of a life he had lived long centuries agone, on some other planet. This phantasm was the Reality. Standing beneath the wind-vane, his eyes fixed on the polar skies, he could not bring himself to realize that the Southland really existed, that at that very moment it was a-roar with life and action. There was no Southland, no men being born of women, no giving and taking in marriage. Beyond his bleak sky-line there stretched

vast solitudes, and beyond these still vaster solitudes. There were no lands of sunshine, heavy with the perfume of flowers. Such things were only old dreams of paradise. The sunlands of the West and the spicelands of the East, the smiling Arcadias and blissful Islands of the Blest,—ha! ha! His laughter split the void and shocked him with its unwonted sound. There was no sun. This was the Universe, dead and cold and dark, and he its only citizen. Weatherbee? At such moments Weatherbee did not count. He was a Caliban, a monstrous phantom, fettered to him for untold ages, the penalty of some forgotten crime.

He lived with Death among the dead, emasculated by the sense of his own insignificance, crushed by the passive mastery of the slumbering ages. The magnitude of all things appalled him. Everything partook of the superlative save himself—the perfect cessation of wind and motion, the immensity of the snow-covered wilderness, the height of the sky and the depth of the silence. That wind-vane,—if it would only move. If a thunderbolt would fall, or the forest flare up in flame. The rolling up of the heavens as a scroll, the crash of Doom—anything, anything! But no, nothing moved; the Silence crowded in, and the Fear of the North laid icy fingers on his heart.

Once, like another Crusoe, by the edge of the river he came upon a track,—the faint tracery of a snowshoe rabbit on the delicate snow-crust. It was a revelation. There was life in the Northland. He would follow it, look upon it, gloat over it. He forgot his swollen muscles, plunging through the deep snow in an ecstasy of anticipation. The forest swallowed him up, and the brief midday twilight vanished; but he pursued his quest till exhausted nature asserted itself and laid him helpless in the snow. There he groaned and cursed his folly, and knew the track to be the fancy of his brain; and late that night he dragged himself into the cabin on hands and knees, his cheeks frozen and a strange numbness about his feet. Weatherbee grinned malevolently, but made no offer to help him. He thrust needles into his toes and thawed them out by the stove. A week later mortification set in.

But the clerk had his own troubles. The dead men came out of their graves more frequently now, and rarely left him, waking or sleeping. He grew to wait and dread their coming, never passing the twin cairns without a shudder. One night they came to him in his sleep and led him forth to an appointed task. Frightened into inarticulate horror, he awoke between the heaps of stones and fled wildly to the cabin. But he had lain there for some time, for his feet and cheeks were also frozen.

Sometimes he became frantic at their insistent presence, and danced about the cabin, cutting the empty air with an axe, and smashing everything within reach. During these ghostly encounters, Cuthfert huddled into his blankets and followed the madman about with a cocked revolver, ready to shoot him if he came too near. But, recovering from one of these spells, the clerk noticed the weapon trained upon him. His suspicions were aroused, and thenceforth he, too, lived in fear of his life. They watched each other closely after that, and faced about in startled fright whenever either passed behind the other's back. This apprehensiveness became a mania which controlled them even in their sleep. Through mutual fear they tacitly let the slush-lamp burn all night, and saw to a plentiful supply of bacon-grease before retiring. The slightest movement on the part of one was sufficient to arouse the other, and many a still watch their gazes countered as they shook beneath their blankets with fingers on the trigger-guards.

What with the Fear of the North, the mental strain, and the ravages of the disease, they lost all semblance of humanity, taking on the appearance of wild beasts, hunted and desperate. Their cheeks and noses, as an aftermath of the freezing, had turned black. Their frozen toes had begun to drop away at the first and second joints. Every movement brought pain, but the fire box was insatiable, wringing a ransom of torture from their miserable bodies. Day in, day out, it demanded its food,—a veritable pound of flesh,—and they dragged themselves into the forest to chop wood on their knees. Once, crawling thus in search of dry sticks, unknown to each other they entered a thicket from opposite sides. Suddenly, without warning, two peering death's-heads confronted each other. Suffering had so transformed them that recognition was impossible. They sprang to their feet, shrieking with terror, and dashed away on their mangled stumps; and falling at the cabin door, they clawed and scratched like demons till they discovered their mistake.

Occasionally they lapsed normal, and during one of these sane intervals, the chief bone of contention, the sugar, had been divided equally between them. They guarded their separate sacks, stored up in the cache, with jealous eyes; for there were but a few cupfuls left, and they were totally devoid of faith in each other. But one day Cuthfert made a mistake. Hardly able to move, sick with pain, with his head swimming and eyes blinded, he crept into the cache, sugar canister in hand, and mistook Weatherbee's sack for his own.

January had been born but a few days when this occurred. The sun had some time since passed its lowest southern declination, and at meridian now threw flaunting streaks of yellow light upon the northern sky. On the day following his mistake with the sugar-bag, Cuthfert found himself feeling better, both in body and in spirit. As noontime drew near and the day brightened, he dragged himself outside to feast on the evanescent glow, which was to him an earnest of the sun's future intentions. Weatherbee was also feeling somewhat better, and crawled out beside him. They propped themselves in the snow beneath the moveless wind-vane, and waited.

The stillness of death was about them. In other climes, when nature falls into such moods, there is a subdued air of expectancy, a waiting for some small voice to take up the broken strain. Not so in the North. The two men had lived seeming æons in this ghostly peace. They could remember no song of the past; they could conjure no song of the future. This unearthly calm had always been,—the tranquil silence of eternity.

Their eyes were fixed upon the north. Unseen, behind their backs, behind the towering mountains to the south, the sun swept toward the zenith of another sky than theirs. Sole spectators of the mighty canvas, they watched the false dawn slowly grow. A faint flame began to glow and smoulder. It deepened in intensity, ringing the changes of reddish-yellow, purple, and saffron. So bright did it become that Cuthfert thought the sun must surely be behind it,—a miracle, the sun rising in the north! Suddenly, without warning and without fading, the canvas was swept clean. There was no color in the sky. The light had gone out of the day. They caught their breaths in half-sobs. But lo! the air was a-glint with particles of scintillating frost, and there, to the north, the wind-vane lay in vague outline on the snow. A shadow! A shadow! It was exactly midday. They jerked their heads hurriedly to the south. A golden rim peeped over the mountain's snowy shoulder, smiled upon them an instant, then dipped from sight again.

There were tears in their eyes as they sought each other. A strange softening came over them. They felt irresistibly drawn toward each other. The sun was coming back again. It would be with them to-morrow, and the next day, and the next. And it would stay longer every visit, and a time would come when it would ride their heaven day and night, never once dropping below the skyline. There would be no night. The ice-locked winter would be broken; the winds would blow and the forests answer; the land would bathe in the blessed sunshine, and life renew. Hand in hand, they would quit this horrid dream and journey back to the Southland. They lurched blindly forward, and their

hands met,—their poor maimed hands, swollen and distorted beneath their mittens.

But the promise was destined to remain unfulfilled. The Northland is the Northland, and men work out their souls by strange rules, which other men, who have not journeyed into far countries, cannot come to understand.

*

An hour later, Cuthfert put a pan of bread into the oven, and fell to speculating on what the surgeons could do with his feet when he got back. Home did not seem so very far away now. Weatherbee was rummaging in the cache. Of a sudden, he raised a whirlwind of blasphemy, which in turn ceased with startling abruptness. The other man had robbed his sugar-sack. Still, things might have happened differently, had not the two dead men come out from under the stones and hushed the hot words in his throat. They led him quite gently from the cache, which he forgot to close. That consummation was reached; that something they had whispered to him in his dreams was about to happen. They guided him gently, very gently, to the woodpile, where they put the axe in his hands. Then they helped him shove open the cabin door, and he felt sure they shut it after him,—at least he heard it slam and the latch fall sharply into place. And he knew they were waiting just without, waiting for him to do his task.

"Carter! I say, Carter!"

Percy Cuthfert was frightened at the look on the clerk's face, and he made haste to put the table between them.

Carter Weatherbee followed, without haste and without enthusiasm. There was neither pity nor passion in his face, but rather the patient, stolid look of one who has certain work to do and goes about it methodically.

"I say, what's the matter?"

The clerk dodged back, cutting off his retreat to the door, but never opening his mouth.

"I say, Carter, I say; let's talk. There's a good chap."

The master of arts was thinking rapidly, now, shaping a skillful flank movement on the bed where his Smith & Wesson lay. Keeping his eyes on the madman, he rolled backward on the bunk, at the same time clutching the pistol.

"Carter!"

The powder flashed full in Weatherbee's face, but he swung his weapon and leaped forward. The axe bit deeply at the base of the spine, and Percy

Cuthfert felt all consciousness of his lower limbs leave him. Then the clerk fell heavily upon him, clutching him by the throat with feeble fingers. The sharp bite of the axe had caused Cuthfert to drop the pistol, and as his lungs panted for release, he fumbled aimlessly for it among the blankets. Then he remembered. He slid a hand up the clerk's belt to the sheath-knife; and they drew very close to each other in that last clinch.

Percy Cuthfert felt his strength leave him. The lower portion of his body was useless. The inert weight of Weatherbee crushed him,—crushed him and pinned him there like a bear under a trap. The cabin became filled with a familiar odor, and he knew the bread to be burning. Yet what did it matter? He would never need it. And there were all of six cupfuls of sugar in the cache,—if he had foreseen this he would not have been so saving the last several days. Would the wind-vane ever move? It might even be veering now. Why not? Had he not seen the sun to-day? He would go and see. No; it was impossible to move. He had not thought the clerk so heavy a man.

How quickly the cabin cooled! The fire must be out. The cold was forcing in. It must be below zero already, and the ice creeping up the inside of the door. He could not see it, but his past experience enabled him to gauge its progress by the cabin's temperature. The lower hinge must be white ere now. Would the tale of this ever reach the world? How would his friends take it? They would read it over their coffee, most likely, and talk it over at the clubs. He could see them very clearly. "Poor Old Cuthfert," they murmured; "not such a bad sort of a chap, after all." He smiled at their eulogies, and passed on in search of a Turkish bath. It was the same old crowd upon the streets. Strange, they did not notice his moosehide moccasins and tattered German socks! He would take a cab. And after the bath a shave would not be bad. No; he would eat first. Steak, and potatoes, and green things,—how fresh it all was! And what was that? Squares of honey, streaming liquid amber! But why did they bring so much? Ha! ha! he could never eat it all. Shine! Why certainly. He put his foot on the box. The bootblack looked curiously up at him, and he remembered his moosehide moccasins and went away hastily.

Hark! The wind-vane must be surely spinning. No; a mere singing in his ears. That was all,—a mere singing. The ice must have passed the latch by now. More likely the upper hinge was covered. Between the moss-chinked roof-poles, little points of frost began to appear. How slowly they grew! No; not so slowly. There was a new one, and there another. Two—three—four; they were coming too fast to count. There were two growing together. And

there, a third had joined them. Why, there were no more spots. They had run together and formed a sheet.

Well, he would have company. If Gabriel ever broke the silence of the North, they would stand together, hand in hand, before the great White Throne. And God would judge them, God would judge them!

Then Percy Cuthfert closed his eyes and dropped off to sleep.

LAND HO!

from *In the Land of White Death* (1917)

Valerian Albanov

In 1912, Valerian Albanov, a Russian navigator, signed on to an expedition to search out new hunting grounds. Two hundred miles east of Novaya Zemlya, the expedition's ship, the Santa Anna, *became frozen in. It drifted for a year and a half, carried nearly 2,500 miles by the moving ice. Finally, Albanov and thirteen other crew members left the* Santa Anna, *and set off on foot for Franz Josef Land. Only he and one other man completed the journey. In the excerpts below, Albanov recounts his departure from the ship, in January 1914, and his first sighting of land, six months later.*

LATE IN THE EVENING the lieutenant called me once more into his cabin to give me a list of items we would be taking with us and which I must, if possible, return to him at a later date. Here is that list as it was entered into the ship's record: 2 Remington rifles, 1 Norwegian hunting rifle, 1 double-barreled shotgun, 2 repeating rifles, 1 ship's log transformed into a pedometer for measuring distances covered, 2 harpoons, 2 axes, 1 saw, 2 compasses, 14 pairs of skis, 1 first-quality malitsa, 12 second-quality malitsi, 1 sleeping bag, 1 chronometer, 1 sextant, 14 rucksacks, and 1 small pair of binoculars.*

Brusilov asked me if he had forgotten to list anything. His pettiness astounded me. It was as if he thought there were horses waiting at the gangway to take those of us who would be leaving to the nearest railway station or steamship terminal. Had the lieutenant forgotten that we were about to set

---

* Malitsi are heavy, sacklike, Samoyed garments sewn from reindeer hide, with the fur on the inside. Slipped over the head, they have crude openings to accommodate the arms and the face. Thirteen of the men in Albanov's party used malitsi in lieu of sleeping bags at night.

off on foot on a daunting trek across drifting ice, in order to search for an unknown landmass, and this under worse conditions than any men who had gone before us? Did he have no greater concerns on this last evening than toting up rucksacks, axes, a defective ship's log, a saw, and harpoons? If truth be told, even as he read the list to me, I felt myself succumbing to a familiar rage. I experienced the sensation of strangling as my throat constricted in anger. But I controlled myself and reminded Brusilov that he had forgotten to list the tent, the kayaks, the sledges, a mug, cups, and a galvanized bucket. He immediately wrote down the tent, but decided not to mention the dishes. "I will not list the kayaks or sledges, either," he offered. "In all probability they will be badly damaged by the end of your trip, and the freight to ship them from Svalbard would cost more than they are worth. But if you succeed in getting them to Alexandrovsk, deliver them to the local police for safekeeping." I told Brusilov I was in agreement with this.

I left the lieutenant's cabin very upset, and went below. On the way to my cabin, Denisov stopped me to ask where I would open the packet of ship's mail and post the letters—in Norway or Russia? That was the last straw, and I could not contain my emotions any longer. I exploded and threatened to dump not only the mail, but also the rucksacks, the cups, and the mugs into the first open lead we came to, because I had serious doubts that we would ever reach a mail train in Norway, Russia, or anywhere else. But then I quickly regained my composure and promised Denisov that, wherever we landed, I would make every effort to see that the ship's mail reached its destination.

Denisov went on his way, reassured. The ship was dark. Everyone had gone to bed. I was dismayed and depressed. It was as if I were already wandering across the endless, icy wastes, without any hope of returning to the ship, and with only the unknown lying ahead.

On that gloomy, decisive night prior to my departure from the *Saint Anna*, filled with anxiety, I wondered about each of the men who would be accompanying me. I already possessed grave doubts about their health and stamina. One was fifty-six years old and all of them complained of sore feet; not one of them was really fit. One man had open sores on his legs, another had a hernia, a third had been suffering from pains in his chest for a long time, and all, without exception, had asthma and palpitations.

In short, these were the dark thoughts that assailed and disheartened me that evening. Was this a premonition of some great misfortune that I was heading for, with no hope of escape?

June 9. The wind swings back and forth between the northwest and the west-northwest. Despite the overcast skies, I was able to determine that with no effort on our part we had reached 80° 52' north and 40° 20' east of Greenwich. But I cannot guarantee the exactness of the longitude.

As I have often done, at around nine in the evening I climbed onto a high ice formation to study the horizon. Ordinarily I saw what looked like islands in every direction, but which on closer examination turned out to be either icebergs or clouds. This time, I sighted something quite different on the shimmering horizon. I was so staggered that I sat down on the ice to clean the lenses of my binoculars and rub my eyes. My pulse was racing in great anticipation, and when I fixed my apprehensive gaze once more on the vision that held such promise, I could discern a pale, silver strip with sinuous contours running along the horizon and then disappearing to the left. The right-hand side of this phenomenon was outlined with unusual clarity against the azure of the sky. This whole formation, including its gradations of color, reminded me of a phase of the moon. The left edge seemed to grow slowly paler while the right stood out even more distinctly, like a yellowish line traced along the blue horizon. Four days earlier I had observed a similar phenomenon; but the bad light led me to think that it was a cloud. During the night I returned five times to check on my strange discovery, and each time my original impression was more or less clearly confirmed; the main features of shape and color had certainly not changed. So far, nobody else had noticed this wonderful sight. I had to restrain myself severely from dashing back to the tent and shouting with excitement: "Wake up, everyone, come and see that our prayers have been answered at last and we are about to reach land!" I was then convinced that it was land that I could see, but I wanted to keep my discovery secret, so I contented myself with thinking: "If you others want to see this miracle, you will have to open your eyes." But my companions were as oblivious as ever, and had not even noticed my ill-concealed excitement. Instead of going out and inspecting the horizon, the only way of evaluating our immediate prospects, they either went back to sleep or started to hunt for "game"—as we have named the lice that are regular guests in our malitsi. That seems to be more important to them!

June 10. The morning was beautiful. My hypothetical land stood out even more clearly, its yellowish hue increasingly extraordinary. Its shape was totally different from what I had been expecting as I scanned the horizon over the past two months. Now I could also see, to my left, a few isolated headlands, set quite far back, however, and between them seemed to be glaciers. I wondered

idly how far away we were, for my eyes were not at all used to judging such distances. I estimated that there must be fifty or sixty nautical miles to the most distant peaks; how far we might be from the shore could not even be roughly determined: twenty to thirty-five nautical miles, perhaps more, perhaps less. The only certainty was that we were now closer to being rescued than we had been for the last two years. I silently offered up my thanks; but how on earth could we get there?

At around noon I managed to fix our position from the sun. We were crossing latitude 80° 52'. Wind from the south. We ate quickly, packed up our belongings, and decided to head for land. By nine o'clock we had covered between two and three miles and made the decision not to pitch the tent until we reached land. Could we do it? The ice floes were in perpetual motion; it was almost impossible to advance without resorting to the kayaks. We spotted quite a few bear tracks; we also succeeded in shooting a seal.

Evening has arrived. We sit together in the tent with mixed feelings, for not only have we failed to reach the island, we are now even farther away from it than this morning. The weather is very gloomy; it is snowing and raining, with wind from the south. The surface of the ice was dreadful; my companions call it "glutinous." It was impossible to make any sort of progress today, either on foot or by kayak. Exhausted, soaked through, and famished, we decided to stop and pitch the tent. South wind still blowing. Major efforts have brought us no more than two miles at the most. But we managed to kill a seal, which we are cooking; we have brewed up a very nourishing broth with the seal's blood. Once we really start cooking we do not skimp on the size of the portions. Today we had a good, solid breakfast; at midday a bucketful of soup and just as much tea; in the evening, a pound of meat each, washed down with more tea. Our food supply is ample, for in addition to what I have just mentioned, each man receives a pound of ship's biscuits per day. Our appetites are wolfish! In gloomy moments we are struck by the thought that such voraciousness normally occurs in cases of severe starvation. God protect us from that!

Yesterday I noticed that seven pounds of biscuit had disappeared. This unfortunate discovery forced me to call my companions together and inform them that if it happened again, I would hold all of them responsible and reduce their rations; and if I managed to catch the ignominious thief red-handed, I would shoot him on the spot. However bitter it seems, I must

admit there are three or four men in the group with whom I have nothing in common.

Only someone who has experienced such an ordeal can fully understand how impatient I was to reach the island where our two-year odyssey through the Arctic wastes would finally end. Once we reached our landfall, our situation would improve dramatically. We would be able to capture hosts of birds and walruses and we would also be able to take a bath. We have not washed now for two months. Catching a chance glimpse of my face in the sextant's mirror the other day gave me a terrible fright. I am so disfigured that I am unrecognizable, covered as I am with a thick layer of filth. And we all look like this. We have tried to rub off some of this dirt, but without much success. As a result we look even more frightening, almost as if we were tattooed! Our underclothes and outer garments are unspeakable. And since these rags are swarming with "game," I am sure that if we put one of our infested jerseys on the ground, it would crawl away all by itself!

Here is a glimpse of life inside the tent: Everyone is squatting in a circle on the ground; with grim expressions, they are silently absorbed in some serious-looking task. What can these men be doing? Hunting lice! This "pastime" is always reserved for the evening. It is the only possible form of hygiene, since we have neither soap nor water for proper ablutions. And even if we had some water, the fearful cold would prevent us from washing. All too often we have not even had enough water to quench our thirst.

Some of us had originally taken a vow not to wash until we reached land. Who would have suspected that it would be two months before we sighted land? No wonder we all felt the need to indulge in our nightly "hunt." This communal activity united us in a remarkable fashion, and all the squabbles usually ceased during those hours.

In the afternoon I went out with three men on a reconnaissance. Beyond the four leads we will have to cross tomorrow morning, we shall find better going. The ice blocks are unusually dark and dirty, with algae, sand, and even rocks sticking to them. We took a couple of small stones, seaweed, and two small pieces of wood with us, as our first gift from the land—an olive branch, so to speak.

We found a lot of bear tracks. The weather, as usual, is damp and foggy. There is wet snow falling, almost rain. Wind from the south.

June 11 A satisfactory day's march. We covered four miles. Toward evening, we pitched camp on a little ice floe surrounded by pools and brash ice. The morning's northeasterly had by evening become a chilly northerly. The current

has pushed us away toward the east, and now our island of salvation appears to be farther south. Good hunting: one seal and a duck. Our eyes are very painful again.

June 12. The wind is still blowing from the north, but the weather is warm and clear. Only the kayak crossings were difficult: We covered scarcely more than a mile. Seven of the men, including myself, are suffering from serious eye inflammation. While crossing one of the open leads we had the serious misfortune of dropping one of our two remaining Remingtons into the sea. It was Lunayev who dropped it, with Smirennikov's assistance. Such negligence made me so angry that I lost my temper and struck out at anybody who crossed my path. This is the second rifle we have lost because of heedless behavior, and anyone who can picture himself in my shoes would surely understand my frustration with such unforgivable carelessness. Now we have just one rifle for which there is abundant ammunition. The smaller repeating rifle is hardly of any use, since there are only eighty cartridges left for it. We still have shells for the shotgun, but it is almost useless against bears, which may be lurking behind every block of ice.

I would have liked to take a sun shot with the sextant, but my eyes were not up to it. The sun seemed to be misty and indistinct and I could not see the horizon at all. According to my companions who can still see clearly, our island is particularly visible today: One can even make out a few details. We saw many eiders in flight that must have come from the island. As our supply of seal meat has run out, for lunch we cooked the bear meat we dried the other day, and in the evening we prepared a soup from the same meat. There is no more sugar, and the tea will last only a few more days.

We are still making little headway trekking over the ice. But we have thought up a new strategy: We work out our course from the top of a rise—that is, we identify the places ahead of time where our kayaks or skis are most likely to get through. Often we are forced to skirt along the edge of a channel on our skis, dragging the sledge-laden kayak behind us on the water. But chunks of disintegrating icebergs, called growlers, often obstruct the boats, and it is not a simple affair to get them moving again. From time to time during our backbreaking toil, one of us sinks through the ice, and that is when we see who can move the fastest. It is imperative to leap out of the icy water, remove one's boots that are rapidly filling up, empty them, and get back to work, all in a matter of seconds!

June 13. The wind has shifted, coming now from the south-southwest. We set off at eight o'clock and traveled, with only an hour's break, until six-

thirty in the evening. The end result: about five miles. We had to cross more extensive pack ice that had been eroded by the wave action and covered in deep snow. Crossing a channel we were startled when a bearded seal suddenly bounded out of the water. We also saw a great many ordinary seals but were unable to shoot one.

When the horizon grew lighter, those of us who were not suffering from snow blindness were able to see the island to the southeast. From now on, the tides will probably swirl growlers and brash ice continually along the shores, and we shall be confronted with this repulsive stuff, this ice porridge, all the way to our landfall. Toward evening, the wind from the south-southwest picked up, bringing with it fine hail.

June 14. The same wind persists, with cold, dark weather. We did two and a half miles this morning. On very thin ice, Konrad suddenly broke through a seal's breathing hole that had been drifted over. Totally submerged, he became tangled in his hauling line while the sledge slid forward and covered the hole. We all rushed to his rescue, cut the hauling line, dragged the sledge aside, and pulled Konrad out. He was soaked to the skin and had swallowed some water. We had to pitch the tent right away and light a fire to warm him up.

Our supplies are dwindling. We have only 120 pounds of biscuits left, and our reserves of meat are finished; for lunch we had nothing but biscuit soup, to which we added our last can of condensed milk. The dire state of our supplies forced us to take some quick action, and we decided on some long-term plans that included abandoning the tent and continuing in our nearly empty kayaks. We would be sorry to leave behind nearly all our belongings: axes, harpoons, ski poles, spare skis, warm clothing, footgear, and empty cans. These represent a considerable load, but at the same time how indispensable all such things will be if we have to winter over on these islands. And in all probability, we will not be spared a wintering.

No sooner had we set off again than we came upon some seals and shot two of them. Fortune had smiled upon us once again during our hour of need. This lucky event restored our courage to such a degree that we went back for the tent. The route was dreadful and required great caution; we barely covered one mile.

ICELANDIC PIONEER

from *Independent People* (1934)

Halldór Laxness

The Nobel Prize–winning author Halldór Laxness was born in Iceland in 1902. Over his long career—he died at the age of 95—he wrote fifty-one novels. Independent People, *Laxness's most celebrated work, was published in Icelandic in 1934. It tells the story of Bjartur of Summerhouses, a stiff-necked sheep farmer who recites traditional ballads while enduring unspeakable miseries. In the excerpt below, Bjartur has gone looking for a lost sheep from his flock.*

It is one of the peculiarities of life that the most unlikely accident, rather than the best-laid plan, may on occasion determine the place of a man's lodging; and thus it fared for Bjartur of Summerhouses now. Just as he was about to cross one of the many gullies that cleave the sides of the valley all the way down to the river, he saw some animals leap lightly down a watercourse not far ahead of him and come to a halt well out on the river bank. He saw immediately that they were reindeer, one bull and three cows. They tripped about on the bank for a little while, the bull next the river and the cows seeking shelter in his lee, all with their antlers in the weather and their hindquarters facing the man, for the wind was blowing from across the river.

Halting in the gulley, Bjartur eyed the animals for some moments. They kept up a continual shifting about, but always so that they were turned away from him. They were fine beasts, probably just in their prime, so it was no wonder that it occurred to Bjartur that he was in luck's way tonight, for it would be no mean catch if he could trap only one of them even. The bull especially looked as if it would make an excellent carcass, judging by its size, and he had not forgotten that reindeer venison is one of the tastiest dishes that ever graced a nobleman's table. Bjartur felt that even if he did not find

the ewe, the trip would have proved well worth while if he managed now to capture a reindeer. But supposing that he caught the bull, how was he to kill it so that its blood did not run to waste?—for from reindeer blood may be made really first-class sandwich meat. The best plan, if he could only manage it, would be to take it back home alive, and with this intention in his mind he searched his pockets for those two articles which are most indispensable to a man on a journey, a knife and some string, and found both, a nice hank of string and his pocket-knife. He thought: "I'll make a rush at him now and get him down. Then I'll stick the point of my knife through his nose, thread the string through the hole, and make a lead of it. In that way I ought to be able to lead him most of the way over the moors, or at least till I come to some easily remembered spot where I can tether him and keep him till I go down to the farms and fetch men and materials." Summerhouses was, of course, easily a day's journey for a man travelling on foot. When Bjartur had completed his plan of attack he stole half-bent down the gully till he was opposite the reindeer, where they stood with their horns in the wind on the strip between the gully and the river. He stole cautiously over the runnels, crept silently up the bank, and, peeping over the edge, saw that he was no more than twelve feet from the buck. His muscles began to taughten with the thrill of the hunt and he felt a certain amount of palpitation. Inch by inch he pulled himself higher over the brink, until he was standing on the bank; slowly, very slowly he stole up to the bull, half a pace alongside—and the next instant had leaped at him and gripped him by one of the antlers, low down near the head. At the man's unexpected attack, the animals gave a sudden bound, flung up their heads, and pricked their ears, and the cows were off immediately, running lightly down the river through the drizzling snow. At first the bull had intended making off with Bjartur holding on to its head as if he made no difference at all, but Bjartur hung on and the bull could not get free, and though it tossed its head repeatedly, it was none the freer for it. But Bjartur soon found that his hold on the antler was uncertain, there being something on it like smooth bark that kept on slipping in his grip, and the creature too lively to allow a secure purchase anywhere else. He saw too, when it came to the point, that he would have to abandon his hope of getting under the animal's neck and gripping it with a wrestling hold, for its horns were of the sharpest and the prospect of having them plunged into his bowels not particularly attractive. For a while they continued their tug of war, the reindeer gradually gaining ground, till it had reached a tolerable speed and had dragged Bjartur quite a distance down the river. Then involuntarily there

flashed across Bjartur's mind the trick he had been taught from childhood to use with wild horses: try to get alongside them, then jump on their backs. It succeeded. Next instant he was sitting astride the reindeer's back holding on to its antlers—and said later that though this animal species seemed light enough on its feet, a bull reindeer was as rough a ride as he had ever come across, and, indeed, it took him all his time to hang on. But the jaunt was not to be a long one. For when the bull had hopped a few lengths with this undesirable burden on his back without managing to shake it off, he saw quickly that desperate measures would have to be taken and, making a sudden leap at right angles to his previous course, shot straight into Glacier River and was immediately churning the water out of his depth.

Well, well. Bjartur had set out on a trip after sheep right enough, but this was becoming something more in the nature of a voyage. Here he was sitting neither more nor less than up to the waist in Glacier River, and that on no ordinary steed, but on the only steed that is considered suitable for the most renowned of adventures. But was Bjartur really proud of this romantic progress? No, far from it. He had at the moment no leisure to study either the distinctive features of his exploit or the rarity of its occurrence, for he had as much as he could do to hold his balance on the reindeer's back. Desperately he hung on to its horns, his legs glued to its flanks, gasping for breath, a black mist before his eyes. The rush of the water swept the animal downstream for a while, and for a long time it seemed as if it intended making no effort to land. Across the river the banks, which rose high and steep out of the water, showed intermittently through the snow, but in spite of the nearness of land Bjartur felt himself as unhappily situated as a man out in mid-ocean in an oarless boat. Sometimes the cross-currents caught the bull, forcing it under, and then the water, so unbearably cold that it made his head reel, came up to the man's neck and he was not sure which would happen first, whether he would lose consciousness or the deer would take a dive that would be the end of him. In this fashion they were carried down Glacier River for some time.

✳

At long last it began to look as if the bull was thinking of landing. Bjartur suddenly realized that they had neared the eastern bank of the river and were now not more than a yard or two from the jagged fringe of ice that formed the only shore. They were carried downstream along by the ice for a while longer, but as the banks rose everywhere with equal steepness from the ice edge, the

matter of effecting a landing remained a most unattractive project. Bjartur nevertheless felt that his best course, if the bull neared the land sufficiently, would be to seize the opportunity and throw himself overboard, then try to haul himself up on the ice, for this stay in cold water was becoming more than he could stand. He realized, of course, that it would be a death-jump that could only end in one of two ways. Finally there came a time when the bull swam for a few yards not more than half an arm's length from the ice, and the man watched his chance, let go of the antlers, heaved himself out of the water, and swung the upper part of his body on to the ice; and there Bjartur parted from the bull, never to set eyes on it again, and with a permanent dislike for the whole of that animal species.

There occurred moments, both then and later, when it struck Bjartur that the bull reindeer was no other than the devil Kolumkilli in person.

The ice was thin and broke immediately under the man's weight, so that he was near to being carried away with the fragments; but as his days were not yet numbered he managed somehow to hang on to the unbroken ice, and succeeded finally in wriggling his lower limbs also out of the water. He was shaking from head to foot with the cold, his teeth chattering, not a single dry stitch in all his clothes. But he did not feel particularly safe on this narrow fringe of ice and began now to tackle the ascent of the river bank. This in itself was a sufficiently hazardous undertaking, for the bank was not only precipitous, but also covered with icicles formed by the rising of the river, and there could only be one end to a fall if hand or foot should lose its grip. As he was fatigued after his exploit in the water, it took him longer to work his way up to the top than it would otherwise have done, but finally the moment arrived when he was standing safe and sound on the eastern bank of Glacier River—on the far pastures of another county. He took off his jacket and wrung it out, then rolled about in the snow to dry himself, and considered the snow warm in comparison with the glacier water. At intervals he stood up and swung his arms vigorously to rid himself of his shivering. It was, of course, not long before he realized to the full what a trick the bull had played on him by ferrying him over Glacier River. In the first place he had cheated him of the quarters he had proposed to use for the night, the shepherds' hut on the western side of the river. But that actually was only a trifle. Altogether more serious was to find himself suddenly switched to the eastern bank of Glacier River, for the river flowed north-east, whereas Bjartur's direction home lay a trifle west of north-west. To cross the river he would therefore be forced to make a detour in an opposite direction to Summerhouses, all the way down to the aerial ferry in the farming districts, and this was

not less than a twenty hours' walk, even at a good speed, for the nearest farm in Glacierdale was at least fifteen hours away. Though he were to travel day and night this adventure of his would thus delay him almost forty-eight hours—and that in weather like this, and his lambs still out.

He was pretty well worn out, though loath to admit it to himself, and his wet clothes would be a poor protection if he decided to bury himself in the snow in this hardening frost. The snowflakes grew smaller and keener; no sooner had they fallen than the wind lifted them again and chased them along the ground in a spuming, knee-deep smother. His underclothes remained unaffected by the frost as long as he was on the move, but his outer clothes were frozen hard and his eyelashes and beard stiff with ice. In his knapsack there remained one whole blood pudding, frozen hard as a stone, and half of another; he had lost his stick. The night was as black as pitch, and the darkness seemed solid enough to be cut with a knife. The wind blew from the east, sweeping the blizzard straight into the man's face. Time and time again be tumbled from another and yet another brink into another and yet another hollow where the powdery snow took him up to the groin and flew about him like ash. One consolation only there was: happen what might, he could not lose his way, for on his left he had Glacier River with its heavy, sullen roar.

He swore repeatedly, ever the more violently the unsteadier his legs became, but to steel his senses he kept his mind fixed persistently on the world-famous battles of the rhymes. He recited the most powerful passages one after another over and over again, dwelling especially on the description of the devilish heroes, Grimur Ægir and Andri. It was Grimur he was fighting now, he thought; Grimur, that least attractive of all fiends, that foul-mouthed demon in the form of a troll, who had been his antagonist all along; but now an end would be put to the deadly feud, for now the stage was set for the final struggle. In mental vision he pursued Grimur the length of his monstrous career, right from the moment when Groa the Sibyl found him on the foreshore, yellow and stuffed with treachery; and again and again he depicted the monster in the poet's words, bellowing, wading in the earth up to the thighs, filled with devilish hate and sorcery, fire spouting from his grinning mouth, by human strength more than invincible:

The monster lived on moor or fen;
The sea was in his power.
He'd shamelessly drink the blood of men,
The steaming flesh devour.

The crags before him split apart,
The rivers ran in spate;
He cleft the rocks by magic art,
His cunning was so great.

For this fiend there was not a shred of mercy in Bjartur. No matter how often he sprawled headlong down the gullies, he was up again undaunted and with redoubled fury making yet another attack, grinding his teeth and hurling curses at the demon's gnashing jaws, determined not to call a halt before Grimur's evil spirit had been hounded to the remotest corners of hell and the naked brand had pierced him through and his death-throes had begun in a ring-dance of land and sea.

Again and again he imagined that he had made an end of Grimur and sent him howling to hell in the poet's immortal words, but still the blizzard assailed him with undiminished fury when he reached the top of the next ridge, clawed at his eyes and the roots of his beard, howled vindictively in his ears, and tried to hurl him to the ground—the struggle was by no means over, he was still fighting at close quarters with the poison-spewing thanes of hell, who came storming over the earth in raging malice till the vault of heaven shook to the echo of their rush.

His loathsome head aloft he reared,
With hellish hate he roared.
His slavering lips with froth were smeared,
Vilely his curses poured.

And so on, over and over again.

Never, never did these thanes of hell escape their just deserts. No one ever heard of Harekur or Gongu-Hrolfur or Bernotus being worsted in the final struggle. In the same way no one will be able to say that Bjartur of Summerhouses ever got the worst of it in his world war with the country's spectres, no matter how often he might tumble over a precipice or roll head over heels down a gully—"while there's a breath left in my nostrils, it will never keep me down, however hard it blows." Finally he stood still, leaning against the blizzard as against a wall; and neither could push the other back. He then resolved to house himself in the snow and began looking for a sheltered spot in a deep gully. With his hands he scooped out a cave in a snowdrift, trying to

arrange it so that he could sit inside on his haunches to pile up the snow at the mouth, but the snow, loose and airy, refused to stick together, and as the man was without implements, the cave simply fell in again. He had not rested long in the snowdrift before the cold began to penetrate him; a stiffness and a torpor crept up his limbs, all the way to his groin, but what was worse was the drowsiness that was threatening him, the seductive sleep of the snow, which makes it so pleasant to die in a blizzard; nothing is so important as to be able to strike aside this tempting hand which beckons so voluptuously into realms of warmth and rest. To keep the oblivion of the snow at bay it was his custom to recite or, preferably, sing at the top of his voice all the obscene verse he had picked up since childhood, but such surroundings were never very conducive to song and on this occasion his voice persisted in breaking; and the drowsiness continued to envelop his consciousness in its mists, till now there swam before his inner eye pictures of men and events, both from life and from the Ballads—horse-meat steaming on a great platter, flocks of sheep bleating in the fold, Bernotus Borneyarkappi in disguise, clergymen's wanton daughters wearing real silk stockings; and finally, by unsensed degrees, he assumed another personality and discovered himself in the character of Grimur the Noble, brother of Ulfar the Strong, when the visit was paid to his bedchamber. Matters stood thus, that the King, father of the brothers, had taken in marriage a young woman, who, since the King was well advanced in years, found a sad lack of entertainment in the marriage bed and became a prey to melancholy. But eventually her eyes fell on the King's son, Grimur the Noble, who far outshone all other men in that kingdom, and the young Queen fell so deeply in love with this princely figure that she could neither eat nor sleep and resolved finally to go to him at night in his chamber. Of the aged King, his father, she spoke in the most derisive of terms:

Of what use to red-blood maid
Sap of such a withered blade?
Or to one so sore in need,
Spine of such a broken reed?

Grimur, however, found this visit displeasing and relished even less such shameless talk, but for some time he retreated in courtly evasion of the issue. But

No refusals ought availed,
Words of reason here had failed.

All intent on lustful play
Softly on the bed she lay.

And before Grimur the Noble had time to marshal his defences, there occurred the following:

In her arms she clasped him tight,
Warm with promise of delight;
Honey-seeming was her kiss,
All her movements soft with bliss.

But at this moment there dawned upon Grimur the Noble the full iniquity of what was taking place, and springing to his feet in a fury, he turned upon the shameless wanton:

Up the hero rose apace,
Smote her sharply on the face;
Scornful of such shameful deed,
Thrust her to the floor with speed.

Angrily the hero cried,
Whilst she lay, bereft of pride:
"Lustful art thou as a swine,
Little honour can be thine."

"To hell with me, then," cried Bjartur, who was now standing in the snow after repulsing the seductive bed-blandishments of the lecherous Queen. Did the heroes of the rhymes ever allow themselves to be beguiled into a life of adultery, debauchery, and that cowardice in battle which characterizes those who are the greatest heroes in a woman's embrace? Never should it be said of Bjartur of Summerhouses that on the field of battle he turned his back on his foes to go and lie with a trollopy slut of a queen. He was in a passion now. He floundered madly about in the snow, thumping himself with all his might, and did not sit down again till he had overcome all those feelings of the body that cry for rest and comfort, everything that argues for surrender and hearkens to the persuasion of faint-hearted gods. When he had fought thus for some time, he stuck the frozen sausages inside his trousers and warmed them on his flesh, then gnawed them from his fist in

the darkness of this relentless winter night and ate the driving snow as savoury.

This was rather a long night. Seldom had he recited so much poetry in any one night; he had recited all his father's poetry, all the ballads he could remember, all his own palindromes backwards and forwards in forty-eight different ways, whole processions of dirty poems, one hymn that he had learned from his mother, and all the lampoons that had been known in the Fourthing from time immemorial about bailiffs, merchants, and sheriffs. At intervals he struggled up out of the snow and thumped himself from top to toe till he was out of breath.

Finally his fear of frost-bite became so great that he felt it would be courting disaster to remain quietly in this spot any longer, and as it must also be wearing on towards morning and he did not relish the idea of spending a whole day without food in a snowdrift miles from any habitation, he now decided to forsake his shelter and leave the consequences to take care of themselves. He forced his way at first with lowered head against the storm, but when he reached the ridge above the gully, he could no longer make any headway in this fashion, so he slumped forward on to his hands and knees and made his way through the blizzard on all fours, crawling over stony slopes and ridges like an animal, rolling down the gullies like a peg; barehanded, without feeling.

On the following night, long after the people of Brun, the nearest farm of Glacierdale, had retired to bed—the storm had raged relentlessly now for a full twenty-four hours—it came to pass that the housewife was wakened from her sleep by a hubbub at the window, a groaning, even a hammering. She woke her husband, and they came to the conclusion that some creature gifted with the power of reasoning must surely be afoot and about the house, though on this lonely croft visitors were the last thing to be expected in such a storm—was it man or devil? They huddled on their most necessary garments and went to the door with a light. And when they had opened the door, there toppled in through the drift outside a creature resembling only in some ways a human being; he rolled in through the doorway armoured from head to foot in ice, nose and mouth encrusted, and came to rest in a squatting position with his back against the wall and his head sunk on his chest, as if the monstrous spectre, despairing of maltreating him further, had finally

slung him through the door and up against the wall; the light of the house shone on this visitor. He panted heavily, his chest heaving and groaning, and made an effort to clear his throat and spit, and when the crofter asked him who he was and where he came from, he tried to get to his feet, like an animal trying to stand up on its hind legs, and gave his name—"Bjartur of Summerhouses."

The crofter's son had now risen also, and together he and his father made an attempt to help their visitor into the room, but he refused any such assistance. "I'll walk by myself," he said, "I'll follow the woman with the lamp." He laid himself across the son's bed and for a while made no answer to their questions, but mumbled like a drunkard, rumbled like a bull about to bellow. At last he said:

"I am thirsty."

The woman brought him a three-pint basin of milk, and he set it to his mouth and drank it off, and said as he passed her the basin: "Thanks for the drink, mother." With her warm hands she helped to thaw the clots of ice in his beard and eyebrows, then drew off his frozen clothes and felt with experienced fingers for frost-bite. Fingers and toes were without feeling, his skin smarting with frost, but otherwise he appeared to have taken no hurt. When the crust of ice had been thawed off, he stretched himself out naked in the son's warm bed and had seldom felt so comfortable in all his life. After the housewife had gone to prepare him some food, father and son sat down beside him, their eyes bewildered, as if they did not really believe this phenomenon and did not know quite what to say. In the end it was he who spoke, as he asked in a hoarse voice from under the coverlet:

"Were your lambs in?"

They replied that they were, and asked in turn how it had come about that he had landed here, on the eastern bank of Glacier River, in murderous weather that would kill any man.

"Any man?" he repeated querulously. "What do the men matter? I always thought it was the animals that came first."

They continued to question him.

"Oh, as a matter of fact I was just taking a little walk by myself," he vouchsafed. "I missed a ewe, you see, and took a stroll along the heights there just to soothe my mind."

For a while he was silent, then he added:

"It's been a trifle rough today."

"It wasn't any pleasanter last night either," they said, "a regular hurricane."

"Yes," agreed Bjartur, "it was just a trifle rough last night, too."

They wanted to know where he had put in the night, and he replied: "In the snow." They were particularly curious about how he had managed to cross Glacier River, but he would give no details. "It's a nice thing to have one's lambs out in this," he said mournfully.

They said that in his shoes they wouldn't trouble themselves about lambs tonight, but think themselves lucky to be where they were.

"It's easy to see," he replied, "that you people have found your feet. But I am fighting for my independence. I have worked eighteen years for the little livestock I have, and if they're under snow, it would be better for me to be under snow too."

But when the woman had brought him a meal in bed and he had eaten his fill, he lay down without further discourse and was asleep and snoring loudly.

THE LAND, BREATHING

from *Arctic Dreams* (1985)

Barry Lopez

Barry Lopez's Arctic Dreams *is a far-ranging study of the northern landscape and the creatures—musk oxen, polar bears, narwhals—who inhabit it. It won the National Book Award for Nonfiction. The passage below is from the chapter on migration.*

It was still dark, and I thought it might be raining lightly.

I pushed back the tent flap. A storm-driven sky moving swiftly across the face of a gibbous moon. Perhaps it would clear by dawn. The ticking sound was not rain, only the wind. A storm, bound for somewhere else.

Half awake, I was again aware of the voices. A high-pitched cacophonous barking, like terriers, or the complaint of shoats. The single outcries became a rising cheer, as if in a far-off stadium, that rose and fell away.

Snow geese, their night voices. I saw them flying down the north coast of Alaska once in September, at the end of a working day. The steady intent of their westward passage, that unwavering line, was uplifting. The following year I saw them over Banks Island, migrating north in small flocks of twenty and thirty. And that fall I went to northern California to spend a few days with them on their early wintering ground at Tule Lake in Klamath Basin.

Tule Lake is not widely known in America, but the ducks and geese gather in huge aggregations on this refuge every fall, creating an impression of land in a state of health, of boundless life. On any given day a visitor might look upon a million birds here—pintail, lesser scaup, Barrow's goldeneye, cinnamon teal, mallard, northern shoveler, redhead, and canvasback ducks; Great Basin and cackling varieties of Canada geese, white-fronted geese, Ross's geese, lesser snow geese; and tundra swans. In open fields between the lakes

and marshes where these waterfowl feed and rest are red-winged blackbirds and Savannah sparrows, Brewer's sparrows, tree swallows, and meadowlarks. And lone avian hunters—marsh hawks, red-tailed hawks, bald eagles, the diminutive kestrel.

The Klamath Basin, containing four other national wildlife refuges in addition to Tule Lake, is one of the richest habitats for migratory waterfowl in North America. To the west of Tule Lake is another large, shallow lake called Lower Klamath Lake. To the east, out past the tule marshes, is a low escarpment where barn owls nest and the counting marks of a long-gone aboriginal people are still visible, incised in the rock. To the southwest, the incongruous remains of a Japanese internment camp from World War II. In agricultural fields to the north, east, and south, farmers grow malt barley and winter potatoes in dark volcanic soils.

The night I thought I heard rain and fell asleep again to the cries of snow geese, I also heard the sound of their night flying, a great hammering of the air overhead, a wild creaking of wings. These primitive sounds made the Klamath Basin seem oddly untenanted, the ancestral ground of animals, reclaimed by them each year. In a few days at the periphery of the flocks of geese, however, I did not feel like an interloper. I felt a calmness birds can bring to people; and, quieted, I sensed here the outlines of the oldest mysteries: the nature and extent of space, the fall of light from the heavens, the pooling of time in the present, as if it were water.

There were 250,000 lesser snow geese at Tule Lake. At dawn I would find them floating on the water, close together in a raft three-quarters of a mile long and perhaps 500 yards wide. When a flock begins to rise from the surface of the water, the sound is like a storm squall arriving, a great racket of shaken sheets of corrugated tin. (If you try to separate the individual sounds in your head, they are like dry cotton towels snapping on a windblown clothesline.) Once airborne, they are dazzling on the wing. Flying against broken sunlight, the opaque whiteness of their bodies, a whiteness of water-polished shells, contrasts with grayer whites in their translucent wings and tail feathers. Up close they show the dense, impeccable whites of arctic fox. Against the bluish grays of a storm-laden sky, their whiteness has a surreal glow, a brilliance without shadow.

When they are feeding in the grain fields around Tule Lake, the geese come and go in flocks of five or ten thousand. Sometimes there are forty or fifty thousand in the air at once. They rise from the fields like smoke in great, swirling currents, rising higher and spreading wider in the sky than one's field

of vision can encompass. One fluid, recurved sweep of ten thousand of them passes through the spaces within another, counterflying flock; while beyond them lattice after lattice passes, like sliding Japanese walls, until in the whole sky you lose your depth of field and feel as though you are looking up from the floor of the ocean through shoals of fish.

What absorbs me in these birds, beyond their beautiful whiteness, their astounding numbers, the great vigor of their lives, is how adroitly each bird joins the larger flock or departs from it. And how each bird while it is a part of the flock seems part of something larger than itself. Another animal. Never did I see a single goose move to accommodate one that was landing, nor geese on the water ever disturbed by another taking off, no matter how closely bunched they seemed to be. I never saw two birds so much as brush wingtips in the air, though surely they must. They roll up into a headwind together in a seamless movement that brings thousands of them gently to the ground like falling leaves in but a few seconds. Their movements are endlessly attractive to the eye because of a tension they create between the extended parabolic lines of their flight and their abrupt but adroit movements, all of it in three dimensions.

And there is something else that draws you in. They come from the ends of the earth and find this small lake every year with unfailing accuracy. They arrive from breeding grounds on the northern edge of the continent in Canada, and from the river valleys of Wrangel Island in the Russian Arctic. Their ancient corridors of migration, across Bering Strait and down the Pacific coast, down the east flank of the Rockies, are older than the nations they fly from. The lives of many animals are constrained by the schemes of men, but the determination in these lives, their traditional pattern of movement, are a calming reminder of a more fundamental order. The company of these birds in the field is guileless. It is easy to feel transcendent when camped among them.

Birds tug at the mind and heart with a strange intensity. Their ability to flock elegantly as the snow goose does, where individual birds turn into something larger, and their ability to navigate over great stretches of what is for us featureless space, are mysterious, sophisticated skills. Their flight, even a burst of sparrows across a city plaza, pleases us. In the Arctic, one can see birds in enormous numbers, and these feelings of awe and elation are enhanced. In spring in the Gulf of Anadyr, off the Russian coast, the surface of the water flashes silver with schools of Pacific herring, and flocks of puffins fly straight into the water after them, like a hail of gravel. They return with

the herring to steep cliffs, where the broken shells of their offspring fall on gusts of wind into the sea by the thousands, like snow. On August 6, 1973, the ornithologist David Nettleship rounded Skruis Point on the north coast of Devon Island and came face to face with a "lost" breeding colony of black guillemots. It stretched southeast before him for 14 miles. On the Great Plain of the Koukdjuak on Bamn Island today, a traveler, crossing the rivers and wading through the ponds and braided streams that exhaust and finally defeat the predatory fox, will come on great windrows of feathers from molting geese, feathers that can be taken in handfuls and thrown up in the air to drift downward like chaff. From the cliffs of Digges Island and adjacent Cape Wolstenholme in Hudson Strait, 2 million thick-billed murres will swim away across the water, headed for their winter grounds on the Grand Banks.

Such enormous concentrations of life in the Arctic are, as I have suggested, temporary and misleading. Between these arctic oases stretch hundreds of miles of coastal cliffs, marshes, and riverine valleys where no waterfowl, no seabird, nests. And the flocks of migratory geese and ducks come and go quickly, laying their eggs, molting, and getting their young into the air in five or six weeks. What one witnesses in the great breeding colonies is a kind of paradox. For a time the snow and ice disappear, allowing life to flourish and birds both to find food and retrieve it. Protected from terrestrial predators on their island refuges and on nesting grounds deep within flooded coastal plains, birds can molt all their flight feathers at once, without fearing the fact that this form of escape will be lost to them for a few weeks. And, for a while, food is plentiful enough to more than serve their daily needs; it provides the additional energy needed for the molt, and for a buildup of fat reserves for the southward journey.

For the birds, these fleeting weeks of advantage are crucial. If the weather is fair and their timing has been good, they arrive on their winter grounds with a strange, primal air of achievement. When the snow geese land at Tule Lake in October, it is not necessary in order to appreciate them to picture precisely the line and shading of those few faraway places where every one of them was born—Egg River on Banks Island, the mouth of the Anderson River in the Northwest Territories, the Tundovaya River Valley on Wrangel Island. Merely knowing that each one began its life, took first breath, on those intemperate arctic edges and that it alights here now for the first or fifth or tenth time is enough. Their success urges one to wonder at such a life, stretched out over so many thousands of miles, and moving on every four or five weeks, always moving on. Food and light running out behind in the fall, looming ahead in the spring.

I would watch the geese lift off the lake in the morning, spiral up white into the blue California sky and head for fields of two-row barley to feed, able only to wonder what this kind of nomadic life meant, how their lives fit in the flow of time and made clearer the extent of space between ground and sky, between here and the Far North. They flew beautifully each morning in the directions they intended, movements of desire, arabesques in the long sweep south from Tundovaya Valley and Egg River. At that hour of the day their lives seemed flush with yearning.

*

One is not long in the field before sensing that the scale of time and distance for most animals is different from one's own. Their overall size, their methods of locomotion, the nature of the obstacles they face, the media they move through, and the length of a full life are all different. Formerly, because of the ready analogy with human migration and a tendency to think only on a human scale, biologists treated migratory behavior as a special event in the lives of animals. They stressed the great distances involved or remarkable feats of navigation. The practice today is not to differentiate so sharply between migration and other forms of animal (and plant) movement. The maple seed spiraling down toward the forest floor, the butterfly zigzagging across a summer meadow, and the arctic tern outward bound on its 12,000-mile fall journey are all after the same thing: an environment more conducive to their continued growth and survival. Further, scientists now understand animal movements in terms of navigational senses we are still unfamiliar with, such as an ability to detect an electromagnetic field or to use sound echos or differences in air pressure as guides.

In discussing large-scale migration like that of snow geese, biologists posit a "familiar area" for each animal and then speak of its "home range" within that area, which includes its winter and summer ranges, its breeding range, and any migratory corridors. The familiar area takes in the whole of the landscape an animal has any notion of, an understanding it gains largely through exploration of territory adjacent to its home range during adolescence. Intense adolescent exploration, as far as we know, is common to all animals. Science's speculation is that such exploring ensures the survival of a group of animals by familiarizing them with alternatives to their home ranges, which they can turn to in an emergency.

A question that arises about the utilization of a home range is: how do

animals find their way to portions of the home range they have never seen? And how do they know when going there would be beneficial? The answers to these questions still elude us, but the response to them is what we call migration, and we have some idea about how animals manage those journeys. Many animals, even primitive creatures like anemones, possess a spatial memory of some sort and use it to find their way in the world. Part of this memory is apparently genetically based, and part of it is learned during travel with parents and in exploring alone. We know animals use a considerable range of senses to navigate from one place to another, to locate themselves in space, and actually to learn an environment, but which senses in which combinations are used, and precisely what information is stored—so far we can only speculate.

The vision most of us have of migration is of movements on a large scale, of birds arriving on their wintering grounds, of spawning salmon moving upstream, or of wildebeest, zebra, and gazelle trekking over the plains of East Africa. The movements of these latter animals coincide with a pattern of rainfall in the Serengeti-Mara ecosystem; and their annual, roughly circular migration in the wake of the rains reveals a marvelous and intricate network of benefits to all the organisms involved—grazers, grasses, and predators. The timing of these events—the heading of grasses in seed, the dropping of manure, the arrival of the rains, the birth of the young—seems perfectly fortuitous, a melding of needs and satisfactions that caused those who first examined the events to speak of a divine plan.

The dependable arrival of swallows at the mission of San Juan Capistrano, the appearance of gray whales off the Oregon coast in March, and the movement of animals like elk from higher to lower ranges in Wyoming in the fall are other examples of migration familiar in North America. I first went into the Arctic with no other ideas than these, somewhat outsize events to guide me. They opened my mind sufficiently, however, to a prodigious and diverse movement of life through the Arctic; they also prompted a realization of how intricate these seemingly simple natural events are. And as I watched the movement of whales and birds and caribou, I thought I discerned the ground from which some people have derived so much of their metaphorical understanding of symmetry, cadence, and harmony in the universe.

Several different kinds of migration are going on in the Arctic at the same time, not all of them keyed to the earth's annual cycle. Animals are still adjusting to the retreat of the Pleistocene glaciers, which began about 20,000 years ago. Some temperate-zone species are moving gradually but steadily

northward, altering their behavior or, like the collared lemming and the arctic fox, growing heavier coats of fur as required.

Climatic fluctuations measured over a much shorter period of time—on the order of several hundred years—are responsible for cyclic shifts of some animal populations north and south during these periods. Over the last fifty years, for example, cod and several species of bird have been moving farther north up the west coast of Greenland, while populations of red fox have been establishing themselves farther north on the North American tundra.* As animals long resident in the Arctic respond to certain kinds of short-term ecological disaster, as was the case with muskoxen in the winter of 1973–74, or to violent fluctuations in their population, as with lemmings, they reinhabit, over time, former landscapes and abandon others.

To cope with annual cycles—the drop in temperature, the loss of light, the presence of snow cover, and a reduction in the amount of food available—arctic animals have evolved several strategies. Lemmings move under the snow; bumblebees hibernate; and arctic foxes move out onto the sea ice. Many other animals, including caribou, walrus, whales, and birds, migrate over quite significant distances. Arctic terns, for example, fly to the Antarctic Ocean at the end of the arctic summer, an annual circuit on which they see fewer hours of darkness than probably any animal on earth. Other migratory birds that head out to sea change their ecological niche. The long-tailed jaeger, a rodent hunter on the summer tundra, becomes a pelagic scavenger on the high seas in winter.

On a scale smaller yet than these annual cycles are the migrations of animals during a season, like the movement of muskoxen; and the regular patterns of localized movement keyed to an animal's diurnal rhythms, like the habit among some wolf packs of leaving a den each evening to hunt. (Arctic animals, as mentioned earlier, maintain a diurnal pattern in spite of the presence of continuous daylight in summer.)

When one considers all these comings and goings, and that an animal like the muskox might be involved simultaneously in several of these cycles, or that when the lemming population crashes, snowy owls must fly off in the direction of an alternative food supply, and when one adds to it the movement

* American robins have moved as far north as Baffin Island in recent years. The Eskimos around Pond Inlet and Arctic Bay, who recognized the bird from stories white travelers told them about it, first saw them around 1942. Eskimos say the robin came that far north then because there was "a lot of fighting in the south" at that time.

of animals to the floe edges in spring, or the insects that rise in such stupendous numbers on the summer tundra, a vast and complex pattern of animal movement in the Arctic begins to emerge. Also to be considered are the release of fish and primitive arthropods with the melting of lake and ground ice. And the peregrinations of bears. And a final, wondrous image—the great ocean of aerial plankton, that almost separate universe of ballooning spiders and delicate larval creatures that drifts over the land in the summer.

The extent of all this movement is difficult to hold in the mind. Deepening the complications for anyone who would try to fix this order in time is that within the rough outlines of their traditional behaviors, animals are always testing the landscape. They are always setting off in response to hints and admonitions not evident to us.

The movement of animals in the Arctic is especially compelling because the events are compressed into but a few short months. Migratory animals like the bowhead whale and the snow goose often arrive on the last breath of winter. They feed and rest, bear their young, and prepare for their southward journey in that window of light before freeze-up and the first fall snowstorms. They come north in staggering numbers, travel hundreds or even thousands of miles to be here during those few weeks when life swirls in the water and on the tundra and in the balmy air. Standing there on the ground, you can feel the land filling up, feel something physical rising in it under the influence of the light, an embrace or exaltation. Watching the animals come and go, and feeling the land swell up to meet them and then feeling it grow still at their departure, I came to think of the migrations as breath, as the land breathing. In spring a great inhalation of light and animals. The long-bated breath of summer. And an exhalation that propelled them all south in the fall.

[. . .]

For years scientists have been aware of different rhythms of life in the Arctic, though they are not particularly arctic rhythms. Tundra soil cores examined by fossil-pollen experts have shown that changes in the composition of arctic plant communities have occurred periodically with a change in climate. Borings in the Greenland ice cap have revealed rhythmic fluctuations in average temperature over the centuries. A careful examination of arctic refuse middens by archaeologists, paleobotanists, and paleozoologists has revealed a succession of differently equipped early human cultures, whose entries into the Arctic are also related to periods of climatic change. The animal bones found in their camps confirm parallel fluctuations in the populations of the animals they hunted.

A number of scientists feel all this information should mesh, that in some way the rhythms of human migration, climatic change, and animal population cycles should be interrelated. With a precise enough mathematics even the "nine-year lynx-snowshoe hare cycle" and the "seventy-year caribou cycle" should fit neatly into a basic pattern. Few have sought to rigorously integrate this material, and many don't believe the relationships even exist, except in a general way. Since the 1930s, however, the Danish scientist Christian Vibe has taken the possibility very seriously, and no other body of work has been so clearly linked with the attempt to find a basic period of arctic cycling, a tantalizing bit of information of enormous interest to biologists, historians, and arctic developers.

Climatic change—the advance and retreat of glacial ice in the Northern Hemisphere—is the hallmark of the Pleistocene, the epoch of man's emergence.* Vibe, keeping this in mind, and believing whatever he learned could be applied to understanding the climatic future of Europe and America, posed certain questions for himself. Why, he asked, were seals scarce at Ammassalik on the east coast of Greenland at the turn of the century, while at the same time they were plentiful along Greenland's southwest coast? Why did the caribou population of western Greenland crash suddenly at the end of both the eighteenth and nineteenth centuries? And what accounted for the periodic northward movements of Atlantic herring and cod in the North Atlantic?

Vibe scrutinized the records of the Royal Greenland Trading Company, which took in sealskins and fox skins, narwhal ivory and other indicators, and by comparing these records with annual records of sea-ice movement and annual rainfall and snowfall, Vibe thought he could discern patterns. He checked his findings, to corroborate them further, by going over 232 years of fur-trading records from the Hudson's Bay Company in Canada, and by examining records kept by wool growers in southwest Greenland.

The first pattern to emerge for Vibe was a cycle of sea-ice formation and movement that lasted about 150 years, which records from arctic ships of exploration seemed to support. Vibe regarded as a key insight in this early work the fact that fluctuations in the arctic climate that were responsible for shifts

* Glacial periods are relatively rare in the earth's history. Scientists have discovered only four in the last 600 million years, the last of which is stlll going on. The Holocene, as far as we know, is only an interglacial stage, a reprieve, between the retreat of the Wisconsin ice (or Würm ice in Europe) and the next glacial advance.

of land and sea animals north and south over prolonged periods were tied to a lunar cycle of 18.6 years (the time it takes the moon to intersect the earth's orbit around the sun again at the same spot). Because the length of this lunar cycle is not a whole number, the maximum and minimum effect it has on the earth's tides (and therefore on ice formation and weather) can occur at different seasons of the year, in successive 18.6-year periods. This led Vibe to posit a primary period of 698 years for the Arctic's weather pattern, with secondary periods of 116.3 years, and what Vibe calls a basic "true ecological cycling period" of 11.6 years.

Depending upon your point of view, either Vibe's insights are ingenious and his mathematics elegant, or his system is impossibly broad and complicated and of little help in understanding arctic change. His inquiry might be considered an entirely esoteric and rarefied pursuit, in fact, if it were not for two things. In the Arctic one is constantly aware of sharp oscillation. It is as familiar a pattern of human thought and animal movement to the arctic resident as the pattern of four seasons is to a dweller in the Temperate Zone. In spite of the many manifestations of this rhythm, and the effect of sharp oscillation not only on resident animals but, probably, too, on the cultures that matured in these regions, Vibe's remains the only serious attempt at a description. Second, insofar as Vibe's theories explain oscillation in temperate-zone climate patterns or indicate harbingers of another ice age, they have a significant bearing on our developing patterns of commerce and economics, especially in the Arctic.

It is easy to say that the Arctic is characterized by sharp oscillation, just as it is to say that the airs of a temperate-zone spring are felicitous, but it is difficult to say precisely why. The basal annual rhythm of the North is winter/summer. The weeks during breakup and freeze-up are short, frequently perilous times, when strategies employed by both animals and human hunters to secure food are momentarily disrupted. The long winter and short summer constitute a temporal pattern around which life carefully arranges itself. Preparations for winter show up clearly everywhere in the land. The short-tailed weasel grows its white coat and the collared lemming its long snow claws. Tundra rodents shift from their night-active summer pattern to a day-active winter pattern, with but a few days of irregular rhythm in between. The arctic fox lines lemmings up in neat rows in its winter caches.

A second pattern complements this oscillation—long stillnesses broken by sudden movement. The river you have been traveling over by dogsled every week for eight months, and have come to think of as a solid piece of the

earth, you wake one day to find a heaving jumble of ice. The spring silence is broken by pistol reports of cracking on the river, and then the sound of breaking branches and the whining pop of a falling tree as the careening blocks of ice gouge the riverbanks. A related but far eerier phenomenon occurs in the coastal ice. Suddenly in the middle of winter and without warning a huge piece of sea ice surges hundreds of feet inland, like something alive. The Eskimo call it *ivu*.* The silent arrival of caribou in an otherwise empty landscape is another example. The long wait at a seal hole for prey to surface. Waiting for a lead to close. The Eskimo have a word for this kind of long waiting, prepared for a sudden event: *quinuituq*. Deep patience.

As I moved through the Arctic I thought often about a rhythm indigenous to this land, not one imposed on it. The imposed view, however innocent, always obscures. The evidence that there is a different rhythm of life here seemed inescapably a part of the expression of the animals I encountered, though I cannot say precisely why. A coherent sense of the pervasiveness of such a rhythm is elusive.

The indigenous rhythm, or rhythms, of arctic life is important to discern for more than merely academic reasons. To understand why a region is different, to show an initial deference toward its mysteries, is to guard against a kind of provincialism that vitiates the imagination, that stifles the capacity to envision what is different.

Another reason to wonder which rhythms are innate, and what they might be, is related as well to the survival of the capacity to imagine beyond the familiar. We have long regarded animals as a kind of machinery, and the landscapes they move through as backdrops, as paintings. In recent years this antiquated view has begun to change. Animals are understood as mysterious, within the context of sophisticated Western learning that takes into account such things as biochemistry and genetics. They are changeable, not fixed, entities, predictable in their behavior only to a certain extent. The world of variables they are alert to is astonishingly complex, and their responses are sometimes highly sophisticated. The closer biologists look, the more the individual animal, like the individual human being, seems a reflection of that organization of energy that quantum mechanics predicts for the particles that compose an atom.

---

* Eskimo descriptions of this phenomenon were not taken seriously until 1982, when archaeologists working at Utqiagvik, a prehistoric village site near Barrow, Alaska, discovered a family of five people that had been crushed to death in such an incident.

The animal's environment, the background against which we see it, can be rendered as something like the animal itself—partly unchartable. And to try to understand the animal apart from its background, except as an imaginative exercise, is to risk the collapse of both. To be what they are they require each other.

Spatial perception and the nature of movement, the shape and direction something takes in time, are topics that have been cogently addressed by people like Werner Heisenberg, Erwin Schrödinger, Paul Dirac, and David Bohm, all writing about subatomic phenomena. I believe that similar thoughts, potentially as beautiful in their complexity, arise with a consideration of how animals move in their landscapes—the path of a raven directly up a valley, the meander of grazing caribou, the winter movements of a single bear over the sea ice. We hardly know what these movements are in response to; we choose the dimensions of space and the durations of time we think appropriate to describe them, but we have no assurance that these are relevant. To watch a gyrfalcon and a snowy owl pass each other in the same sky is to wonder how the life of the one affects the other. To sit on a hillside and watch the slow intermingling of two herds of muskoxen feeding in a sedge meadow and to try to discern the logic of it is to grapple with uncertainty. To watch a flock of snow geese roll off a headwind together is to wonder where one animal begins and another ends. Animals confound us not because they are deceptively simple but because they are finally inseparable from the complexities of life. It is precisely these subtleties of fact and conception that comprise particle physics, which passes for the natural philosophy of our age. Animals move more slowly than beta particles, and through a space bewildering larger than that encompassed by a cloud of electrons, but they urge us, if we allow them, toward a consideration of the same questions about the fundamental nature of life, about the relationships that bind forms of energy into recognizable patterns.

ALIBERTI'S RIDE

from *This Cold Heaven* (2001)

Gretel Ehrlich

Gretel Ehrlich made her first trip to Greenland in 1993. Over the next seven years, she returned repeatedly, drawn by the landscape and by a desire to understand native culture. In the excerpt below, she travels by dogsled with a hunter named Aliberti from the tiny island town of Uummannaq to the even tinier town of Niaqornat.

THE SHIT TRUCK'S CHAINED snow tires awakened me. It had taken only a day for all of us to fill up the toilet. The door opened. Two young men slipped in, emptied the bucket, returned it, then disappeared down the narrow street. It was a good job because it paid well and they were finished with their work early, so they could go out hunting in the evening. I thought of the Buddhist monks who asked for the job of cleaning toilets in cities to help "ground" them, bind them to what is actual during a long course of otherworldly practices. But how much grounding does a subsistence hunter need?

The household woke slowly. We dressed for frigid weather and carted loads down the steep hill to the ice where the dogs were staked out—food, sleeping bags, fuel, stoves, warm clothes, presents for friends. Our route would take us from Uummannaq to Ukkusissat, to Illorsuit, north to Nuugaatsiaq, back to Illorsuit, then to Niaqornat, Qaarsut, and home.

The ice at the base of the town was a Coney Island of hunters, sleds, children, and six thousand dogs all yipping, howling, barking, screeching, and yapping. Children played tag between sleds while hunters harnessed their teams. One team was being fed frozen halibut. They yelped and yapped as the chunks were tossed into their mouths while the other teams looked on

silently, knowing there would be no food for them. After a long winter, the dogs had spring fever: females in heat ranged freely among chained-up males, causing yowls of sexual union and longing. Everywhere I looked, dogs were fucking, getting stuck together, crouching, sitting, biting, and yapping. One male grabbed the trace line of the female he wanted, jerked it tight, then jumped on her from the rear. A dog that had been left behind sat up on his hind legs, a lone figure pawing at a world of ice and air.

Our party consisted of sixty-six dogs, six children, and seven adults. Aliberti was the hunter with whom I would travel, as his only passenger. It was his sled and those of two other hunters—Jacob and Unatoq, or "Mr. Warm"—that we had seen coming the night before from Ikerasak. The other adults were Ann, Olejorgen, and an Inuit woman from the Children's House, Louisa.

At fifty-eight, Aliberti's face was weathered, making him look older than his years. He had the small, tight build typical of Inuit hunters and moved like a cat. A cigarette dangled from his mouth. When I laid my duffle and sleeping bag on the sled, he smiled, and the cigarette stuck to his lower lip. He was nearly toothless—just a few black stubs. Standing face to face we could look straight into each other's eyes. He slipped my parka and duffle under caribou skins and lashed them tight.

Seven cigarettes were stubbed out on the ice; we had been waiting. Olejorgen's dogs were tangled and fighting. One female got loose and ran away through the midst of six thousand other dogs chained up on the ice or being harnessed, and a collective howl rose, echoing. The young sled drivers—Ludwig and Aliberti's son—were having troubles too: their trace lines had broken and two more dogs got loose and ran off with another team.

Aliberti looked on coolly. It is not the Inuit way to give help unless there is real danger, otherwise no lessons will be learned. He tightened the lash ropes on our sled again, blew his nose, secured his mittens under the lines—until finally he couldn't stand it anymore. Shaking his head, he hooked the trace lines to the sled. The dogs lurched forward and with one flying leap we landed side by side on the sled. He looked back at the chaos behind us and laughed. It would be some time before the others could follow.

I thought of a story told to Rasmussen by Inaluk in 1902: "There was once an orphan boy who drifted out to sea on an ice-floe, and arrived among strange people. They took him into their service at once and used him for all

their menial work. But he had a brother who was a great magician and who, when the little boy did not come back, began to look for him in soul-flights. . ."

Now, almost one hundred years later, Aliberti had his own story. Three days before Christmas in 1959, Aliberti, then nineteen, went out to get a seal. He lived on the island of Ikerasak. A warm wind, a foehn, blew and the fjord ice began to break up. Before he knew what was happening, Aliberti was adrift with his dogs on an icy slab. He had no food, no stove to melt ice for water, no extra clothes, just his rifle—he hadn't been planning to stay out very long.

Aliberti's ride lasted five days. At night the temperature was 20 below zero. He drifted past the village of Uummannaq, then north up the coast past Niaqornat. He had gotten wet trying to jump to shore and his clothes froze stiff. There was no way to get dry or warm. It was December and completely dark, with no light at all in the sky. That week there wasn't even a moon. Up the coast he went, pushed by currents and pulled by tides. No one saw him drifting.

His parents waited. By Christmas Day he was presumed dead. Perhaps he had slipped under the ice, as so many had done before. His obituary was read on the radio. But he was alive and still drifting. "I remember seeing the candles in people's houses at the villages. Christmas came and went. I could hear singing. No one could see me. I didn't think I would live."

The currents took him north into the Illorsuit Strait along the east coast of Ubekendt Ejland—Unknown Island. He had already drifted seventy miles. "Things got worse. When it's that cold and you have no food or water, *sila* becomes stronger than you are. My dogs died one by one and I pushed them over into the water. I just lay on my sled and waited. On the fourth day I noticed that the ice floe had gotten smaller. Then it split in half and I had to jump from the bad part to what was left of the floe. The ice didn't look like it would hold. I wondered what my family was thinking. They couldn't come look for me because there was too much ice to go out in a boat and not enough to travel by dogsled. My clothes were wet on the inside and frozen on the outside. You get pretty cold. On the fifth day something woke me. I had been drifting in my mind too, but I heard the ice bump against something. I looked up: I had come to a stop a hundred yards from the village of Illorsuit.

"For a moment I thought a miracle had happened. I could see people in their houses with candles burning. They would see me and rescue me and I

would live after all. I lamented the fact that all but one of my dogs had died. I allowed myself to think about eating and drinking. I lifted up on one arm and yelled and waved. But no one came. Another day passed. I was telling time by the rising and setting of the moon. Sometimes I called out with what strength I had left, but the wind was blowing the wrong way. My last dog died.

"A villager came outside to pee and he saw me. He yelled out and I lifted an arm. Then he went back inside his house. He was gone a long time. I gave up hope. I think I fell asleep. The sound of voices jarred me awake. I didn't know if they were human. The man I'd seen on the beach and another hunter were making their way toward me in a skiff, pushing big pieces of ice aside with an oar. When they got close they looked like giants. I wasn't right in the head by then. They took me to shore.

"The hunters thought I was a ghost because my death had already been announced on the radio. My hair was coming out in handfuls and my teeth were loose. When they carried me ashore, people ran the other direction.

"The hunter took me to his parents' house. There I was fed and kept warm. They got through to my parents on the two-way radio and told them that I was alive but my mother didn't believe it. After a few days, the weather improved and the hunters took me home to Ikerasak. And as you can see, I am still alive."

Bright sun, clear skies, a slight breeze, the temperature about zero. The ice was smooth and fast. With our light load, the sled fishtailed and the dogs' panting became the only sound we heard as we slid from the noise of town. The smoke from Aliberti's cigarette snaked back across his cheek as we glided forward. He turned and smiled. There was no need to talk. To be alive and on a dogsled in Greenland was enough and I was happy.

Far out in the middle of the fjord, the dogs slowed to a trot. My whole body worked like an eye, watching the world scroll under and over us. Aliberti and the five others headed straight north, gliding between the uninhabited islands of Abut and Saleq, then veered northwest toward the village of Ukkusissat on the thumb's tip of land that extended from under the ice cap.

Cool, alert, and relaxed, Aliberti sat sideways at the front of the sled with his legs sticking out straight. Seagulls and eider ducks flew overhead in

flocks—a sign of spring. From shoreline—ice line—everything looked mathematical. We glided through its permutations. If water is time's shapeless infinity, then ice is time's body, inhabited by light and shadow, tormented by sun and cold.

A row of stranded icebergs was a giant slalom course leading out to sea. We glided by, staring at their outsize beauty. Icebergs are lessons in geometry: they give and take light at will, changing a shined, chrome side into a dull fastness, then back to a sizzling angle of vitality, with a fine, thin, razor-sharp ridge cutting winter out of the sky.

Jacob and Unatoq came up behind us on their two sleds, carrying Ann on one and Louisa and a boy from the Children's House on the other. Finally the others—Olejorgen, Ludwig, and Aliberti's son—caught up. Ann zinged by, whooping, then Louisa and the boy. They shouted something in Greenlandic I couldn't understand. The teenage girls rode with the teenage boys and they were always last. This was an Eskimo–Danish–Faroe Islander–American laughing, gossiping sled party heading north. First stop, Ukkusissat, where we would celebrate the birthday of a friend.

Between Uummannaq and the mainland the ice was rough. We bumped over small waves, their topsy-turvydom frozen in place, then stopped and waited again for the others. Olejorgen pulled up beside us and whistled his dogs to a stop. With his movie-star looks, he resembled an Eskimo pasha reclining on a bearskin rug and wearing polar bear pants. His hair had grown long and he wore dark glasses. I imagined a remake of *Doctor Zhivago,* set in Greenland rather than Russia. As he tried to settle the dogs, he promptly ran over one of them. The animal screamed in pain. We lifted the sled up and Olejorgen held the injured animal. He wasn't sure what to do next, so Aliberti told him to carry the dog on the front of the sled until we got to the village. The others showed up and we continued on.

Seagulls flew over: there must be open water somewhere. A slow-moving sled pulled by five skinny dogs passed us going the other way. The hunter looked comatose, lying flat on his back and gazing skyward, his dogs wandering. When we stopped for lunch, Jacob said that during his monthlong sled trip to Qaanaaq the previous spring he had developed stomach pains. He stopped in Upernavik and the next day he was operated on there. While the others continued north, he rested. When they returned three weeks later, Jacob drove his own dogsled home.

On the way, Jacob was lucky and got a polar bear—he was in need of new

pants for the next winter. "Not only did he get an ice bear," Olejorgen exclaimed, "the bear fell over dead on Jacob's sled, almost killing him!" Laughter.

Jacob quipped: "We call this story 'How I almost died twice in one springtime hunting trip.' "

Rounding a bend, we saw a long arm of rock pushing out into the fjord. On its tip was Ukkusissat. As we approached, litter blew across multiple sled tracks—beer bottles and plastic bags. Ann insisted on stopping and picking up every piece, indignant for her beloved (though not native) Greenland. "We must not be so dirty," she exclaimed to no one in particular. The hunters smiled. Her garrulousness amused them, but she also held their respect for the fine job she did with the children.

Ukkusissat consisted of a few small houses set in stepping-stone fashion up a hill. Behind the last building a towering wall of basalt lay crumbled in ruin, as if an older village had once been sited there. Birthe ran down the hill to greet us. Fine-boned, strong-willed, and skittish as a colt, she was joyous at having company. She was the only Dane in this village of one hundred people and she drove her own dogsled. "I just barely know how to do it," she said apologetically. "But it's transportation between here and my friends in Uummannaq." After the dogs were unharnessed and fed and the duffles and sleeping bags had been unloaded, we sat on the empty sleds and basked in the evening sun.

An old man came to inspect the injured dog, who had been laid on the snow. The dog's back was broken. The man told Olejorgen: "After you have gone, I will shoot him for you."

The birthday celebration began at Birthe's house that evening. We cooked seal ribs, potatoes, and onions, and drank warm Tuborg beer. How quickly and effortlessly food for twelve was prepared. At midnight the sun hovered, then began to slide behind the mountains into the northern part of the sky. One by one the others went to bed. Ann and Olejorgen appropriated Birthe's bed, so she and I stayed up late, talking.

There was no night, no darkness. I trudged down to the ice. Aliberti had pitched a canvas tent over his sled, blue canvas at one end to keep the sunlight out. Inside, a Primus stove was lit to keep him warm. I unrolled my sleeping bag next to his on the sled and slept. At three in the morning gold and purple stripes of light lay across the ice. The sounds from the village subsided, but the dogs, staked out everywhere in a vast slumber party, talked for the rest of the night.

When we woke there was no sun and the dog whose back was broken was

still alive. He lay unattended in the snow covered with rime frost. A fog had come in and sealed itself to the ice. The dogs sat hunched and waiting. As the fog lifted, an eerie yellow light shone through.

I helped Birthe move her dogs from behind her house to the ice. She had decided to accompany us to the next village. One by one they pulled us downhill—that's all they know how to do. After, we carried the toilet bucket from her house and poured the contents into a pipe that emptied into the fjord. City living, she said with a grin, because on hunting trips any piece of rough ice served as a bathroom.

Aliberti broke camp, harnessed the dogs, and loaded the sled. The flying *S* of his whip snapped: we were on our way to Illorsuit, the village where he was saved after drifting for five days, where Rockwell Kent lived with Salamina, where I spent a summer with Marie Louisa. Again we took off before the others were ready. All was lost behind us in fog. We drove in a white shroud, a white darkness, continually breaking through a crust of ice that had melted and refrozen.

"Sumiippa?" I asked. Where are we? Aliberti smiled but said nothing. We were dressed in skins: sealskin pants, sealskin anoraks with fox-fur ruffs around the hood, sealskin kamiks, and sealskin mittens with dog-hair ruffs at the wrist. His whip straightened out above the dogs' heads like a thought coming apart. Sun shone behind mist; now we traveled in a brightening cloud. *"Uatsi,"* I said. Please wait. I got off the sled to pee. When I looked down, I found that I'd unexpectedly gotten my period.

In the Arctic there is no privacy: Aliberti walked over, looked at the blood on the ice, and laughed. "This is good! It looks like we killed a seal! They will think I am a very good hunter to find a seal in this fog. And also, they will be able to find us."

Soon Jacob, Louisa, and the little boy pulled up and together we waited for the others. Hot tea and a bag of frozen shrimp were passed around. The boy started a game of tag. He was short for his age, as though stunted, and had come to the Children's House from a family where sexual abuse was epidemic. He ran around the circle of adults, then tagged me so hard that I fell, stood, and fell again. We were five people slipping and sliding, playing tag 700 miles from the North Pole in the middle of a frozen ocean with no land in sight. When Louisa was tagged and turned to tag the boy, he vanished into thick mist.

Another hour went by. The horizon's seamless wall of white frayed. I sat on the edge of the sled as if it were the edge of the world, with my hands over

my eyes, trying to locate myself; my body was still, but my eyes tumbled around; my head was a goblet of ice. Is ice a form of indifference or is its intent to obscure? How absurd we must have seemed to the marine mammals swimming below, to the walrus with its colossal appetite and backward stroke, feasting orgiastically on shellfish, and the narwhal with its mysterious white tusk needling the ocean floor for food.

Aliberti dozed on the sled. We were not really lost, just living under the thumb of weather and it was pressing down on us hard. Good time to sleep, he said. But I kept trying to see. My eyes wandered . . . I was looking at a world that had overflowed its outlines, where everything had grown into invisibility.

Finally the others came. Birthe's dogs had gotten loose in the village and she had to start the harnessing process all over again. By tacit agreement, Aliberti took the lead once again. He was still honored in Ikerasak for having survived his five-day ordeal on the drift ice. *"Meeeeuuww, meeeeuuuw,"* he called out, and we took off, seven sleds abreast amid dogfights, tangled trace lines, and laughter. Then the dog-pant, dogtrot rhythm began again.

We drove through a flock of thousands of Arctic gulls. One bird's foot had frozen to the ice. As we approached it struggled to escape. The other gulls, seeing an easy meal, attacked and killed him. Aliberti watched impassively as we slid by, then turned to see if I'd seen. My eyes went from the cannibalized bird to him. He flashed a look that said, yes, it's tough up here. To the north, icebergs sprouted clouds, then fog closed in again, more tightly than ever, hiding my trail of blood.

Mist constricted and freed us simultaneously. Our slip-and-slide dealings with our psyches were never more evident: we long for solitude, but as soon as we have it, we are desperate for friends. I contemplated the shape of Aliberti's head, wondering what a PET scan might reveal about his ability to draw perfect maps of this coast from memory. Plato thought that vision was "a stream of fire or light that issues from the observing eye and coalesces with sunlight." But there was no sun and a hard, glasslike crust slowed down the dogs.

Gliding blind, I lost my bearings, wondering what life must be like for my mother, who was going blind. "The worst part of it is not being able to drive," she told me. Olejorgen's sled passed us, then fell back. I blinked: a rock wall shot up in front of us spangled with ice crystals and new snow. *"Takurngaqtuq!"* I yelled. "I feel as if I'm seeing [something icy] for the first time."

Aliberti turned and smiled. "It is called Apaat . . . this rock."

The way to Illorsuit was silvered, the top sift of snow blowing across our path as if swept by a giant broom. We cut through long-tailed drifts that pointed north like arrows toward the open end of the fjord, which gave onto an entire ocean paved with ice. Where snow had blown off stranded icebergs, exposed walls had been fired by sun and refrozen into glass—all glint and glaze and broken crust like crème brûlée. The snowed-on, fog-sealed ice in front of the sled revealed only the toes of things—icebergs whose drift toward the sea had been postponed by winter.

A headland loomed topped with crosses: Illorsuit's cemetery, which doubled as the heliport. Around a bend the village appeared, the same ramshackle houses lying in the arc of a half-moon bay. Seeing and smelling other dogs excited our dogs. Our seven sleds raced each other to the edge of the village. I took a deep seat and hung on to the lash ropes: we flew through the uneven middle of a decapitated iceberg and lurched down onto flat ice. Aliberti was in the lead, but at the last moment, when we had to go around a piece of rough ice, he was beaten by Mr. Warm.

I saw a young girl weaving between sleds and dogs lying in soft snow, noses tucked under tails, others yowling while being fed, and children pulling handmade toys on a string. She walked toward me, then sat on a sled nearby and waited. *Kina una?* I asked tentatively. Who are you? She smiled. I was still puzzled. *Kina una?* I asked again. Then I knew it was Marie Louisa.

She had grown. Her black hair hung down in a long braid, her cheeks had fattened, and she was four inches taller than she had been two years ago. Then someone grabbed me from behind and spun me around: Hans. We all laughed and hugged. He was grayer and thinner and immediately launched into a litany of grievances about village politics: "Last week they passed a bill allowing snow scooters in this district. Nothing will be the same. Now they say we need a truck in the village. All these hundreds of years we've walked and used a wheelbarrow and in winter a sled to transport our supplies. Now we will have to worry about our children and dogs getting run over."

"Where is your sled?" I asked as he loaded my duffle onto the back of the town's one snowmobile.

He gave a disgruntled shrug. "Last week they began selling Coca-Cola in

Greenland. The helicopter brought it and it was sold out within half an hour." He told me to get on the snow scooter.

"This is against my religion," I said.

He laughed ruefully. "Yes. Those are my feelings too."

"Takuus!" I yelled to Aliberti as we sped off. See you soon. The others dispersed: Ann and Olejorgen went to the schoolteacher's house, the children and Louisa to the clinic (used for extra housing for visitors), the hunters to the house of the family who had saved Aliberti's life forty years ago.

Hans's house had not changed. The walls were pale blue and yellow, the rug was red, the clock still gave the time as noon, and the photographs from his hippie days in Christiania were still on the wall. Snow sifted through the broken skylight and left a perfect white pyramid on the floor. Because there was no word in Greenlandic for pyramid, we referred to the invading snow as "the igloo of Egyptians." The living room was still empty except for the foam pad—my bed—on the floor. "I haven't moved it since you left two years ago," Hans said.

Only one thing had changed: Arnnannguaq was not there. "I told her she was not to come to the house when she was drinking," Hans said. What had been a minor problem had escalated. Now she was drinking all the time. Hendrik and Marie Louisa burst in the door and jumped on the pad, burrowing under the covers. We hugged and wrestled and spoke our usual jumble of Greenlandic and English which only we understood.

In the middle of the night I found myself at the small window where I had kneeled sleepless so many times before. The view across the fjord was marked indelibly on my mind: the rock-faced mountain with two glaciers spilling out on either side like grand stairways, and the gleaming fortress of the ice cap looming above. Dogs howled, the diesel generator chugged, and the weather worsened. Blue sky fused with white, its porcelain fracturing into frostfall.

The children woke and joined me in the living room. We watched the sun droop in the west and bounce back up to circle the northern sky. Despite the sun, the living room started to feel frosty. We looked: the heater was off and the thermometer had dropped to 16 below. The children huddled under my sleeping bag until Hans fixed the heater again.

Much later Marie Louisa awoke and was upset. *"Anaana, Anaana!"* Mother, Mother, she cried out, nuzzling my breasts like a baby. But she was eight years old. I held her tight. Things had not been good for her here.

We slept huddled together during the two hours of twilight in the middle of the night. As soon as the sky lightened again, it began to snow. Marie Louisa looked up and asked if the world was really round. I said yes, but realized that

I was only repeating received information, that I hadn't actually laid my hands on the rounded fender of this planet. I asked how she thought the world was made and she said, "Big flat pieces of ice pushed together." She still didn't understand how we stayed attached to a sphere. I said, "Gravity—a kind of glue."

In the morning Hans played "Mr. Tambourine Man" while I made pancakes for the children. Marie Louisa played the electric piano that Hans had given her for Christmas and tried to keep up with the song. Later, when the children went off to play, Hans told me his troubles with A.: "She is drunk most of the time now; she doesn't give love when the children need it, only when she wants it. This is no good. But what can I do? Where can I go? I'm fifty years old with an unfinished education and it's too late for me to go home to Denmark and get a job. I didn't think about these things when I was young. Now the children need more education, which means I have to go somewhere else. The law isn't in my favor because we are not married, the law will want to give her the children. But I have been feeding and dressing and taking the children to school every day for years. But how can I prove that to the Greenlandic social worker in Uummannaq? Who will believe me?"

✳

When Aliberti, Olejorgen, Ann, Jacob, Louisa, Unatoq, and the children packed up the sleds and continued north to the village of Nuugaatsiaq, I stayed behind to help Hans. A foot of new snow fell during the week, then another. Arnnannguaq came home, sober. I helped her get water. Two years earlier we had collected water from a spring behind their house. Now water was fetched from a huge red tank on the hill. "Progress," Hans said.

The Arctic geometries began to soften. Sky and ice turned powder blue and the glaciers on the far side of the fjord bent up into a white nothingness. All afternoon the north-facing village rested in shadow, but at night it was bathed in bright sun. Kristian Moller, who had brought me to Illorsuit on his boat two summers before, stood on the beach and looked at his dogs while Marie Louisa and I played catch on the ice with a pink ball. Later, I watched her pounding up and down the hill behind the village on short red skis. I told Hans that I would take her to live with me and go to school if he found he couldn't leave the village. She trusted me and I loved her and I would bring her home to him for the summers. He agreed. The next night she slept curled up against me. Her face shone in the midnight sun. A tremendous wind howled.

We climbed the steep mountain behind the village, kicking footholds in ice and snow. Near the top we stopped to rest. From above I could see that the way was swept clean. The grooves the glacier had made on the rock were free of snow. Across the fjord two glaciers flanked an immense rock wall, their gashed roofs carrying debris like dark hats to the sea. Above the crumbling snout and the stretched *seracs,* the castellated masses of the glacier, the ice mound was aswirl in blowing snow.

We descended, following a track made by a rolling stone that led us under the thin spray of a frozen waterfall where, in summer, we had bathed daily, then came down to the shadowed crescent where the houses were. The beach was all gravel laced with ice. The lines made fast to skiffs went slack and taut, slapping the open water and leaving marks that looked like arrows pointing toward another season.

Our heels had pushed down through snow, then scree. We jumped over the boggy grass around a spring and landed on the beach where chunks of ice caught on pebbles. In the summer, Marie Louisa liked to take off her clothes and swim in the frigid Arctic water. Now it was a white floor on which she danced, tossing the pink ball, then skate-skied as fast as she could away from where I was standing.

The fjord is six miles wide but looked narrower. The brown wall we had descended was a soft amphitheater into which the sounds of collapsing icebergs would soon be gathered. Behind us the village lay in shadow. Out on the ice Marie Louisa grew smaller and smaller. Come back to me, I wanted to yell, but did not. I suddenly sensed how forbidding solitude could be here, how effortlessly death could occur—just a slip through the ice. Then I did yell, and Marie Louisa turned laughing, then went farther until she was no bigger than an ant. Finally she skied toward me.

✳

The next day Olejorgen, Ann, and the Uummannaq bunch returned in snow that was so deep, it came up to the middle of the dogs' chests. They could barely move; the hunters had to walk in front to break trail. While they rested for a few days, the wind began to blow. Snow stuck to brown rock behind the long string of houses. The light shifted and the sky came apart as frostfall, glitter from the Land of Day.

The last night the children climbed into my sleeping bag with me—a

slumber party, I said—but they didn't understand. Framed in the window was a pyramid-shaped iceberg. An eerie twilight filled the room. Toward morning, a ring circled the sun, which meant bad weather.

The sleds were packed. As Aliberti took off in the lead, Marie Louisa ran after us calling *"Anaana."* The sky cleared and it was deeply cold. We pulled our sealskin anoraks on. From the numbness beginning in my feet, I knew it must be zero.

On our seven-sled procession out of Illorsuit, sharp bits of ice flew in our faces. We bowed our heads in the face of Sila as the dogs' fur grew thick with snow. "Sometimes spring feels colder than winter," Aliberti said, pulling his hood up tight around his face. We were dressed head-to-toe in sealskin pants and anoraks with fox-fur ruffs. "We eat the inside and wear the outside," Ole-jorgen said, passing us as we left the village. Marie Louisa stood in the ice storm watching. I wanted to take her with me right then, but she was still in school. Hans said he would call me in Uummannaq. I knelt backwards on the sled, waving until she was out of sight.

Where pieces of ice stuck up like gravestones, snow had drifted in long inverted V's and the wind broke these apart into scudding curds that slid on ice. The rest of the ice was clear. Beneath us a tormented ocean heaved, while continually pushing up against the lid of ice, the topside was smooth and mirrorlike, reflecting only calm. The poet Muso Soseki wrote "No clarity can flatten torment, no fragment can undo clarity. . ."

Aliberti cracked the whip in a circular underhand motion and the dogs ran. A cloud bank grayed the horizon and sun lit the place where the fjord gave way to the whole of Baffin Bay. We veered southwest toward the village of Niaqornat. Wind howled out of the west and blew from distant headlands all the way across the frozen sea. When we passed behind an iceberg, we were shielded from wind, but the snow grew suddenly deeper where it had drifted, almost bringing the dogs to a standstill. Between stranded floes, the ice was blown clear. The dogs speeded up and the sled fishtailed on what looked like a thousand-mile-long mirror.

Halfway down the flank of Unknown Island, the weather worsened: a ground blizzard stirred. Bits of ice, like tiny continents, blasted our eyes. Aliberti motioned to me to lie down behind him on the sled. He laid his right hand on my hip and I draped my left arm over his shoulder. The sled tipped and bumped: we held each other on. Wind-driven ice tore at the dogs. Their feet and legs were bleeding and their muzzles were encrusted. The fog lifted

but still we couldn't see. I looked down at my body and Aliberti's. Dressed like seals we slid over seals: thousands of pinnipeds, with only a thin sheaf of ice separating us. If they could see us, would they think we were seals?

Ice flew. We slid closer together and pulled our hoods down almost over our eyes. Once in a while I saw one of the other sleds in the distance, then it disappeared. Where the ice under us was smooth, we could almost doze. Then we hit a rough patch and had to hold on. Ice cut us; snow blinded us. So much in the Arctic attempts to obstruct vision: fog, snow, darkness, ice. But each element has its built-in clarity, an opaque shine. Another ancient theory of vision went like this: "An eye obviously has fire within, for when one is struck, this fire flashes out."

We bumped along lying close. Seen another way, we were fake seals, like decoys, trying to attract the seals under the ice, as we made gestures of intimacy, my feet in his hands, his back against my chest, my hand on his sealskin shoulder: seal love with no thought of possession. All I knew was that the seal body in front of me was blessedly blocking the wind.

We traveled for eight hours without talk, chastely intimate in a bond of blood, snow, and fur. Yet I knew that what I was seeing was transient: glaciers, human love, sea ice, dogs, humans, and my own perishable body and his—Aliberti's. Too often we confuse what is happening in the moment with notions of permanence. The intimacy would not last.

My feet hurt with the cold. Even on a simple trip such as this one, things could go wrong. Jorgen Brönlund, who had traveled with Rasmussen up this strait, wrote in his diary of the 1908 Denmark Expedition: "Perished on 79th Latitude North after attempt home—journey round the inland ice, in November I came here in a waning moon and could not continue because of frost-bitten feet and the darkness. Hagen died on 15 November and Mylius about 10 days later."

A few hours later we came to a hut. Two sleds from our party had reached it before us. Aliberti stopped the dogs. We stood up and looked at each other: we were both blasted white with snow. I stumbled when I walked: my feet were completely numb, but I said nothing since I hadn't brought an extra pair of kamiks. Inside, tea and coffee were being brewed. I passed around nuts, raisins, and figs from California. When Aliberti saw me wriggling my toes, he quietly removed my boots and socks and began massaging my toes.

Years ago in a Wyoming winter, two of my toes were frostbitten; a neighbor who had lived through seventy winters helped thaw them out. Now they were white and painful again. Aliberti held them between his knees while he

drank coffee. Nothing was said. Then he signaled to one of the young boys and said something in Greenlandic. A moment later he was slipping a different pair of boots on my feet.

It was time to go. Aliberti motioned to me and I followed. We were two seals who had sprouted legs. Untangling the traces, he hooked the dogs in and the lines pulled tight. We flew off the lip of a cornice and slammed down onto the ice, hooked together on the sled. My feet were warm again as we clattered, bumped, tipped, and shuddered across the frozen sea toward the village of Niaqornat.

*

Sunday morning. Complacencies of the anorak and the doubletime contrapuntal panting of dogs. Nothing moved but my eyes. All beauty stayed behind only to give way to more—not the green freedom of the cockatoo, but the liberation of the diamondlike ice and Arctic sun that, after ten days of traveling, no longer went down, but only lingered at the eastern and western extremities of the sky, enticing us forward.

When the dogs' paws bled because they were trotting through glass, we stopped to put sealskin booties on their feet. A northwest wind poured across the fjord like cold water. Earlier, the Greenlandic radio played Marilyn Monroe singing "Diamonds Are a Girl's Best Friend." We drove over wind-drifted humps of ice blown free of snow, a rocky road paved with shining, beveled diamonds, then stopped to hack off a piece of young ice to be melted for drinking water. Its faceted interior was a blue wilderness. At its base, droplets of dog blood stained the snow.

Snowy polka dots dappled the sled track in front of us. They looked like eyes. Who was watching? Could Sila see? Fog was snow-flecked. Was it possible to draw eyeballs on chaos?

When the storm cleared two ravens flew in front of the sled, one on top of the other. Aliberti pointed at the double-decker birds and made a quick gesture to indicate "fucking." He laughed and his black teeth showed. A dog got tangled in the loose trace lines. Aliberti jumped off the moving sled, ran alongside, plunged into the middle of the dogs, snapped a line, tied a knot in another, jumped back out, and leaped back on.

We traveled close to the brown flank of Ubekendt Ejland. The island was brown, copper, and slate—all mudstones piled vertically and opening out at the bottom in fluted vaginas cut in half by ice. Ahead the way was silvered, the

top sift of snow blowing across our path and sweeping it clean, then refilling it again.

Where Illorsuit Strait flooded into Baffin Bay, a sweep of ice filled my eye, stopped by the hammered silver at the horizon. We turned and headed into Uummannaq Fjord. I dozed until Aliberti shook me awake to show me an eider duck flying by. Then we lay on the sled as we had become accustomed to doing, hooked together: we were two seals moving over a world of seals with only a thin sliver of ice between us. I felt the weight of his hand on my thigh; my hand rested on his shoulder. Palisades of rock strobed by. We slipped across sea ice that had no end.

UNEXPECTED POISONS

from *Silent Snow* (2005)

Marla Cone

In 1996, Marla Cone, an environmental reporter for the Los Angeles Times, *was researching a series of articles on chemicals that suppress the immune system when she learned that some of the most heavily exposed populations on earth live in the remotest regions. This realization became the impetus for her book* Silent Snow, *which documents how the Arctic and its people are disproportionately burdened with toxins like DDT and PCBs.*

BACK IN THE early 1970s, Tom Smith knew more about seals than any white man in Holman, a treeless expanse of frozen tundra that is one of the northernmost outposts in all of Canada. He knew when the seals had pups, how long they lactated, how big they grew, what they ate. Most of what he learned as a field scientist he gathered with the help of Inuit hunters, who camped on the sea ice of Canada's Northwest Territories every winter, enduring twenty-below Fahrenheit temperatures as they waited for ringed seals to pop through their breathing holes. The few hundred people of Holman, who hunted seals for food, skins, and oil, still lived like their ancestors did and had little contact with the rest of the world. Sometimes Smith would persuade the hunters to share their prey with him, and he used the blubber to explore whether Arctic seals were healthy and well-nourished.

One day Smith read a piece in an obscure scientific journal written by a chemist in Nova Scotia who had tested harp seals in the waters off Québec's urbanized coast. The chemist, Richard Addison, had discovered toxic pesticides and industrial compounds in the urban seals' blubber. It was 1973, an era of extraordinary pollution, when DDT, PCBs, and other compounds were building up in animals throughout North America and Europe. Many

birds—eagles, robins, pelicans—were vanishing as the chemicals destroyed their eggs. More curious than concerned about his seals, Smith thought that the specimens he had been collecting in Holman could be a treasure trove for a chemist since no one had ever tested the seals before. So he called Addison and asked: Would you like some Arctic blubber?

Addison had never been to Holman. In fact, he had never been to the Arctic at all. A few years earlier, in 1969, a Scottish fisheries scientist, Alan Holden, had found residue of DDT and PCBs in a few Arctic seals from Norway's northern coast and Canada's Baffin Island. The amounts were minute, almost undetectable, and Holden had declared them "substantially free of contamination in all but the 'background' sense." With the Holman seals inhabiting such remote waters of the Beaufort Sea, Addison thought it was likely that they, too, would essentially be "blanks," carrying no toxic substances or mere traces. Nevertheless, on a whim, Addison told Smith: Sure, send them and I'll take a look.

Soon afterward, blubber from about forty Arctic seals arrived at Addison's lab in Dartmouth, Nova Scotia, wrapped in foil and still frozen. It offered the first tangible clue to toxic detectives that chemicals were invading the far North.

Addison knew a lot about animal fat. Growing up Belfast, Northern Ireland, he had gotten his doctorate in agriculture, specifically studying the fat intake of poultry. He was most interested in applying chemistry to the real world, and when he was offered a job in Halifax, Canada, working in the "lipids group" for the government's fisheries research board, he accepted. He knew nothing about fish but he thought it would be a chance to help Canada figure out some commercial uses for blubber and fish oil. He started work there in 1966—when the environmental age was in its infancy. Chemical crises were just beginning to unfold around the world. PCBs were being detected in Swedish fish and Great Lakes birds. DDT was building up in California seabirds. As a chemist working with fish, Addison soon found himself completely drawn into Canada's emerging environmental problems. In January 1969, a large detergent factory had opened along the shore of Newfoundland, and within two months, local fishermen noticed thousands of dead herring. Divers were sent down and reported that everything in the harbor was dead. It was an ecological and economic disaster—the first pollution event in

Canada attracting national attention. The government directed Addison to investigate. It was a time when environmental chemistry didn't even exist, when environmental science of all types was in its infancy and there were few laws governing industry's handling of chemicals. Addison saw huge amounts of effluent flowing from the plant, a brew of unknown chemicals. Developing new detection techniques for seawater, he identified the culprit in the fish kill—phosphorus from detergents. The plant was shut down and the harbor floor dredged and paved over.

After that crisis, Canada decided to open up a pollution lab in Dartmouth, and Addison, hooked by the new science, took a job there in 1971, with the goal of developing ways to measure the new "bad boys" of pollution, organochlorines—chlorinated chemicals such as DDT and PCBs that had just begun showing up in wildlife. Addison knew he needed to push the limits of old-fashioned detectors in searching the environment for these chlorinated compounds, but he wasn't sure which medium to measure. Most scientists had been sampling water. He could have chosen fish or plankton but he decided on something he knew—fat. He needed large reservoirs of it for the sampling. What could be better than a plump seal with its thick layer of blubber? Canada certainly had lots of seals, and they were fairly easy to catch. He set out to turn seals into a sentinel species, an environmental monitor for the health of the whole oceanic ecosystem. He started in the foul waters of the Gulf of St. Lawrence off Québec.

Science is often serendipity, and in this case, Addison recalls, "the Arctic connection was purely accidental." He never would have thought to sample Arctic animals if Tom Smith hadn't seen his report and happened to pick up the phone. Addison was intrigued but not alarmed by what he detected in the blubber Smith sent: The levels of DDT and PCBs seemed pretty benign, an order of magnitude lower than animals in urban environments such as the Great Lakes. In 1974, Addison published a concise, three-page report documenting that the males were more than twice as contaminated as the females, a sign that the mothers were offloading the chemicals to their pups in their milk. A year later, two Canadian scientists reported PCBs in another Arctic species, polar bears.

At their labs in Scotland and Canada, Holden and Addison suspected that the chemicals they found in the Arctic were coming from distant, urban lands. How else could pesticides and PCBs be turning up in northern latitudes, and even in Antarctica? And what else could explain why DDT was decreasing rapidly in urban seals but not in Arctic ones? As early as 1969, Holden

sounded a warning, the first of many to come: "One further aspect . . . should be emphasized," he wrote at the conclusion of a paper written for a meeting of European marine scientists. "The global distribution of the organochlorine contaminants in the marine environment has been demonstrated. These contaminants, particularly the PCBs, are chemically very stable and presumably other substances of similar stability will also be globally distributed." This contamination, Holden wrote, "could be potentially damaging" to Arctic seals and other animals, just as it has been to urban birds.

It was a prophetic warning—that large volumes of chemicals were spreading globally—but no one in the scientific community or the public heeded it at the time. The early discoveries about one of the most isolated places on Earth were promptly forgotten. No one bothered to follow up on them for nearly a decade. More pressing environmental crises were mounting in cities around the world in the 1970s. In the United States, oil gushed from rigs, rivers caught fire, skies were blackened by soot and smog, songbirds dropped dead, and many species were on the verge of extinction. The Great Lakes and Europe's Baltic and North seas were far more contaminated than the Arctic Ocean. Scientists were busy testing for chemicals in cow's milk and beef and chicken and butter and eggs and fish. Why should anyone care about a few seals or bears near the North Pole? It didn't dawn on them that there was another creature precariously perched at the very top of the food chain, eating marine mammals and passing the chemicals to its young. No one gave a second thought to the Arctic's human hunters.

[. . .]

Eric Dewailly was a teenager about to graduate from high school in northern France when he visited Africa's poverty-stricken Ivory Coast. His father was a doctor, a gastroenterologist, but Dewailly had no interest whatsoever in medicine. Instead, he was fascinated by the forces that made life in isolated Third World places so difficult, and he expected to become a sociologist. By chance, while there, he met a physician handling infectious diseases and visited a clinic, watching the doctor take preventive steps that actually saved people's lives—administering vaccines, cleaning up sewage, finding clean drinking water. He was so impressed that he decided to enroll in medical school, and he knew immediately that protecting public health was his calling.

Dewailly thought tropical medicine would be his specialty, but instead, because of an exchange program between France and the French-speaking Canadian province of Québec, he wound up in the opposite hemisphere. When he was asked in 1983 to return to Québec City to start an environmental health

program at Laval University, he jumped at the chance. At the time, chemicals were being discovered in the breast milk of women in the United States and Europe, but little was known about Canada. The Québec provincial government asked Dewailly to survey women and, in 1986, he chose women giving birth at twenty-two hospitals, mostly from around Montreal, which he assumed would have the worst contaminant levels.

As his project began, he happened to meet Johanne Gagnon, a midwife from East Hudson Bay, at a public health meeting in June of 1986. Gagnon asked him if he would like to include women in Nunavik, the Arctic region of Québec, home to about eight thousand Inuit. At first, he had little interest. Too many logistical nightmares. And the milk of women so far from industries would most certainly be pristine. Nevertheless, he agreed, thinking a few samples might be useful as blanks so he could compare an unexposed population to an urban one.

About a year later, in the fall of 1987, the first batch of samples from Nunavik—glass vials holding a half-cup of frozen milk from each of twenty-four women—arrived via air mail at the laboratory in Québec City. Lab technician Evelyne Pelletier removed a sample from a walk-in refrigerator and began the daylong preparations to analyze it. She first extracted the chemicals by adding a solvent compound to the breast milk and shaking it, then mixing in an acid to destroy the fat so it wouldn't plug up the instruments. She poured the organochlorine mixture into a narrow glass column, removed the impurities, and spun it in a centrifuge so that the liquids evaporated and only the highly concentrated chemicals were left. The next day, using a gas chromatograph, Pelletier screened the extract for twenty-two chemicals—ten insecticides and twelve PCB compounds. She stood at the chart recorder as the machine spit out reams of data, one chemical at a time. Within minutes, Pelletier knew something was wrong. The concentrations of chemicals were off the charts—literally. In a normal test, technicians find individual, needlelike peaks, like those on an electrocardiogram. Instead, the peaks had overloaded the lab's equipment, running off the page. Pelletier showed the charts to the lab director, Jean-Philippe Weber. He had never seen his lab's equipment overloaded by a sample. The concentrations were about thirty times higher than anything he had ever seen before. When he saw that the samples were the milk of Arctic women, making the results even more improbable, he called Dewailly. "We have a problem here," he said. Something was wrong with the Arctic milk. He thought it might have been tainted in transit with some type of solvent.

They decided to test the sample again, diluting it this time to get a more accurate reading, and then tried another batch of Arctic milk, and another. "We knew then that this was not accidental contamination," Weber said. The chemicals were real. They were the same contaminants found in the milk of women in the south—PCBs and pesticides—but the milk of the Arctic mothers had up to ten times more than that of the mothers in Canada's biggest cities.

To Dewailly, who grew up near the North Sea, in one of Europe's most polluted regions, it belied all logic—until he began to search ecological journals and unearth data about PCBs and DDT, including Addison's long-forgotten 1974 report about seals. It became clear that the seals of Holman had been an unrecognized omen. Dewailly knew that the Inuit ate marine mammals but, like most doctors, he had no idea that toxic substances were building up in Arctic animals, as the data were not published in the medical journals he read, only in ecological journals he had never even heard of. Dewailly contacted the World Health Organization in Geneva, where an expert in chemical safety told him that the PCB levels were the highest he had ever seen. Those women, the expert said, should stop breast-feeding their babies—immediately.

Dewailly hung up the phone, his mind reeling. He knew that no food is more nutritious than mother's milk and that Nunavik is so remote that mothers had nothing else to feed their infants. As a doctor, he couldn't, in good conscience, tell them to stop breast-feeding. But he couldn't hide the problem either. "Breast milk is supposed to be a gift," Dewailly says. "It isn't supposed to be a poison." At the same time, elsewhere in Canada, in a small Inuit community on Baffin Island, other medical researchers were finding similar levels of contamination in breast milk there. The news about Nunavik and Baffin Island spread to the highest levels of government in Canada in 1988, frightening and angering Inuit leaders and triggering an international investigation into the health of all Arctic inhabitants.

Dewailly, teaming with a doctor in Greenland, soon discovered that the bodies of some Inuit there carried such extraordinary loads of chemicals that their bodies and breast milk could be classified as hazardous waste. Over the next decade and a half, Dewailly led a team investigating the effects on the babies of Nunavik. He discovered that the Inuit's traditional diet of seal meat, beluga blubber, and walrus is part tonic, part poison: Rich in nutritious fatty acids, the foods protect the Inuit from cancer and heart disease but the research suggests that they also make babies more susceptible to infectious

diseases and damage their developing brains. Nevertheless, Dewailly still firmly believes that the Inuit should keep nursing their babies and eating their traditional foods. Even today, almost two decades later, Canada remains embroiled in a debate over how to protect the health of its aboriginal people from the extreme levels of contaminants.

At about the time Dewailly began testing breast milk in Nunavik, Pál Weihe, across the Atlantic, was wondering about the childen of his own homeland. The son of a harbormaster, Weihe was born in the tiny seafaring village of Sørvágur in the Faroe Islands, a Danish territory in the middle of the North Atlantic, south of the Arctic Circle. In 1969, when he graduated high school, he left for Copenhagen to become a doctor—a surgeon, he thought. But upon studying occupational medicine, he learned about the dangers of chemicals and found this field more intriguing than his surgical studies. Surgery carries few surprises, he decided, but determining the risks of chemical exposure was so full of uncertainty, so mysterious, yet so vital to public health. In 1985, soon after his second child was born, Weihe heard about high levels of mercury in whales of the North Atlantic. The Faroese people are Nordic, not Inuit, but one thing separates them from the rest of their Danish compatriots: They eat pilot whales, in a tradition dating back centuries, perhaps to the days of the Vikings. Sometimes called "black torpedoes," pilot whales migrate long distances along the shores of the Atlantic, accumulating excessive levels of mercury from a variety of sources, including emissions spewed by coal-burning power plants thousands of miles from the Faroe Islands. The evidence that mercury is a neurotoxin that scrambles the brain dates back at least two centuries. "Mad as a hatter," a phrase made famous in *Alice in Wonderland,* originated from the tremors, confused speech, and hallucinations of nineteenth-century hatters poisoned by mercury used to cure felt. Yet it wasn't until the 1950s when the dangers to an infant's developing brain became apparent. At Japan's Minamata Bay, where a chemical factory dumped tons of mercury, thousands of people died or suffered various degrees of brain damage from eating fish, and an unexpected impact surfaced with the next generation: Thousands of children were born with mental retardation, deformed limbs, and other severe problems. Iraqi children suffered a similar fate in the 1970s when grain was contaminated with mercury.

Weihe knew Faroese babies were not exposed to enough mercury to cause

retardation as in Minamata or Iraq. Nevertheless, he wondered if there were more subtle neurological effects at the lower exposures of his fellow islanders. Weihe approached Philippe Grandjean, a Danish environmental epidemiologist known at the time for his studies of another heavy metal, lead, which damages the brains of babies and children. Weihe told him about the pilot whales. Grandjean's response reinforced his concerns. "I'm afraid mercury could very well be like lead," Grandjean said.

Weihe returned to the Faroes as medical director at its hospital system and began to collect blood samples from adults. Sure enough, their bodies contained large amounts of mercury. Weihe and Grandjean mapped out an ambitious plan: Even though they knew it would mean a lifetime of work, they decided to assemble a group of subjects, called a cohort, and follow them from birth. Their first grant proposal was denied but the Danish government gave them $15,000 in support, enough to get started. They asked pregnant women throughout the Faroe Islands to participate. Very few said no. They wound up recruiting 1,023 pregnant women—80 percent of the women who gave birth on the islands in 1986 and 1987. Their umbilical cord blood was stored for mercury analysis.

When the babies were born, Weihe and Grandjean saw no indication of any immediate damage to the infants. But they didn't expect to. Maturation of the brain is what's at stake with lower levels of mercury exposure. Weihe and Grandjean decided to test the children when they reached seven years, the age school begins in the Faroe Islands. In the spring of 1993, the children underwent an extensive series of psychological and neurological tests designed to see whether the mercury impaired any of their mental skills. The results: a measurable delay in transmission of signals from the ears of the most highly exposed children to their brains, a subtle slowing of a key neurological function. It was the "eureka" moment, Grandjean recalls. Other tests on the children found impaired vocabulary, memory, and attention span at what had previously been considered a low and safe level of mercury in pregnant women. Grandjean and Weihe wrote up their first findings in 1994 but the scientific paper came back three times from the publisher for more review because the findings were so worrisome for seafood eaters around the world. The first results were finally published in 1997. Years later, in 2004, results of tests on the children when they reached age fourteen were published, suggesting that at least some of the neurological impacts of mercury are long-lasting, perhaps permanent.

After having tested nearly 2,000 Faroese children, Weihe now is convinced

that the effects are real. As a scientist, he finds the research, which has had international repercussions for setting health standards, exhilarating, but as a doctor with strong ties to the people of his homeland, he is dismayed. Unlike the Inuit of Canada and Greenland, women in the Faroe Islands are now advised, based on Weihe's recommendations as head doctor of the hospital system, to stop eating the whale meat and blubber that have been important to their culture for centuries. It doesn't make Weihe, now fifty-five years old, the most popular person in the islands. But almost 2,000 mothers have come to trust Weihe, the "mercury doctor," with the well-being of their children, returning year after year to his small clinic in a residential neighborhood of Tórsham to undergo the neurological tests.

"My first assumption, back in 1985, was that we would not find any effects, that we have adapted to our diet over hundreds of years," Weihe says. Now he knows he was wrong.

Jim Estes was trying to solve a mystery of his own on another string of sub-Arctic islands half a world away, between the North Pacific and the Bering Sea. A marine biologist at the U.S. Geological Survey (USGS) in Santa Cruz, California, Estes for years had devoted his career to studying sea otters inhabiting the ocean off central California, trying to figure out why they had such a high death rate. Could pollutants like DDT and PCBs be to blame? After all, everyone knew California's waters were contaminated. For the sake of comparison, in the summers of 1991 and 1992, he traveled to the place where everyone assumed sea otters were clean, healthy, and thriving: Alaska's Aleutian Islands. He returned to Santa Cruz with ice chests containing samples of blood and fat from otters on Adak Island, and asked Walter Jarman, a pollutants expert then at the University of Utah, to take a look at their chemical content. The day the lab results came back, Jarman scanned the columns of numbers. No way, he thought. The Aleutian otters were supposed to be the uncontaminated ones, but he had never seen PCB numbers so high. How could otters inhabiting these remote Alaskan islands contain twice as much of these industrial compounds as otters off urban California? They carried 309 parts per billion of total PCBs—thirty-eight times more than otters in southeast Alaska. At the same time, another USGS scientist, Robert Anthony of Oregon, was finding surprisingly high concentrations of DDT and mercury in the eggs of bald eagles nesting in western parts of the Aleutians. Most

baffling of all, neither the otters nor the eagles ever migrated. They were somehow picking up the chemicals without ever leaving the islands.

Within a few years, by 1997, Estes had an even bigger mystery on his hands. The ecosystem of the Aleutians had collapsed. Tens of thousands of the Aleutians' otters disappeared within a matter of years, bringing them perilously close to extinction. There were no bodies to dissect, no clues to decipher. The otters weren't starving. They weren't sick. They simply vanished from the Aleutians without a trace, along with other sea mammals, particularly seals and sea lions, that had begun their descent in the 1980s. Estes believes that contaminants are not the main culprit, although they may play some minor role. Instead, he and his team have collected clues suggesting that various other human impacts, dating all the way back to commercial whaling a half-century ago, have triggered a series of events that upset the region's ecological equilibrium, upending its balance of predator and prey.

Piece by piece, Estes and other scientists think they are cobbling together this intricate puzzle, although the answers are more disturbing than satisfying, more elusive than conclusive. Estes predicts that the Aleutians' otters, in all likelihood, will be extinct in ten years, and their loss will reverberate throughout this entire ecosystem. It seems the ocean's chain of life is actually a fragile silk web. If you remove a strand, the whole thing unravels. And it may never be whole again.

✳

In the late 1980s, scientists worried that the effects of contaminants in the Arctic were too subtle to see in its wildlife but that, slowly, over time, its populations could be decimated. They realized that they needed to probe the bodies of animals for minute biochemical changes. By the 1990s, wildlife researchers, using new techniques honed in part by AIDS researchers, had succeeded: They developed technology to look at specific parameters—such as immune cells, antibodies, or testosterone levels—and see if they changed with concentrations of toxic chemicals in the bodies of animals. By searching for such connections, they could finally answer the question: Were chemicals harming Arctic wildlife in ways impossible to detect with the naked eye?

The answer came on Svalbard, a Norwegian archipelago where Canadian Andrew Derocher was pursuing his dream of studying animals in remote, pristine lands. Coming of age in the heyday of the environmental movement in the early 1970s, Derocher was the only one in his family interested in the

outdoors. He was raised along the lush banks of British Columbia's Fraser River, where he collected bird eggs and garter snakes and fished for salmon fry, trying to keep them alive in jars. After high school, he took a job as a seasonal game warden in British Columbia's provincial parks, spending his days fly-fishing for trout as bears ambled to the banks of the river. Then he studied wildlife ecology, focusing on large mammals, particularly bears. His father, a telephone repairman who worked his way up to upper management of a telephone company, was skeptical of his son's chosen career. His other son was a doctor and his daughter was a nurse, while his middle child was "mucking around with bears." When he took a research job at the Norwegian Polar Institute in 1996, Derocher thought he had found polar paradise. His longtime dream as a biologist was to study polar bears in their purest form, to find a population protected from human contact. Hunting of Svalbard's bears dates back to the sixteenth century—several hundred were trapped and shot for their fur each year—but since 1973, the archipelago has been a revered national refuge where hunting is banned. When Derocher arrived a quarter-century later, the population should have been fully recovered.

It wasn't long before he knew something was wrong. "Things just don't appear right," Derocher told his colleagues. It was as if these bears were still being hunted. Why weren't there more bears? Where were the older ones? Why were there so few females over the age of fifteen bearing cubs? Were they dying? Were they infertile? "Within the first year, it became pretty darned clear that I wasn't working with an unperturbed population," he says. In his second season on Svalbard, Derocher checked the sex of one bear as he routinely did, and found both a vagina and a penis-like knob. "What the hell is this?" he thought. He had examined more polar bears than just about anyone on Earth, yet he had never seen that before. Then he started finding more—three or four out of every one hundred examined. Derocher immediately suspected that chemicals were to blame.

By then, it was indisputable that Svalbard's bears had extraordinarily high concentrations of PCBs. In living animals, worse doses had been found only a handful of times: in Pacific Northwest orcas, European seals, and St. Lawrence River belugas. A few years later, in 2004, Derocher and Norwegian scientists published some groundbreaking findings, documenting an array of effects in Svalbard's polar bears they linked to the PCBs. Included are altered sex hormones—reductions in testosterone, increases in progesterone—as well as depleted thyroid hormones, which regulate brain development of a fetus. The bears also suffer suppressed immune cells and antibodies; altered cortisol,

which is important for managing stress and crucial body functions; lower retinol (vitamin A), which controls growth; and even osteoporosis. Such biochemical changes could impair the bears' ability to fight off disease and give birth to healthy cubs. Polar bear scientists theorize that the chemicals are culling older bears and weakening or killing cubs, perhaps leaving a missing generation of mother bears. Only 11 percent of Svalbard's bears with cubs are over fifteen years old, compared with 48 percent in Canada. When it comes to the most dramatic discovery—the pseudohermaphroditic polar bears with female and partial male genitalia—some scientists now suspect that they are natural occurrences, unrelated to the contaminants. Derocher and others, however, say that contaminants are a more plausible explanation. Essentially, no one really knows.

Nevertheless, Derocher is now virtually certain that there is some connection between the low numbers of bears in Svalbard and the toxic chemicals. "Could you realistically put two hundred to five hundred foreign compounds into an organism and expect them to have absolutely no effect?" he says. "I would be happier if I could find no evidence of pollution affecting polar bears, but so far, the data suggest otherwise."

*

Today, more than thirty years after the first traces of DDT were found in the Canadian seals, the evidence is overwhelming that toxic substances have spread throughout the Arctic, harming animals and people of the far North. An international body of scientists called the Arctic Monitoring and Assessment Programme (AMAP) concluded in a 2002 report that the contamination raises "fundamental questions of cultural survival, for it threatens to drive a wedge of fear between people and the land that sustains them."

Several generations have passed since chemicals first hitched a ride to the Arctic around World War II. The hunters who ate the seals sampled by Tom Smith are now likely to be grandparents, and the infants who drank the breast milk sent to Eric Dewailly's laboratory in Québec are teenagers, about to bestow on their children the chemical load amassing in their bodies from their consumption of marine life. Yet little has changed to protect the next generation of Arctic children.

After a half-century of research, scientists now ponder why it took them so long to make the connection between the Inuit and their prey, and to realize that toxic chemicals can wreak subtle damage on animals and people. Arriving

in the 1940s and discovered in the early 1970s, Arctic contamination was ignored until the late 1980s, and by then it was too late. It had reached extraordinary levels throughout the circumpolar north. "We knew that there were contaminants of concern in the Arctic as early as the 1970s. Why it took twenty years for the other shoe to drop is a puzzle," says Rob Macdonald of Canada's Institute of Ocean Sciences. "There are many we could fault—maybe scientists, maybe politicians, maybe society. We could point fingers in lots of directions. But I think it's probably more instructive to say that sometimes we don't pay attention to things that we should, that we don't connect things well."

Making matters worse, contaminants aren't the only environmental threat to the Arctic. It faces a triple whammy of human influences—contaminants, climate change, and commercial development—that the United Nations Environment Programme says is likely to inflict drastic changes on its natural resources and way of life this century. Seemingly hearty, the Arctic is, in fact, fragile.

What does this portend for the health of the world's children and animals, particularly in the Arctic, with its extraordinary exposure and vulnerability to contaminants? Unfortunately, no one really knows yet. The seeds of the next generation have already been planted in the Arctic, but the answers could still be generations away. The circumpolar north has been transformed into an immense living laboratory, where scientists are gradually unraveling the fate of contaminants on Earth and their effects on all its inhabitants, from pole to pole. Someone once said that ecology isn't rocket science—it's much harder. Although scientists have made great progress, most answers elude them.

Tom Smith still prowls the Arctic, evaluating the health of its seals. Richard Addison retired after a career directing a team of Canadian scientists known for breaking new ground in studying contaminants in seals and whales. Andrew Derocher left Svalbard and returned to Canada to continue his dangerous fieldwork, warning that polar bears, jeopardized by melting ice as well as PCBs, might not survive this century. Pál Weihe and Philippe Grandjean are still testing the youngsters of the Faroe Islands—those born when the experiments began are now teenagers—and it endures as one of the longest-lasting human experiments ever conducted. They, along with Eric Dewailly, are now among the world's leading experts on the human health effects of exposure to industrial poisons and pesticides, and their findings guide world regulators in determining how much tainted fish and other seafood is safe to eat. Recognizing a global need for analyzing contaminated foods, Dewailly hopes to soon take his show on the road, developing a mobile

laboratory to explore the symbiotic relationship between the oceans and human health and seek a balance between the benefits and risks of seafood around the world.

As with most environmental crises, there are no quick and easy solutions to the Arctic's dilemma. Even if the flow of all pesticides, PCBs, and other compounds is halted by every nation today, the tons already in the Arctic cannot be swept away or cleaned up. They are too ubiquitous, too persistent, too deeply embedded in the biota. Old PCBs, DDT, and other chemicals will remain there as long as it takes for nature to cleanse itself. And perhaps the most ominous discovery of all is that new chemicals are continually joining them.

SHISHMAREF, ALASKA

from *Field Notes from a Catastrophe* (2006)

Elizabeth Kolbert

In the spring of 2004, I traveled to Greenland, Iceland, and northern Alaska to observe the effects of global warming. The passage excerpted below discusses the retreat of the Arctic sea ice and the thawing of permafrost in the area around Prudhoe Bay.

The Alaskan village of Shishmaref sits on an island known as Sarichef, five miles off the coast of the Seward Peninsula. Sarichef is a small island—no more than a quarter of a mile across and two and a half miles long—and Shishmaref is basically the only thing on it. To the north is the Chukchi Sea, and in every other direction lies the Bering Land Bridge National Preserve, which probably ranks as one of the least visited national parks in the country. During the last ice age, the land bridge—exposed by a drop in sea levels of more than three hundred feet—grew to be nearly a thousand miles wide. The preserve occupies that part of it which, after more than ten thousand years of warmth, still remains above water.

Shishmaref (population 591) is an Inupiat village, and it has been inhabited, at least on a seasonal basis, for several centuries. As in many native villages in Alaska, life there combines—often disconcertingly—the very ancient and the totally modern. Almost everyone in Shishmaref still lives off subsistence hunting, primarily for bearded seals but also for walrus, moose, rabbits, and migrating birds. When I visited the village one day in April, the spring thaw was under way, and the seal-hunting season was about to begin. (Wandering around, I almost tripped over the remnants of the previous year's catch emerging from storage under the snow.) At noon, the village's transportation planner, Tony Weyiouanna, invited me to his house for lunch. In the living

room, an enormous television set tuned to the local public-access station was playing a rock soundtrack. Messages like "Happy Birthday to the following elders . . ." kept scrolling across the screen.

Traditionally, the men in Shishmaref hunted for seals by driving out over the sea ice with dogsleds or, more recently, on snowmobiles. After they hauled the seals back to the village, the women would skin and cure them, a process that takes several weeks. In the early 1990s, the hunters began to notice that the sea ice was changing. (Although the claim that the Eskimos have hundreds of words for snow is an exaggeration, the Inupiat make distinctions among many different types of ice, including *sikuliaq,* "young ice," *sarri,* "pack ice," and *tuvaq,* "landlocked ice.") The ice was starting to form later in the fall, and also to break up earlier in the spring. Once, it had been possible to drive out twenty miles; now, by the time the seals arrived, the ice was mushy half that distance from shore. Weyiouanna described it as having the consistency of a "slush puppy." When you encounter it, he said, "your hair starts sticking up. Your eyes are wide open. You can't even blink." It became too dangerous to hunt using snowmobiles, and the men switched to boats.

Soon, the changes in the sea ice brought other problems. At its highest point, Shishmaref is only twenty-two feet above sea level, and the houses, most of which were built by the U.S. government, are small, boxy, and not particularly sturdy-looking. When the Chukchi Sea froze early, the layer of ice protected the village, the way a tarp prevents a swimming pool from getting roiled by the wind. When the sea started to freeze later, Shishmaref became more vulnerable to storm surges. A storm in October 1997 scoured away a hundred-and-twenty-five-foot-wide strip from the town's northern edge; several houses were destroyed, and more than a dozen had to be relocated. During another storm, in October 2001, the village was threatened by twelve-foot waves. In the summer of 2002, residents of Shishmaref voted, a hundred and sixty-one to twenty, to move the entire village to the mainland. In 2004, the U.S. Army Corps of Engineers completed a survey of possible sites. Most of the spots that are being considered for a new village are in areas nearly as remote as Sarichef with no roads or nearby cities or even settlements. It is estimated that a full relocation would cost the U.S. government $180 million.

People I spoke to in Shishmaref expressed divided emotions about the proposed move. Some worried that, by leaving the tiny island, they would give up their connection to the sea and become lost. "It makes me feel lonely," one woman said. Others seemed excited by the prospect of gaining

certain conveniences, like running water, that Shishmaref lacks. Everyone seemed to agree, though, that the village's situation, already dire, was only going to get worse.

Morris Kiyutelluk, who is sixty-five, has lived in Shishmaref almost all his life. (His last name, he told me, means "without a wooden spoon.") I spoke to him while I was hanging around the basement of the village church, which also serves as the unofficial headquarters for a group called the Shishmaref Erosion and Relocation Coalition. "The first time I heard about global warming, I thought, I don't believe those Japanese," Kiyutelluk told me. "Well, they had some good scientists, and it's become true."

✳

The National Academy of Sciences undertook its first major study of global warming in 1979. At that point, climate modeling was still in its infancy, and only a few groups, one led by Syukuro Manabe at the National Oceanic and Atmospheric Administration and another by James Hansen at NASA's Goddard Institute for Space Studies, had considered in any detail the effects of adding carbon dioxide to the atmosphere. Still, the results of their work were alarming enough that President Jimmy Carter called on the academy to investigate. A nine–member panel was appointed. It was led by the distinguished meteorologist Jule Charney, of MIT, who, in the 1940s, had been the first meteorologist to demonstrate that numerical weather forecasting was feasible.

The Ad Hoc Study Group on Carbon Dioxide and Climate, or the Charney panel, as it became known, met for five days at the National Academy of Sciences' summer study center, in Woods Hole, Massachusetts. Its conclusions were unequivocal. Panel members had looked for flaws in the modelers' work but had been unable to find any. "If carbon dioxide continues to increase, the study group finds no reason to doubt that climate changes will result and no reason to believe that these changes will be negligible," the scientists wrote. For a doubling of CO_2 from preindustrial levels, they put the likely global temperature rise at between two and a half and eight degrees Fahrenheit. The panel members weren't sure how long it would take for changes already set in motion to become manifest, mainly because the climate system has a built-in time delay. The effect of adding CO_2 to the atmosphere is to throw the earth out of "energy balance." In order for balance to be restored—as, according to the laws of physics, it eventually must be—the entire planet has to heat up, including

the oceans, a process, the Charney panel noted, that could take "several decades." Thus, what might seem like the most conservative approach—waiting for evidence of warming to make sure the models were accurate—actually amounted to the riskiest possible strategy: "We may not be given a warning until the CO_2 loading is such that an appreciable climate change is inevitable."

It is now more than twenty-five years since the Charney panel issued its report, and, in that period, Americans have been alerted to the dangers of global warming so many times that reproducing even a small fraction of these warnings would fill several volumes; indeed, entire books have been written just on the history of efforts to draw attention to the problem. (Since the Charney report, the National Academy of Sciences alone has produced nearly two hundred more studies on the subject, including, to name just a few, "Radiative Forcing of Climate Change," "Understanding Climate Change Feedbacks," and "Policy Implications of Greenhouse Warming.") During this same period, worldwide carbon-dioxide emissions have continued to increase, from five billion to seven billion metric tons a year, and the earth's temperature, much as predicted by Manabe's and Hansen's models, has steadily risen. The year 1990 was the warmest year on record until 1991, which was equally hot. Almost every subsequent year has been warmer still. As of this writing, 1998 ranks as the hottest year since the instrumental temperature record began, but it is closely followed by 2002 and 2003, which are tied for second; 2001, which is third; and 2004, which is fourth. Since climate is innately changeable, it's difficult to say when, exactly, in this sequence natural variation could be ruled out as the sole cause. The American Geophysical Union, one of the nation's largest and most respected scientific organizations, decided in 2003 that the matter had been settled. At the group's annual meeting that year, it issued a consensus statement declaring, "Natural influences cannot explain the rapid increase in global near-surface temperatures." As best as can be determined, the world is now warmer than it has been at any point in the last two millennia, and, if current trends continue, by the end of the century it will likely be hotter than at any point in the last two million years.

In the same way that global warming has gradually ceased to be merely a theory, so, too, its impacts are no longer just hypothetical. Nearly every major glacier in the world is shrinking; those in Glacier National Park are retreating so quickly it has been estimated that they will vanish entirely by 2030. The oceans are becoming not just warmer but more acidic; the difference between daytime and nighttime temperatures is diminishing; animals are shifting their

ranges poleward; and plants are blooming days, and in some cases weeks, earlier than they used to. These are the warning signs that the Charney panel cautioned against waiting for, and while in many parts of the globe they are still subtle enough to be overlooked, in others they can no longer be ignored. As it happens, the most dramatic changes are occurring in those places, like Shishmaref, where the fewest people tend to live. This disproportionate effect of global warming in the far north was also predicted by early climate models, which forecast, in column after column of FORTRAN-generated figures, what today can be measured and observed directly: the Arctic is melting.

Most of the land in the Arctic, and nearly a quarter of all the land in the Northern Hemisphere—some five and a half billion acres—is underlaid by zones of permafrost. A few months after I visited Shishmaref, I went back to Alaska to take a trip through the interior of the state with Vladimir Romanovsky, a geophysicist and permafrost expert. I flew into Fairbanks—Romanovsky teaches at the University of Alaska, which has its main campus there—and when I arrived, the whole city was enveloped in a dense haze that looked like fog but smelled like burning rubber. People kept telling me that I was lucky I hadn't come a couple of weeks earlier, when it had been much worse. "Even the dogs were wearing masks," one woman I met said. I must have smiled. "I am not joking," she told me.

Fairbanks, Alaska's second-largest city, is surrounded on all sides by forest, and virtually every summer lightning sets off fires in these forests, which fill the air with smoke for a few days or, in bad years, weeks. In the summer of 2004, the fires started early, in June, and were still burning two and a half months later; by the time of my visit, in late August, a record 6.3 million acres—an area roughly the size of New Hampshire—had been incinerated. The severity of the fires was clearly linked to the weather, which had been exceptionally hot and dry; the average summertime temperature in Fairbanks was the highest on record, and the amount of rainfall was the third lowest.

On my second day in Fairbanks, Romanovsky picked me up at my hotel for an underground tour of the city. Like most permafrost experts, he is from Russia. (The Soviets more or less invented the study of permafrost when they decided to build their gulags in Siberia.) A broad man with shaggy brown

hair and a square jaw, Romanovsky as a student had had to choose between playing professional hockey and becoming a geophysicist. He had opted for the latter, he told me, because "I was little bit better scientist than hockey player." He went on to earn two master's degrees and two Ph.D.s. Romanovsky came to get me at ten A.M.; owing to all the smoke, it looked like dawn.

Any piece of ground that has remained frozen for at least two years is, by definition, permafrost. In some places, like eastern Siberia, permafrost runs nearly a mile deep; in Alaska, it varies from a couple of hundred feet to a couple of thousand feet deep. Fairbanks, which is just below the Arctic Circle, is situated in a region of discontinuous permafrost, meaning that the city is pocked with regions of frozen ground. One of the first stops on Romanovsky's tour was a hole that had opened up in a patch of permafrost not far from his house. It was about six feet wide and five feet deep. Nearby were the outlines of other, even bigger holes, which, Romanovsky told me, had been filled with gravel by the local public-works department. The holes, known as thermokarsts, had appeared suddenly when the permafrost gave way, like a rotting floorboard. (The technical term for thawed permafrost is "talik," from a Russian word meaning "not frozen.") Across the road, Romanovsky pointed out a long trench running into the woods. The trench, he explained, had been formed when a wedge of underground ice had melted. The spruce trees that had been growing next to it, or perhaps on top of it, were now listing at odd angles, as if in a gale. Locally, such trees are called "drunken." A few of the spruces had fallen over. "These are very drunk," Romanovsky said.

In Alaska, the ground is riddled with ice wedges that were created during the last glaciation, when the cold earth cracked and the cracks filled with water. The wedges, which can be dozens or even hundreds of feet deep, tended to form in networks, so when they melt, they leave behind connecting diamond- or hexagon-shaped depressions. A few blocks beyond the drunken forest, we came to a house where the front yard showed clear signs of ice-wedge melt-off. The owner, trying to make the best of things, had turned the yard into a miniature-golf course. Around the corner, Romanovsky pointed out a house—no longer occupied—that basically had split in two; the main part was leaning to the right and the garage toward the left. The house had been built in the sixties or early seventies; it had survived until almost a decade ago, when the permafrost under it started to degrade. Romanovsky's mother-in-law used to own two houses on the same block. He had urged her

to sell them both. He pointed out one, now under new ownership; its roof had developed an ominous-looking ripple. (When Romanovsky went to buy his own house, he looked only in permafrost-free areas.)

"Ten years ago, nobody cared about permafrost," he told me. "Now everybody wants to know." Measurements that Romanovsky and his colleagues at the University of Alaska have made around Fairbanks show that the temperature of the permafrost in many places has risen to the point where it is now less than one degree below freezing. In places where the permafrost has been disturbed, by roads or houses or lawns, much of it is already thawing. Romanovsky has also been monitoring the permafrost on the North Slope and has found that there, too, are regions where the permafrost is very nearly thirty-two degrees Fahrenheit. While thermokarsts in the roadbeds and talik under the basement are the sort of problems that really only affect the people right near—or above—them, warming permafrost is significant in ways that go far beyond local real estate losses. For one thing, permafrost represents a unique record of long-term temperature trends. For another, it acts, in effect, as a repository for greenhouse gases. As the climate warms, there is a good chance that these gases will be released into the atmosphere, further contributing to global warming. Although the age of permafrost is difficult to determine, Romanovsky estimates that most of it in Alaska probably dates back to the beginning of the last glacial cycle. This means that if it thaws, it will be doing so for the first time in more than a hundred and twenty thousand years. "It's really a very interesting time," Romanovsky told me.

*

The next morning, Romanovsky picked me up at seven. We were going to drive from Fairbanks nearly five hundred miles north to the town of Deadhorse, on Prudhoe Bay. Romanovsky makes the trip at least once a year, to collect data from the many electronic monitoring stations he has set up. Since the way was largely unpaved, he had rented a truck for the occasion. Its windshield was cracked in several places. When I suggested this could be a problem, Romanovsky assured me that it was "typical Alaska." For provisions, he had brought along an oversize bag of Tostitos.

The road that we traveled along—the Dalton Highway—had been built for Alaskan oil, and the pipeline followed it, sometimes to the left, sometimes to the right. (Because of the permafrost, the pipeline runs mostly aboveground,

on pilings that contain ammonia, which acts as a refrigerant.) Trucks kept passing us, some with severed caribou heads strapped to their roofs, others belonging to the Alyeska Pipeline Service Company. The Alyeska trucks were painted with the disconcerting motto "Nobody Gets Hurt." About two hours outside Fairbanks, we started to pass through tracts of forest that had recently burned, then tracts that were still smoldering, and, finally, tracts that were still, intermittently, in flames. The scene was part Dante, part *Apocalypse Now.* We crawled along through the smoke. After another few hours, we reached Coldfoot, named, supposedly, for some gold prospectors who arrived at the spot in 1900, then got "cold feet" and turned around. We stopped to have lunch at a truck stop, which made up pretty much the entire town. Just beyond Coldfoot, we passed the tree line. An evergreen was marked with a plaque that read "Farthest North Spruce Tree on the Alaska Pipeline: Do Not Cut." Predictably, someone had taken a knife to it. A deep gouge around the trunk was bound with duct tape. "I think it will die," Romanovsky told me.

Finally, at around five P.M., we reached the turnoff for the first monitoring station. By now we were traveling along the edge of the Brooks Range and the mountains were purple in the afternoon light. Because one of Romanovsky's colleagues had nursed dreams—never realized—of traveling to the station by plane, it was situated near a small airstrip, on the far side of a quickly flowing river. We pulled on rubber boots and forded the river, which, owing to the lack of rain, was running low. The site consisted of a few posts sunk into the tundra, a solar panel, a two-hundred-foot-deep borehole with heavy-gauge wire sticking out of it, and a white container, resembling an ice chest, that held computer equipment. The solar panel, which the previous summer had been mounted a few feet off the ground, was now resting on the scrub. At first, Romanovsky speculated that this was a result of vandalism, but after inspecting things more closely, he decided that it was the work of a bear. While he hooked up a laptop computer to one of the monitors inside the white container, my job was to keep an eye out for wildlife.

For the same reason that it is sweaty in a coal mine—heat flux from the center of the earth—permafrost gets warmer the farther down you go. Under equilibrium conditions—which is to say, when the climate is stable—the very warmest temperatures in a borehole will be found at the bottom and temperatures will decrease steadily as you go higher. In these circumstances, the lowest temperature will be found at the permafrost's surface, so that, plotted on a graph, the results will be a tilted line. In recent decades, though, the temperature profile of Alaska's permafrost has drooped. Now, instead of a straight line, what

you get is shaped more like a sickle. The permafrost is still warmest at the very bottom, but instead of being coldest at the top, it is coldest somewhere in the middle, and warmer again toward the surface. This is a sign—and an unambiguous one—that the climate is heating up.

"It's very difficult to look at trends in air temperature, because it's so variable," Romanovsky explained after we were back in the truck, bouncing along toward Deadhorse. It turned out that he had brought the Tostitos to stave off not hunger but fatigue—the crunching, he said, kept him awake—and by now the enormous bag was more than half empty. "So one year you have around Fairbanks a mean annual temperature of zero"—thirty-two degrees Fahrenheit—"and you say, 'Oh yeah, it's warming,' and other years you have mean annual temperature of minus six"—twenty-one degrees Fahrenheit—"and everybody says, 'Where? Where is your global warming?' " In the air temperature, the signal is very small compared to noise. What permafrost does is it works as low-pass filter. That's why we can see trends much easier in permafrost temperatures than we can see them in atmosphere." In most parts of Alaska, the permafrost has warmed by three degrees since the early 1980s. In some parts of the state, it has warmed by nearly six degrees.

✳

When you walk around in the Arctic, you are stepping not on permafrost but on something called the "active layer." The active layer, which can be anywhere from a few inches to a few feet deep, freezes in the winter but thaws over the summer, and it is what supports the growth of plants—large spruce trees in places where conditions are favorable enough and, where they aren't, shrubs and, finally, just lichen. Life in the active layer proceeds much as it does in more temperate regions, with one critical difference. Temperatures are so low that when trees and grasses die they do not fully decompose. New plants grow on top of the half-rotted old ones, and when these plants die the same thing happens all over again. Eventually, through a process known as cryoturbation, organic matter is pushed down beneath the active layer into the permafrost, where it can sit for thousands of years in a botanical version of suspended animation. (In Fairbanks, grass that is still green has been found in permafrost dating back to the middle of the last ice age.) This is the reason that permafrost, much like a peat bog or, for that matter, a coal deposit, acts as a storage unit for accumulated carbon.

One of the risks of rising temperatures is that the storage process can start

to run in reverse. Under the right conditions, organic material that has been frozen for millennia will begin to break down, giving off carbon dioxide or methane, which is an even more powerful (though more short-lived) greenhouse gas. In parts of the Arctic, this process is already under way. Researchers in Sweden, for example, have been measuring the methane output of a bog known as the Stordalen mire, near the town of Abisko, nine hundred miles north of Stockholm, for almost thirty-five years. As the permafrost in the area has warmed, methane releases have increased, in some spots by as much as 60 percent. Thawing permafrost could make the active layer more hospitable to plants, which are a sink for carbon. Even this, though, wouldn't be enough to offset the release of greenhouse gases. No one knows exactly how much carbon is stored in the world's permafrost, but estimates run as high as 450 billion metric tons.

"It's like ready-use mix—just a little heat, and it will start cooking," Romanovsky told me. It was the day after we had arrived in Deadhorse, and we were driving through a steady drizzle out to another monitoring site. "I think it's just a time bomb, just waiting for a little warmer conditions." Romanovsky was wearing a rain suit over his canvas work clothes. I put on a rain suit that he had brought along for me. He pulled a tarp out of the back of the truck.

Whenever he has had funding, Romanovsky has added new monitoring sites to his network. There are now sixty of them, and while we were on the North Slope he spent all day and also part of the night—it stayed light until nearly eleven—rushing from one to the next. At each site, the routine was more or less the same. First, Romanovsky would hook up his computer to the data logger, which had been recording permafrost temperatures on an hourly basis since the previous summer. When it was raining, Romanovsky would perform this first step hunched under the tarp. Then he would take out a metal probe shaped like a "T" and poke it into the ground at regular intervals, measuring the depth of the active layer. The probe was a meter long, which, it turned out, was no longer quite long enough. The summer had been so warm that almost everywhere the active layer had grown deeper, in some spots by just a few centimeters, in other spots by more than that. In places where the active layer was particularly deep, Romanovsky had had to work out a new way of measuring it using the probe and a wooden ruler. (I helped out by recording the results of this exercise in his waterproof field notebook.) Eventually, he explained, the heat that had gone into increasing the depth of the active layer would work its way downward, bringing the permafrost that much closer to the thawing point. "Come back next year," he advised me.

On the last day I spent on the North Slope, a friend of Romanovsky's, Nicolai Panikov, a microbiologist at the Stevens Institute of Technology, in New Jersey, arrived. He was planning on collecting cold-loving microorganisms known as psychrophiles, which he would take back to New Jersey to study. Panikov's goal was to determine whether the organisms could have functioned in the sort of conditions that, it is believed, were once found on Mars. He told me that he was quite convinced that Martian life existed—or, at least, had existed. Romanovsky expressed his opinion on this by rolling his eyes; nevertheless, he had agreed to help Panikov dig up some permafrost.

That same day, I flew with Romanovsky by helicopter to a small island in the Arctic Ocean, where he had set up yet another monitoring site. The island, just north of the seventieth parallel, was a bleak expanse of mud dotted with little clumps of yellowing vegetation. It was filled with ice wedges that were starting to melt, creating a network of polygonal depressions. The weather was cold and wet, so while Romanovsky hunched under his tarp I stayed in the helicopter and chatted with the pilot. He had lived in Alaska since 1967. "It's definitely gotten warmer since I've been here," he told me. "I have really noticed that."

When Romanovsky emerged, we took a walk around the island. Apparently, in the spring it had been a nesting site for birds, because everywhere we went there were bits of eggshell and piles of droppings. The island was only about ten feet above sea level, and at the edges it dropped off sharply into the water. Romanovsky pointed out a spot along the shore where the previous summer a series of ice wedges had been exposed. They had since melted, and the ground behind them had given way in a cascade of black mud. In a few years, he said, he expected more ice wedges would be exposed, and then these would melt, causing further erosion. Although the process was different in its mechanics from what was going on in Shishmaref, it had much the same cause and, according to Romanovsky, was likely to have the same result. "Another disappearing island," he said, gesturing toward some freshly exposed bluffs. "It's moving very, very fast."

*

On September 18, 1997, the *Des Groseilliers,* a three-hundred-and-eighteen-foot-long icebreaker with a bright-red hull, set out from the town of Tuktoyaktuk, on the Beaufort Sea, and headed north under overcast skies. Normally, the *Des Groseilliers,* which is based in Québec City, is used by the Canadian

Coast Guard, but for this particular journey it was carrying a group of American geophysicists, who were planning to jam it into an ice floe. The scientists were hoping to conduct a series of experiments as they and the ship and the ice floe all drifted, as one, around the Arctic Ocean. The expedition had taken several years to prepare for, and during the planning phase its organizers had carefully consulted the findings of a previous Arctic expedition, which had taken place back in 1975. The researchers aboard the *Des Groseilliers* were aware that the Arctic sea ice was retreating; that was, in fact, precisely the phenomenon they were hoping to study. Still, they were caught off guard. Based on the data from the 1975 expedition, they had decided to look for a floe averaging nine feet thick. When they reached the area where they planned to overwinter—at seventy-five degrees north latitude—not only were there no floes nine feet thick, there were barely any that reached six feet. One of the scientists on board recalled the reaction on the *Des Groseilliers* this way: "It was like 'Here we are, all dressed up and nowhere to go.' We imagined calling the sponsors at the National Science Foundation and saying, 'Well, you know, we can't find any ice.' "

Sea ice in the Arctic comes in two varieties. There is seasonal ice, which forms in the winter and then melts in the summer, and perennial ice, which persists year-round. To the untrained eye, all of it looks pretty much the same, but by licking it you can get a good idea of how long a particular piece has been floating around. When ice begins to form in seawater, it forces out the salt, which has no place in the crystal structure. As the ice thickens, the rejected salt collects in tiny pockets of brine too highly concentrated to freeze. If you suck on a piece of first-year ice, it will taste salty. Eventually, if the ice stays frozen long enough, these pockets of brine drain out through fine, veinlike channels, and the ice becomes fresher. Multiyear ice is so fresh that if you melt it, you can drink it.

The most precise measurements of Arctic sea ice have been made by NASA, using satellites equipped with microwave sensors. In 1979, the satellite data show, perennial sea ice covered 1.7 billion acres, or an area nearly the size of the continental United States. The ice's extent varies from year to year, but since then the overall trend has been strongly downward. The losses have been particularly great in the Beaufort and Chukchi Seas, and also considerable in the Siberian and Laptev Seas. During this same period, an atmospheric circulation pattern known as the Arctic Oscillation has mostly been in what climatologists call a "positive" mode. The positive Arctic Oscillation

is marked by low pressure over the Arctic Ocean, and it tends to produce strong winds and higher temperatures in the far north. No one really knows whether the recent behavior of the Arctic Oscilliation is independent of global warming or a product of it. By now, though, the perennial sea ice has shrunk by roughly 250 million acres, an area the size of New York, Georgia, and Texas combined. According to mathematical models, even the extended period of a positive Arctic Oscillation can account for only part of this loss.

At the time the *Des Groseilliers* set off, little information on trends in sea-ice depth was available. A few years later, a limited amount of data on this topic—gathered, for rather different purposes, by nuclear submarines—was declassified. It showed that between the 1960s and the 1990s, sea-ice depth in a large section of the Arctic Ocean declined by nearly 40 percent.

Eventually, the researchers on board the *Des Groseilliers* decided that they would just have to settle for the best ice floe they could find. They picked one that stretched over some thirty square miles. In some spots it was six feet thick, in some spots just three. Tents were set up on the floe to house experiments, and a safety protocol was established: anyone venturing out onto the ice had to travel with a buddy and a radio. (Many also carried a gun, in case of polar-bear problems.) Some of the scientists speculated that, since the ice was abnormally thin, it would grow thicker during the expedition. Just the opposite turned out to be the case. The *Des Groseilliers* spent twelve months frozen into the floe, and, during that time, it drifted some three hundred miles north. Nevertheless, at the end of the year, the average thickness of the ice had declined, in some spots by as much as a third. By August 1998, so many of the scientists had fallen through that a new requirement was added to the protocol: anyone who set foot off the ship had to wear a life jacket.

Donald Perovich has studied sea ice for thirty years, and on a rainy day not long after I got back from Deadhorse, I went to visit him at his office in Hanover, New Hampshire. Perovich works for the Cold Regions Research and Engineering Laboratory, or CRREL (pronounced "crell"). CRREL is a division of the U.S. Army that was established in 1961 in anticipation of a very cold war. (The assumption was that if the Soviets invaded, they would probably do so from the north.) He is a tall man with black hair, very black eyebrows, and an earnest manner. His office is decorated with photographs

from the *Des Groseilliers* expedition, for which he served as the lead scientist; there are shots of the ship, the tents, and, if you look closely enough, the bears. One grainy-looking photo shows someone dressed up as Santa Claus, celebrating Christmas in the darkness out on the ice. "The most fun you could ever have" was how Perovich described the expedition to me.

Perovich's particular area of expertise, in the words of his CRREL biography, is "the interaction of solar radiation with sea ice." During the *Des Groseilliers* expedition, Perovich spent most of his time monitoring conditions on the floe using a device known as a spectroradiometer. Facing toward the sun, a spectroradiometer measures incident light, and facing toward earth, it measures reflected light. By dividing the latter by the former, you get a quantity known as albedo. (The term comes from the Latin word for "whiteness.") During April and May, when conditions on the floe were relatively stable, Perovich took measurements with his spectroradiometer once a week, and during June, July, and August, when they were changing more rapidly, he took measurements every other day. The arrangement allowed him to plot exactly how the albedo varied as the snow on top of the ice turned to slush, and then the slush became puddles, and, finally, some of the puddles melted through to the water below.

An ideal white surface, which reflected all the light that shone on it, would have an albedo of one, and an ideal black surface, which absorbed all the light, would have an albedo of zero. The albedo of the earth, in aggregate, is 0.3, meaning that a little less than a third of the sunlight that strikes it is reflected back out. Anything that changes the earth's albedo changes how much energy the planet absorbs, with potentially dramatic consequences. "I like it because it deals with simple concepts, but it's important," Perovich told me.

At one point, Perovich asked me to imagine that we were looking down at the earth from a spaceship hovering above the North Pole. "It's springtime, and the ice is covered with snow, and it's really bright and white," he said. "It reflects over 80 percent of the incident sunlight. The albedo's around 0.8, 0.9. Now, let's suppose that we melt that ice away and we're left with the ocean. The albedo of the ocean is less than 0.1; it's like 0.07.

"Not only is the albedo of the snow-covered ice high; it's the highest of anything we find on earth," he went on. "And not only is the albedo of water low; it's pretty much as low as anything you can find on earth. So what you're doing is you're replacing the best reflector with the worst reflector." The more open water that's exposed, the more solar energy goes into heating the ocean.

The result is a positive feedback, similar to the one between thawing permafrost and carbon releases, only more direct. This so-called ice-albedo feedback is believed to be a major reason that the Arctic is warming so rapidly.

"As we melt that ice back, we can put more heat into the system, which means we can melt the ice back even more, which means we can put more heat into it, and, you see, it just kind of builds on itself," Perovich said. "It takes a small nudge to the climate system and amplifies it into a big change."

APPENDIX

A BRIEF CHRONOLOGY OF ARCTIC EXPLORATION

Third century BC — Pytheas, a Greek merchant and explorer from Massilia, now Marseille, writes of visiting an island six days' sail from Great Britain, which he refers to as Thule. Many candidates have been proposed for Thule, including Iceland, Greenland, and the Faroe Islands. It has also been speculated that Pytheas chronicled a fictional journey.

985 AD — Having been evicted from Iceland, Eric the Red arrives in Greenland, accompanied by nearly seven hundred followers. They establish two settlements, the Eastern Settlement, which is actually in the south, and the Western Settlement, which is to the north.

1576–1578 — Martin Frobisher, a British sailor, makes three voyages in search of the Northwest Passage. On the second, he finds what he thinks is gold, and carries 1,500 tons back to Britain, at which point he is informed that it is iron pyrite. He gets as far as Frobisher Bay, an inlet of the Labrador Sea.

1610 — Henry Hudson and the crew of the *Discovery* reach what is now Hudson Bay. That winter, the ship gets iced in in James Bay. The following spring, Hudson's crew mutinies and sets him adrift in a small boat. He is never heard from again.

1831 — James Clark Ross locates the North Magnetic Pole on the west coast of the Boothia Peninsula.

1845 Sir John Franklin and crew of more than 120 set off in search of the Northwest Passage on two ships, the *Terror* and the *Erebus.* The ships are last sighted by Europeans in Melville Bay. After two years with no word from the expedition, several rescue ships are dispatched. (More men would be lost looking for members of the Franklin expedition than set out on it to begin with.) No survivors are ever found.

1853 Elisha Kent Kane sets off on an American expedition in search of Franklin and his crew.

1854 In the course of searching for survivors of the Franklin expedition, the British explorer Robert McClure and the crew of the *Investigator* become the first Europeans to traverse the Northwest Passage, in part by boat and in part by sled. Upon returning to England, McClure is knighted.

1871 Charles Hall, an American businessman-turned-explorer sets off in search of the North Pole, dies under mysterious circumstances.

1882–83 First International Polar Year

1888 The Norwegian explorer Fridtjof Nansen completes the first crossing of Greenland.

1895 Fridtjof Nansen reaches 86° 14′ N, the highest latitude then attained.

1897 Jack London joins the Klondike Gold Rush.
The Swedish explorer Salomon August Andrée attempts a balloon flight over the North Pole; he and his two companions perish.

1906 The Norwegian explorer Roald Amundsen completes the first expedition to successfully traverse the Northwest Passage by ship. The route he travels is not commercially viable.

1909 The American explorer Robert Peary claims to have reached the North Pole.

1924 Knud Rasmussen completes the Fifth Thule Expedition, a 20,000-mile trek, via dogsled, from Greenland to Siberia.

1926 The American aviator Richard Byrd claims to have flown over the North Pole; this claim is later cast into doubt.

1932–33 Second International Polar Year

1955 Halldór Laxness is awarded the Nobel Prize for Literature.

1957–58 Third International Polar Year

1958 The USS *Nautilus*, the world's first operational nuclear submarine, makes a transit underneath the geographic North Pole. She surfaces northeast of Greenland.

1959 The U.S. Army constructs Camp Century, a research station built below the surface of the Greenland ice sheet. The camp is powered by a portable nuclear reactor.

1993 The first full-length core of the Greenland ice is completed. The core is ten thousand feet long, and contains a climatological data going back more than a hundred thousand years.

2004 The Arctic Climate Impact Assessment concludes that "climate change presents a major and growing challenge to the Arctic and the world as a whole."

2005 Arctic sea ice extent reaches a record low.

2006 Studies predict that the Arctic Ocean could be ice-free in summer by 2040.

2007–9 Fourth International Polar Year

ACKNOWLEDGMENTS

Various publishers and estates have generously given permission to use extracts from the following works:

"See the Esquimaux" by Andrea Barrett from *The Voyage of the Narwhal.* Copyright © 1998 by Andrea Barrett. Used by permission of W.W. Norton & Company and HarperCollins Publishers, Ltd. "Andrée's Second Diary" by Salomon August Andrée from *Andrée's Story* by S.A. Andrée and Nils Strindberg and K. Fraenkel, translated by Edward Adams-Ray, copyright © 1930 by Albert Bonniers Forlag & Hearst Enterprises Inc. © 1930 Viking Press, Inc., translation. Used by permission of Viking Penguin, a division of Penguin Group (USA) Inc. Copyright "Christmas in Greenland" by Tété-Michel Kpomassie from *An African in Greenland.* Copyright © 1981, Flammarion, Paris, English translation by James Kirkup, copyright © 1983 by Harcourt, Inc and Martin Secker & Warburg Ltd., reprinted by permission of Harcourt, Inc. "Land Ho!" by Valerian Albanov from *In the Land of the White Death.* Copyright © 2000 by Random House, Inc. Used by permission of Modern Library, a division of Random House, Inc. "Icelandic Pioneer" by Halldór Laxness from *Independent People,* translated by J.A. Thompson. Copyright © 1946, published by Harvill, reprinted by permission of the Random House Group Ltd. "The Land, Breathing" by Barry Lopez from *Arctic Dreams,* copyright by Barry Lopez © 1986. Reprinted by permission of SLL/Sterling Lord Literistic, Inc. Published in the UK by Harvill. Reprinted by permission of the Random House Group Ltd. Published by Harvill. Reprinted by permission of the Random House Group Ltd. "Aliberti's Ride" by Gretel Ehrlich from *This Cold Heaven,* copyright © 2001 by Gretel Ehrlich. Used by permission of Pantheon Books, a division of Random House, Inc., and HarperCollins Publishers Ltd. "Unexpected Poisons" by Marla Cone from *Silent Snow,* copyright © 2005 by Marla Cone. Used by permission of Grove/Atlantic, Inc. "Shishmaref, Alaska" by Elizabeth Kolbert from *Field Notes from a Catastrophe,* copyright © 2006 by Elizabeth Kolbert. Map of the Northern Hemisphere and the

Arctic, France, 18th century. Held by the Bibliotheque Nationale, Paris, France. Copyright © Erich Lessing/Art Resource, NY.

Every effort has been made to contact copyright holders of the extracts used in this volume. However, the publisher would be happy to rectify any errors or omissions in future editions.

ABOUT THE EDITOR

Elizabeth Kolbert is the author of *Field Notes from a Catastrophe*, and has been a staff writer for the *New Yorker* since 1999. Prior to that, she was a reporter for the *New York Times*. She lives with her husband and three sons in Williamstown, Massachusetts.